Physiology in Health and Disease

Published on behalf of The American Physiological Society by Springer

Physiology in Health and Disease

This book series is published on behalf of the American Physiological Society (APS) by Springer. Access to APS books published with Springer is free to APS members.
APS publishes three book series in partnership with Springer: *Physiology in Health and Disease* (formerly *Clinical Physiology*), *Methods in Physiology*, and *Perspectives in Physiology* (formerly *People and Ideas*), as well as general titles.

More information about this series at http://www.springer.com/series/11780

Kelly Anne Hyndman • Thomas L. Pannabecker
Editors

Sodium and Water Homeostasis

Comparative, Evolutionary and Genetic Models

Springer

aps
American Physiological Society

Editors
Kelly Anne Hyndman
The University of Alabama
at Birmingham
Birmingham
Alabama
USA

Thomas L. Pannabecker
Department of Physiology
University of Arizona Health Sciences Center
Tucson
Arizona
USA

Physiology in Health and Disease
ISBN 978-1-4939-3212-2 ISBN 978-1-4939-3213-9 (eBook)
DOI 10.1007/978-1-4939-3213-9

Library of Congress Control Number: 2015954944

Springer New York Heidelberg Dordrecht London

Printed on acid-free paper

Springer Science+Business Media LLC New York is part of Springer Science+Business Media (www.springer.com)

Contents

Chapter 1
Overview

Kelly Anne Hyndman and Thomas L. Pannabecker

Abstract Natriuresis and diuresis, we all have to do it! Whether you are a euryhaline fish that migrates between fresh and seawater, or a person who just ate a large salty meal, the body has to regulate salts and water to maintain in balance. This book highlights recent advances in the fields of comparative, evolutionary and genetic models of sodium and water homeostasis in an attempt to encourage collaboration and discussion among comparative and biomedical researchers in this field. Seven themes are covered and these focus broadly on a number of molecular, cellular and whole animal studies that involve multiple physiological systems. Each theme is presented as a paired set of chapters; the first chapter consists of a discussion from the viewpoint of the comparative physiologist and the second chapter consists of a discussion from the viewpoint of the biomedical researcher. Although there is often overlap of ideas and concepts between each set of paired chapters, it will be seen that different approaches to understanding the same physiological processes greatly expands our understanding of these fundamental processes. An enhanced interconnectivity of comparative, evolutionary and genetic animal models in basic science and medical science should improve our overall understanding of the mechanisms of sodium and water balance.

Keywords Water channels • Aquaporins • Angiotensin • Hypertension • Sex differences • Endothelin • Comparative genetics • miRNA • Urinary concentrating mechanism • Kidney • Giraffe • Malpighian tubule • Circadian clock • Insect visual system

K.A. Hyndman, Ph.D. (✉)
Division Nephrology, Department of Medicine, The University of Alabama at Birmingham, 1720 2nd Ave S, Kaul 830, Birmingham, AL 35294-0024, USA
e-mail: hyndmank@uab.edu

T.L. Pannabecker, Ph.D.
Department of Physiology, University of Arizona Health Sciences Center, AHSC 4128, 1501 N Campbell Avenue, Tucson, AZ 85724-5051, USA
e-mail: pannabec@email.arizona.edu

K.A. Hyndman, T.L. Pannabecker (eds.), *Sodium and Water Homeostasis*,
Physiology in Health and Disease, DOI 10.1007/978-1-4939-3213-9_1

Sodium and water both are essential intracellular and extracellular molecules of the body. Where goes sodium, water soon follows whether by osmosis or bolus flows. All living organisms have to regulate sodium and water in order to remain in homeostasis and thrive. Depending on the environment, sodium and/or water may be plentiful or limited, and thus the organisms must overcome these challenges to remain in balance. The question of what occurs physiologically for an organism to remain in sodium and water homeostasis spans many disciplines. One goal of this book is to highlight how different approaches to understanding the same physiological processes greatly expand our understanding of these fundamental processes.

For every biological process there is an appropriate species in which investigations of that process may be most insightful, a concept underscored as the "Krogh Principle" and ruminated upon by many others over the years. This is a key principle shared by comparative physiologists, and has lead to a breadth of studies in a variety of species in order to elucidate physiological pathways either conserved or divergent over the course of evolution. The processes by which sodium and water remain in homeostasis in different organisms include many similar molecular and biochemical pathways and physiological events that have followed the course of evolution and animal development. A common genetic makeup underlies these processes.

To best understand common molecular and cellular pathways involved in fluid-electrolyte balance, it is essential to account for the evolutionary history of these mechanisms. The tree of life is ever changing (a new species is discovered, other species go extinct), but on a molecular and cellular level, many of the pathways we study have a deep evolutionary history. For example, as Chaps. 6 and 7 will highlight, the endothelin system that is critical for maintaining fluid-electrolyte balance, was a vertebrate innovation with endothelins first being expressed in fishes. This suggests endothelins have been around for 400–500 million years, and many of the downstream targets are common from fishes to man.

In studies with a focus on biomedical research and to improve human health, a number of unique genetic models have been developed. With the advent of whole body genetic deletions, tissue or cell-specific genetic deletions or insertions into the mouse genome, there has been a flurry of major advances made in a biomedical context. Now, with Cre/Lox, zinc-finger nuclease, and CRISPR technology, mice, rat, and other models for human health can be manipulated at the genomic level, and this is leading to very important medical advances. This information is not only relevant for the betterment of humans, but also is interesting in a comparative and evolutionary context. Likewise, comparative and evolutionary studies can lend important information to biomedical studies. In this digital age, information from different disciplines is very accessible, and as scientists we should try to incorporate this information from different perspectives so that we test well-informed hypotheses.

From euryhaline fishes to desert adapted rodents (Chap. 10) to human diseases such as ANGII hypertension (Chap. 5) and salt-sensitive hypertension, the mechanisms involved in maintaining sodium and water balance can be experimentally tested in comparative, evolutionary and genetic animal models. The physiology and pathophysiology of sodium homeostasis can be understood at multiple levels. For example, hormones regulate sodium movement across plasma membranes and into and out of cells and across epithelia. Flow pathways through an organ can be complex, as in the multi-segmented nephrons and capillary networks of the mammalian kidney, or very simple, as in the blind-ended insect Malpighian tubule. Whether it is a female mosquito, that must excrete copious amounts of sodium and water after she has engorged on the blood of a vertebrate host (Chap. 13), or a human that has just eaten lots of salty foods, and must excrete the excess salt to prevent extracellular fluid expansion, many common mechanisms have evolved in order to maintain homeostasis.

From circadian rhythms of expression of sodium transporters in the in the fly eye (Chap. 14) or sodium transporters in the mammalian kidney (Chap. 15) to miRNAs that bind to RNA and, in so doing, regulate expression of sodium transporters in the zebrafish kidney (Chap. 8) or miRNA that function similarly in the mammalian kidney (Chap. 9), this book highlights recent advances in the fields of comparative, evolutionary and genetic models of sodium and water homeostasis in an attempt to encourage collaboration and discussion among comparative and biomedical researchers in this field. An enhanced interconnectivity of comparative, evolutionary and genetic animal models in basic science and medical science should improve our overall understanding of the mechanisms of sodium and water balance.

Chapter 2
Comparative and Evolutionary Physiology of Water Channels

Stanley D. Hillyard

Abstract Experiments with a variety of plant and animal tissues over the past 250 years have given rise to concepts of osmosis, osmotic pressure and permeability that are used to this day. The frog skin and mammalian red blood cells proved particularly useful in establishing that water transport across membranes was mediated by pores rather than simple diffusion. Comparative studies identified membrane intrinsic proteins (MIPs) in a variety of tissues that included red blood cells, bovine lens, plant cell membranes, fruit fly brain and microbial membranes. From these studies emerged the concept of water conducting channels called aquaporins (Aqps) that were present in virtually all living organisms. The discovery of Aqps and methodology for identifying them by RT-PCR cloning resulted in an enormous literature that provided mechanistic bases for physiological functions related to ionic and osmotic regulation. Among the vertebrates mammalian Aqps are the best studied due to their importance for biomedical issues such as diabetes insipidus and the availability of knock out (KO) models where specific deficiencies can be studied. From an evolutionary perspective, fish have multiple copies of most of the canonical Aqps that have been characterized in mammals. The development of genomic technologies is beginning to identify episodes of whole genome duplication, beginning in the earliest chordates, that resulted in the variety of Aqps that serve osmoregulatory functions in living vertebrate taxa. Comparative studies are beginning to describe how different Aqps have been coopted to osmoregulatory epithelia but more research is needed to understand the coordination of apical and basolateral membrane function.

Keywords Aquaporin • Evolution • Water • Comparative physiology • Evolution

---

S.D. Hillyard (✉)
Department of Biomedical Sciences, UNLV School of Dental Medicine, 1001 Shadow Lane, Las Vegas, NV 89106, USA
e-mail: stanley.hillyard@unlv.edu

K.A. Hyndman, T.L. Pannabecker (eds.), *Sodium and Water Homeostasis*, Physiology in Health and Disease, DOI 10.1007/978-1-4939-3213-9_2

2.1 Introduction

The study of water transport across biological membranes has a long history of experimentation with plant and animal tissues that established basic principles of osmosis, the role of membrane integral proteins and ultimately the identification of aquaporins (Aqps) as water conducting membrane channels. The first objective of this chapter is to use selected studies to give a linear time line for these events. Techniques for cloning Aqps have generated an enormous literature that makes it difficult if not impossible to address a topic as broad as the chapter title within the time and space limitations for its preparation. Given the context of this volume, the second objective will focus on vertebrate Aqps but even then it is likely there will be omissions that some may find fault with. There is an equally interesting literature on Aqps in plants [summarized by Maurel et al. (2008)] and invertebrates [summarized by Campbell et al. (2008)].

The literature on the structure and function of vertebrate Aqps is dominated by research with mammalian tissues [summarized by Ishibashi et al. (2009)] and the numerical terminology for mammals is applied to Aqps in other vertebrate classes. Of particular interest are recent studies that reveal the expression of homologs to mammalian Aqps in elasmobranch, actinopterigian and sarcopterigian fishes, suggesting the diversity of Aqps associated with mammalian evolution dates from the earliest chordates (Cutler et al. 2012; Cerda and Finn 2010; Konno et al. 2010). The invasion of land is associated with unique amphibian Aqps in the skin and bladder (Suzuki and Tanaka 2009, 2010). Mammalian-like Aqps have been cloned from osmoregulatory tissues of reptiles and birds, including the avian kidney that has nephrons with loops of Henle and modest concentrating capability (Nishimura and Yang 2013). Despite the broad evolutionary aspects of vertebrate Aqps surprisingly few physiological conclusions can be made. The presence of mRNA used for cloning doesn't necessarily indicate the Aqp is translated or even expressed in a membrane. Antibodies used to immunolocalize Aqps often show expression in the apical or basolateral membrane of osmoregulatory epithelia but not both. Furthermore, many Aqps perform functions other than simply water channels including cell adhesion and motility in adult and embryonic tissues (Ishibashi et al. 2011). Phylogenies based on Aqp sequence identities, in combination with comparative physiological studies of ionic and osmotic regulation (Evans 2009), raise interesting questions regarding the role of gene duplication and co-optation of Aqps in the evolution of water balance strategies of many species.

2.2 Historical Perspectives

2.2.1 Osmosis and Water Permeability

The movement of water across biological structures has been studied for many years, using different plant and animal tissues for experiments. Among the earliest was Nollet (1752, 1995 translation) who stretched a pig bladder over the opening of a flask filled with an ethanol solution and placed the bladder into a vase of water. Water was taken into the solution, a process he termed *ebullition*, leading to the conclusion that ethanol exerted a force attracting water across the tissue. Anecdotal reference was made to similar results obtained by another researcher in 1688 which is only a short time after the term "cell" was coined by Hooke (1664) and the description of red blood cells by Malphighi (1666, translated by Forrester 1995). Hewson (1773) observed changes in the shape of red blood cells of many species as they rolled down a tilted microscope slide. He noted that cells bathed in serum were disc shaped and when placed in water they swelled and became spherical. It was suggested that cells were contained within a membrane and that water was taken up from dilute pond water [Reviewed by Kleinzeller (1996)].

The term "osmosis" is attributed to Dutrochet (1827) who used a device similar to that of Nollet (1752) with the caecum of a chicken separating an upper chamber containing a solution from a container of pure water (Fig. 2.1a) Movement from water in the lower chamber into the solution in the upper chamber was termed endosmosis. Over time solute permeated the tissue resulting in an outward movement of water that he called exosmosis. A similar Dutrochet-like osmometer was used by Matteucci and Cima (1845) and Reid (1890) to demonstrate inward water movement across isolated frog skin. The pressure generated by osmotic water movement across a membrane was quantified by the German botanist Wilhelm Pfeffer (1877) who constructed a similar osmometer using an artificial membrane of copper ferrocyanide precipitated on a porous ceramic surface. A mercury manometer was used to measure the pressure generated as water flowed across the membrane into the solution (Fig. 2.1b). The osmotic pressure was taken as the hydrostatic pressure at which osmotic water flow ceased. The relationship between solute concentration and osmotic pressure (denoted as π) was quantified by van't Hoff (1887, 1901) who applied the ideal gas law to this relationship and derived the famous equation:

$$\pi = C_s R\, T \tag{2.1}$$

in which C_s is the concentration of an ideal solute, R the appropriate gas constant and T the absolute temperature. For this synthesis he was awarded the first Nobel Prize in chemistry in 1901.

In practice, the osmolar concentration (C_{osm}) is generally lower than the molar concentration and can be corrected for by an osmotic coefficient for a particular solute. For example the osmotic coefficient for NaCl is assumed to be about 0.93 at

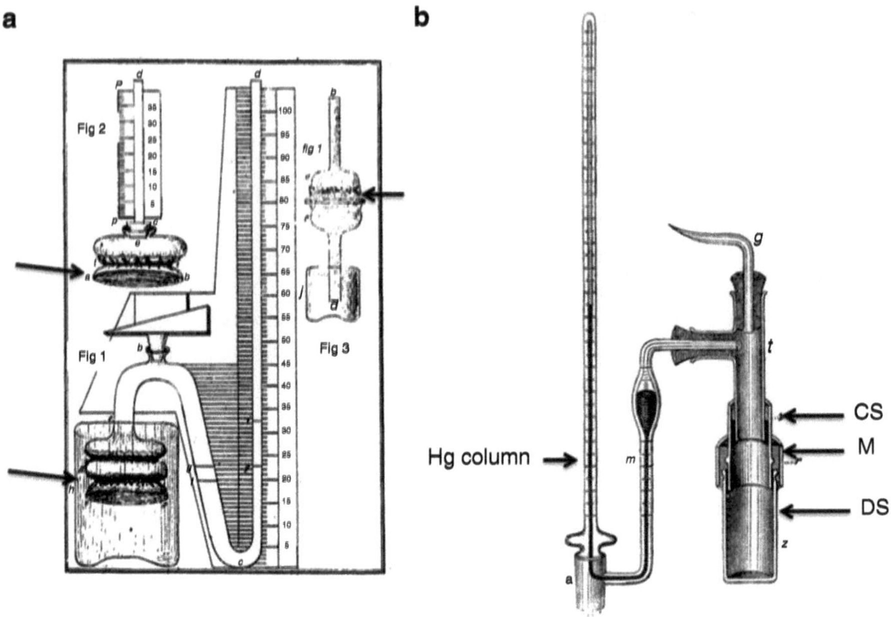

Fig. 2.1 (**a**) Osmometers used by Dutrochet (1827) to measure osmotic water movement across chicken cecum tied over the opening of a flask and immersed into a container of water. (**b**) Osmometer constructed by Pfeffer (1877) in which an artificial membrane made of copper ferrocyanide precipitated on a ceramic plate (r) separated water in the lower chamber (z) from solutions of different concentration in the upper chamber (t). Osmotic flow into the upper chamber displaced a mercury column (m). The pressure measured when osmotic flow was balanced by the mercury column was the osmotic pressure

physiological concentrations (Stadie and Sunderman 1931). Further, many biological membranes may be permeable to the solute as well. This may be compensated for by correcting Eq. (2.1) for solute permeability with the reflection coefficient (δ_s) for a given solute. A reflection coefficient of 1 indicates complete solute impermeability while a value of 0 indicates a solute permeability equal to that of water. The osmotic pressure generated by a concentration difference across a water permeable barrier is then:

$$\pi = \delta_s \, R \, T \, \Delta C_{osm} \tag{2.2}$$

Osmotic pressure is generally expressed as a positive value since it is measured as the force exerted as water moves from a more dilute to a more concentrated solution. The free energy of water can be determined from the equation (Marshall 1978),

$$\pi \, V_w = RT\ln X_a \tag{2.3}$$

where V_w is the partial molal volume of water (18 $cm^3\ mol^{-1}$) and X_a is the mole fraction of water [(moles water)/(moles water + moles solute)]. In this notation the free energy of pure water ($Xa = 1$) is taken to be zero and the addition of solute results in a progressively more negative value ($Xa < 1$) so osmotic water movement occurs down a favorable free energy gradient. Note that corrections from ideality are also required for accurate evaluation of osmotic gradients. A more rigorous description of the thermodynamics of osmotic water movement is provided by Larsen et al. (2014). The osmotic permeability coefficient P_f, (cm s^{-1}) is often used for comparisons among biological structures.

$$P_f = ((R\ T)/V_w)\ L_p \tag{2.4}$$

In this equation L_p is the hydraulic conductivity (the rate of water movement across a given area per unit of osmotic concentration gradient.

2.2.2 The Amphibian Skin and Urinary Bladder

The English naturalist Robert Townson (1795) observed living frogs (*Rana temporaria*) and tree frogs *Hyla* (then *Rana*) *arborea* to absorb water across their skin when water was made available, rapidly regaining water lost by evaporation. He also noted the storage of water in the urinary bladder as a reservoir to offset evaporation. Edwards (1824) made similar observations that frogs dehydrated by 15 % of their hydrated weight recovered 2/3 of the loss in 15 min. These and other studies on amphibian water balance during the nineteenth and twentieth centuries were reviewed by Jørgensen (1997) who noted most of the research was directed at the application of basic principles of water transport to human physiology (e.g. Matteucci and Cima 1845; Reid 1890). Comparative physiology as an independent discipline had not yet been established. Frogs were often used because they were readily available and "*tenacious of life*" (attributed to Claude Bernard). A significant advancement in understanding the regulation of osmotic water permeability was the finding by Brunn (1921) that neurohypophyseal hormones (Pituitrin) were able to stimulate water absorption by living frogs (Fig. 2.2a). The "Brunn effect" or hydroosmotic response established antidiuretic hormone as major factor in the regulation of amphibian water balance and the utility of frog skin for studying mechanisms of regulation. The availability of D_2O as an isotopic tracer for water movement was investigated by Hevesy et al. (1935) who found osmotic flow was three to five times more rapid than the theoretical value calculated from heavy water flux. They concluded: "*The large difference between the observed rate for heavy and ordinary water came as a surprise to us*" and that the discrepancy "*limits the practical applicability of heavy water or any other indicator to comparative experiments.*" These concerns were addressed by Koefoed-Johnsen and Ussing

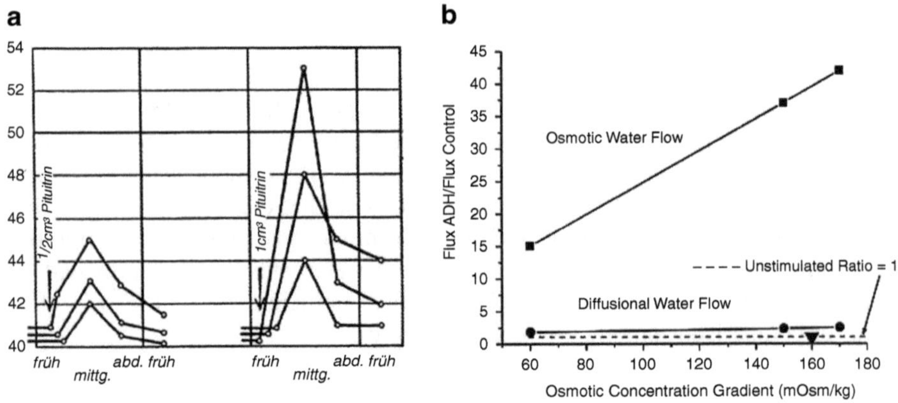

Fig. 2.2 (**a**) The increase of body mass of three frogs each injected with two doses of pituitrin (from Brunn (1921) with permission). (**b**) Osmotic water flow across isolated toad urinary bladder following treatment with antidiuretic hormone, is much greater than diffusional flow measured with tritiated water (plotted from tabular data in Hays and Leaf (1962) in Hillman et al. (2009), with permission)

(1953) who formulated a mathematical analysis of osmotic flow vs diffusion of isotopic water. They observed that the stimulation of osmotic flow by treatment of isolated skin from frogs (*Rana temporaria*) and toads (*Bufo bufo*) with neurohypophyseal extracts was increased by 100–200 % while heavy water diffusion "*...had only slightly increased.*" They concluded, "*The results are consistent with the assumption that neurohypophyseal hormones increase the pore size in some layer of the skin without increasing the total area available for diffusion.*"

The anuran urinary bladder, long known as a storage organ for dilute urine formed when water is available (Steen 1929), also became a valuable tool for studying the effects of neurohypophyseal peptides on epithelial water permeability (Bentley 1958). Similar increases in water permeability could be obtained following treatment either with the mammalian antidiuretic hormone, arginine vasopressin (AVP), or the amphibian antidiuretic hormone, arginine vasotocin (AVT). Hays and Leaf (1962) compared the increase in water flux calculated from D_2O and tritiated water with that determined from osmotic flow and, like Koefoed-Johnsen and Ussing (1953), found a large increase in osmotic flow compared with that calculated from isotopic flux (Fig. 2.2b). Further experiments with the toad bladder showed the increase in water permeability to correspond with an increase in membrane area, measured as capacitance (Stetson et al. 1982) and was inhibited by colchicine, a disruptor of microtubules (Taylor et al. 1973) both suggesting the insertion of vesicles containing putative water channels. This concept was further supported by the identification of clusters of intramembranous particles that could be seen in freeze fracture electron micrographs of toad bladders stimulated by vasopressin or vasotocin (Chevalier et al. 1974; Kachadorian et al. 1975).

2.2.3 Discovery of Aquaporins

In parallel with research on amphibian epithelial tissues, experiments with red blood cells showed a similar discrepancy between osmotic water flow and diffusion of tritiated water (Paganelli and Solomon 1957). This study cited Koefoed-Johnsen and Ussing (1953) and similarly hypothesized water transport to be mediated by pores in the membrane. It was later found that osmotic water permeability of red blood cells was inhibited by $HgCl_2$ and organomercurial compounds (Macey and Farmer 1970). Benga et al. (1986) subsequently utilized (chloromercuri) benzene sulfonate (PCMBS) with ^{130}Hg as a label to identify proteins associated with inhibition of water permeability of red blood cell ghosts. Two labeled bands were identified with gel electrophoresis that the authors concluded "*....have to be considered as playing a role in water transport.*" During this same period, studies on proteins associated with junctional complexes in the bovine lens revealed a "*main intrinsic polypeptide*", abbreviated as MIP, with a molecular weight of about 27 kD (Broekhuyse et al. 1976). Takemoto et al. (1983) were able to purify the MIP with a molecular weight of 26 kDa (called the ***major*** *intrinsic polypeptide,* MIP26) and obtain its amino acid sequence. From this sequence, Gorin et al. (1984) were able to construct an array of 20 mer oligonucleotide probes to screen a cDNA library established for bovine lens. The derived amino acid sequence obtained from the resulting cDNA revealed a protein with six transmembrane domains "*...and that it has at least one amphiphilic transmembrane segment, as expected if the protein were to participate in the formation of an aqueous channel.*" In this regard MIP26 (now called Major Intrinsic Protein) was proposed to be a candidate for a gap junction protein and was identified in the lens of many mammalian species (Kistler and Bullivant 1980).

Alternative approaches also supported the existence of water channel proteins in the kidney that are known to reabsorb a considerable amount of water from the glomerular filtrate. In one study water permeability of brush border vesicles from proximal tubules in rat kidney was evaluated by measuring changes in volume by light scattering and radiation inactivation. Because absorption of radiation is a linear function of molecular weight, the radiation dose that inactivates water permeability could be used to estimate the mass of the water conducting protein as 30 ± 5 kDa (van Hoek et al. 1991). The discovery that mRNA for membrane proteins from mammalian tissues was able to be expressed in *Xenopus* oocytes made it possible to use this assay for water channels. Zhang et al. (1990) micro injected poly A mRNA from rat and rabbit kidney into oocytes and showed a greater increase in water permeability compared with water injected oocytes.

While isolating a 32 kDa Rh protein from red blood cells, Agre and co-workers discovered a novel 28 kDa protein that was also present in rat kidney (Agre et al. 1987; Denker et al. 1988). This protein was further characterized by Smith and Agre (1991) who obtained its amino acid sequence and observed the first 35 NH_2-terminal amino acids had 37 % homology with the 26 kDa MIP described in bovine lens by Gorin et al. (1984). The 28 kDa protein was shown to occur as a tetramer. Oligonucleotide primers corresponding to the N-terminal amino acid sequence were used to screen a cDNA library from erythroid tissue and create a cDNA for the 28 kDa protein (Preston and Agre 1991). The deduced amino acid sequence was highly homologous with other members of the MIP family that had been identified in wide range of organisms (Pao et al. 1991). Because these proteins were collectively proposed to form membrane channels, the term channel-like integral membrane protein of 28 kDa (CHIP28) was applied to this protein. When in vitro transcribed CHIP28 RNA was injected into *Xenopus* oocytes the water permeability was increased from 27.9×10^{-4} cm s^{-1} to 210×10^{-4} cm s^{-1} and inhibited by exposure to $HgCl_2$ (Preston et al. 1992). This was further demonstrated by the incorporation of highly purified CHIP28 protein into proteoliposomes and showing similar increase in water permeability (Zeidel et al. 1994). Thus, a specific protein was identified as a water channel and Peter Agre was awarded the Nobel Prize in chemistry in 2003, a century after the pioneering work of van't Hoff.

2.2.4 Diversity of MIP Proteins

The MIP family of integral membrane transport proteins examined by Pao et al. (1991) included a neurogenic protein in *Drosophila* (Big Brain, Rao et al. 1990), plant tissues including the cell membrane and that of the tonoplast (PIP and TIP, Johnson et al. 1990) and in the inner membrane of a bacterial cell (*Escherichia coli*, glycerol facilitator protein, GlpF) where it formed a pore to facilitate glycerol uptake (Miramatsu and Mizuno 1989). Alignment of the consensus amino acid sequences for these proteins revealed six transmembrane helical domains connected by extracellular and intracellular loops A–E. Domains 1–3 (segment 1) resembled domains 4–6 (segment 2) but their orientation in the membrane (inward vs outward facing) was reversed. Comparison of segments 1 and 2 with corresponding segments 1 and 2 of several species showed a 32 % average identity whereas comparison of segment 1 with segment 2 of these species showed a 23 % average identity. It was hypothesized that MIP proteins arose by internal gene duplication of a three transmembrane protein in an ancestral organism. Preston and Agre (1991) found similar sequence similarities and membrane orientation for CHIP28. Further examination of CHIP 28 revealed four intron sequences with intron-exon boundaries that were identical with human MIP26 despite there being only a 44 % overall sequence identity between the two and different sizes of introns (Moon et al. 1993). The gene for human CHIP28 was localized to chromosome 7p14 and found to exist as a single copy.

Research during the following decade revealed a plethora of MIP-related proteins in virtually all living organisms. Given the similarities among many members of the MIP family, Agre et al. (1993) suggested, "*We believe that other sequence-related water pores will be identified and feel that the functional name, aquaporin, will be useful for communication among scientists...*". With this designation CHIP28 became Aqp1, MIP26 became Aqp0 and an aquaporin that had been recently described in the mammalian collecting duct (WCH-CD, Fushimi et al. 1993) became Aqp2. In prokaryotic cells, Calamita et al. (1995) characterized a water channel in *E. coli* that was termed AqpZ. Phylogenetic comparisons of AqpZ and GlpF indicated an ancient gene divergence. At about the same time, Ishibashi et al. (1994) cloned an aquaporin (Aqp3) from rat kidney that was also permeable to glycerol and urea in addition to water (see also Yamaguchi et al. 1994). Thus, the proposed internal gene duplication that gave rise to an ancestral MIP family and the divergence of water vs glycerol conducting isoforms occurred very early in the evolution of cellular life and persists in extant species. In this context, aquaporins that are also permeable to glycerol and other small organic molecules are termed aquaglyceroporins although the term aquaporin is often applied to both.

The characteristic feature of Aqps and aquaglyceroporins is the presence of corresponding asparagine-proline-alanine (NPA) sequences in intracellular loop B and extracellular loop E (Fig. 2.3a). Comparing osmotic water permeability of CHIP28 (Aqp1) with site-directed mutations in the B and E loops Jung et al. (1994) proposed an "hourglass model" in which loops B and E fold into the membrane in such a way that the NPA sequences align to form a single narrow aqueous pore surrounded by the six transmembrane helices (Fig. 2.3b). Resolution of Aqp1 structure was made at 3.8 Å resolution with electron crystallographic data (Murata et al. 2000) and 2.2 Å resolution with X-ray diffraction data (Sui et al. 2001). Further insight into the structure and function of Aqps was made with molecular dynamic simulation (Fujiyoshi et al. 2002). These studies revealed that loops B and E contain α-helical domains each extending half way across the membrane to form a seventh "broken" half membrane helix with the NPA motifs at the center of the channel (reviewed by Törnroth-Horsefield et al. 2010). The positive dipole of the complimentary helices aligns with that of water molecules as they pass by the NPA motifs and restricts proton movement. A selectivity filter for water permeation is provided by a constriction at the extracellular face of the channel containing an "aromatic arginine" motif in which an arginine residue from the E loop interacts with aromatic groups of phenylalanine and histidine residues in transmembrane helices (Fig. 2.3c). Aquaglyceroporins have a different selectivity filter in which arginine is paired opposite phenylalanine and tryptophan (Wang et al. 2005; Wang and Tajkhorshid 2007). A variety of other amino acid residues have been implicated in the regulation of permeation, selectivity and gating of both Aqps and aquaglyceroporins as has the observation that Aqps invariably occur as homotetramers (Wang and Tajkhorshid 2007; Törnroth-Horsefield et al. 2010).

2.2.5 Phylogenetic Considerations

Evolutionary relationships are often expressed as phylogenetic trees produced by computer based analyses of amino acid sequence identities. Figure 2.4 (from Calamita et al. 1995) provides an example of phylogenetic relationships among Aqps that were known at that time. It can be seen that Aqp1 has 37 % sequence

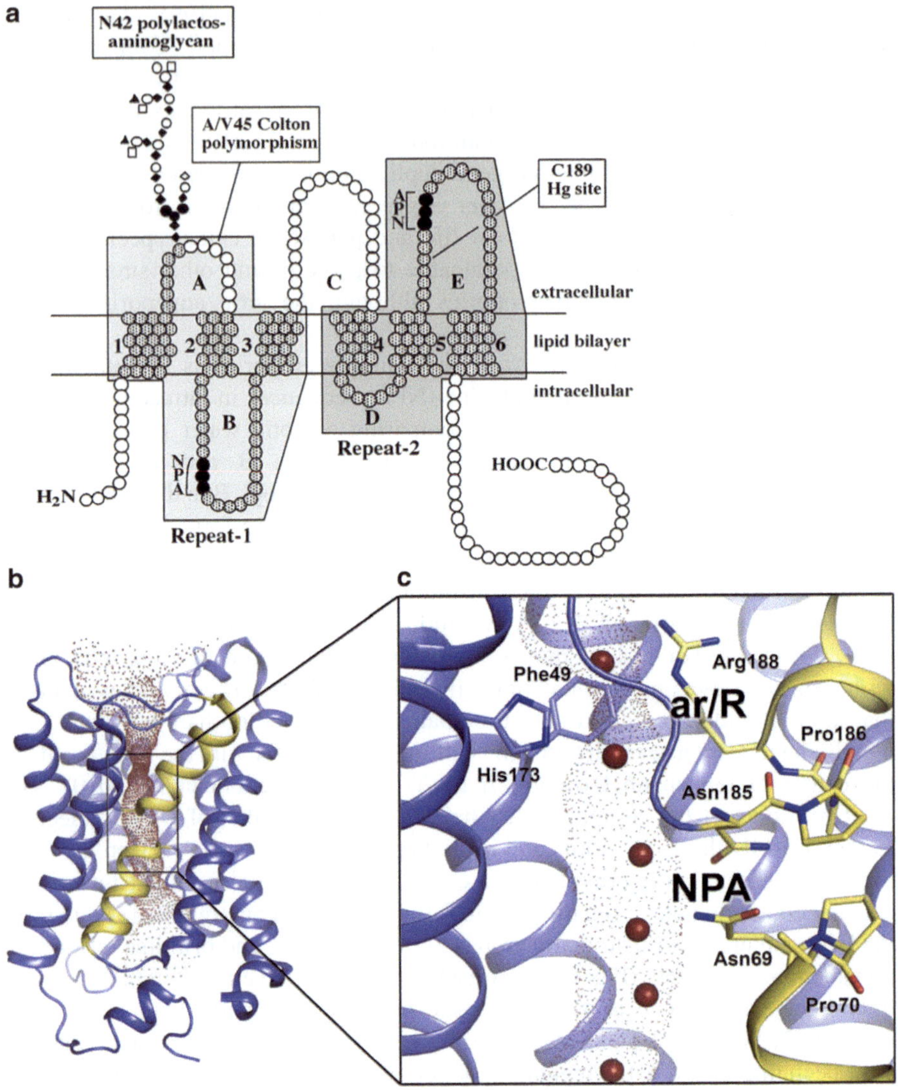

Fig. 2.3 (**a**) The structure of Aqp1 illustrating the two repeats that are inversely oriented in the cell membrane and the NPA sequences of loops B and E that characterize aquaporins and aquaglyceroporins (From Heymann et al. 1998 with permission). Note also a cysteine residue that confers sensitivity to mercury. (**b**) The membrane orientation of the six transmembrane segments and the formation of a water conducting channel by the infolding of loops B and E with helical domains that restrict ion permeation. (**c**) The juxtaposition of arginine with phenylalanine and histidine constitute a size filter for water passing entering the channel (**b** and **c** from Törnroth-Horsefield et al. 2010, with permission)

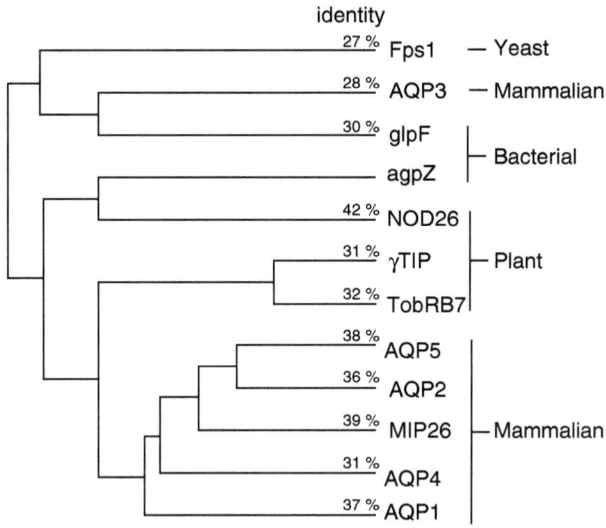

Fig. 2.4 Phylogenetic analysis of AqpZ with other aquaporins and homologous MIPs whose amino acid sequences were known at the time. The numbers refer to the percent sequence identity with AqpZ. Fps1 and Aqp3 are aquaglyceroporins, Nod 26 is a root nodule Aqp, γTIP and TobRB7 are tonoplast Aqps, MIP 26 was the contemporary term for Aqp0 (Calamita et al. 1995)

identity with AqpZ. However, water permeability and selectivity in oocytes is increased to almost 200×10^{-4} cm s^{-1} when injected with AqpZ cRNA, i.e. both homologs are able to stimulate water permeability by a similar amount despite their long established divergence from one another. In contrast, AqpZ lacks a Hg^{2+} sensitive cysteine residue while water permeability resulting from the expression of Aqp1 is inhibited by Hg^{2+} (Yang et al. 1996). It is evident that many functional properties of Aqps and aquaglyceroporins are conserved despite the differences in amino acid sequences used to establish phylogenies and that some, but not all amino acid substitutions result in functional differences. In an examination of the NCBI data base, Zardoya (2005) identified over 450 non-redundant amino acid sequences for members of the MIP family in eubacteria, archaea and eucaryia. Computer programs were used to align amino acid sequences and construct phylogenetic trees. The diversity of Aqps in plants and animals is the result of gene duplication and cooptation for different physiological functions in different tissues. This can be seen in Table 2.1 showing Aqps in mammals, plants, yeast and a bacterium [compiled from Hillyard (2012)]. *E. coli* has a single Aqp and aquaglyceroporin, yeast (*Saccharomyces cervevisiae*) has two copies of each while the model plant (*Arabidopsis thaliana*) has 35 integral membrane proteins including those expressed in the plasma membrane (PIPs), tonoplast (TIP) and root nodules (NIPs). Thirteen mammalian Aqps have been described that include three classes: class 1 are water selective Aqps (Aqp 0, 1, 2, 4, 5, 6, 8), class 2 are

Table 2.1 Aquaporin diversity increases with the evolution of biological complexity

E. Coli	Yeast	Mammals	Plants
AqpZ	Aqy1	Aqp0	PIPs (Plasma membrane intrinsic proteins)
GlpF*	Aqy2	Aqp1	TIPs (Tonoplast intrinsic proteins)***
	GLPY1*	Aqp2	NIPs (Nodulin-like intrinsic proteins)***
	GLPY2*	Aqp3*	
		Aqp4	
		Aqp5	
		Aqp6	
		Aqp7*	
		Aqp8	
		Aqp9*	
		Aqp10*	
		Aqp11**	
		Aqp12**	

Escherichia coli expresses a single aquaporin and aquaglyceroporin * while brewers yeast, *Saccharomyces cerevisiae*, expresses two copies of each. Mammals express 13 aquaporins that includes 4 aquaglyceroporins* and 2 unorthodox or super aquaporins with non-NPA motifs (NPC in Aqp11 and NPT in Aqp12)**. Plants express aquaporins in various tissues including the plasma membrane, tonoplast and root nodules. Some of the TIPs and NIPs are permeable to glycerol***. A total of 35 aquaporins have been described in the model plant, *Arabadopsis thaliana* [Data from Calamita et al. (1995), Ishibashi et al. (2009), Maurel et al. (2008)]

aquaglyceroporins (Aqp 3, 7, 9, 10), class 3 are unorthodox or "superaquaporins" (Aqp 11, 12) characterized by variations in the amino acid sequences of the characteristic NPA sequence in the transmembrane pore.

The diversity of Aqps and aquaglyceroporins can arise from tandem gene duplication or whole genome duplication. Terminology for these patterns of gene duplication is important for interpreting these patterns of Aqp evolution. Differences in sequence identities between homologous Aqps that resulted from a speciation event are termed orthologs while differences that result from gene duplication are termed paralogs. In general (but not always), paralogs assume a different function or become expressed in a different tissue than the parent form. For example, microbial AQPZ and GlpF can be seen as paralogs. Now that genomic sequences are available, it is possible to determine the position of paralogous genes in the same chromosomal regions, a condition termed synteny and such regions are termed paralogons. Patterns of gene duplication and synteny seen in aquaporins provide examples for the evolution of osmoregulatory function at the cellular and organismal level.

2.3 Comparative and Evolutionary Relationships

2.3.1 *Mammals*

Although vertebrate evolution began long before the appearance of mammals the use of the term aquaporin arose in the context of mammalian aquaporins and the numerical assignments of mammalian aquaporins are commonly used for comparisons with other vertebrates (Crane and Goldstein 2007). For this reason it is useful to begin with discussion of the mammalian aquaporins and compare their structure and function with other vertebrates in an evolutionary context. The following description of mammalian aquaporins (with chromosomal locus) is compiled from Ishibashi et al. (2009). This review also notes that the above three classes have characteristic intron-exon boundaries: class 1 are aquaporins with four exons, class 2 are aquaglyceroporins with six exons and class 3 are the unorthodox aquaporins with three exons. Primary literature references from this review are selected to provide additional information.

Aqp0 (12q13) was the first MIP described in the lens in the context of junctions that adhere cells closely together to maintain organization of the cellular structure of the lens (Gorin et al. 1984). Humans with defective Aqp0 and Aqp0 knock out (KO) mice both develop cataracts. These junctions were shown to be alignments of Aqp0 homotetramers of adjacent cells forming an octomer (Gonen et al. 2004). X-ray diffraction studies indicated proteolytic cleavage of a C-terminal peptide segment that made the junctional form impermeable to water while the non-junctional homotetramer expressed in oocytes showed water permeability only one tenth of Aqp1. More recently, Jensen et al. (2008) used molecular dynamic simulation of Aqp0 function in the tetrameric and junctional configuration and found both to have similar low water permeabilities. The low permeability of the junctional form was suggested to allow water movement within cells of the lens as it changes shape in the course of focusing images on the retina. Aqp0 is found in the lens of other vertebrates and, importantly, exists as paralogs that are implicated in the evolution of Aqps 2, 5 and 6 that are syntenic in the amphibian and mammalian genomes.

Aqp1 (7p14) was discovered in red blood cells and is the most characterized of the Aqps. It is widespread in mammalian tissues including the endothelium of non-fenestrated capillaries on the serosal side of epithelia where it may facilitate water absorption and secretion. Humans with defective Aqp1 and Aqp1-KO mice show little or no loss of function for most of the potential tissues where AQP1 has been described although pulmonary edema can be induced in some individuals (Agre 2006). An exception is in the kidney where AQP1 is expressed in the apical and basolateral membrane of epithelial cells of the proximal tubule and descending loop of Henle. Aqp1-null humans and Aqp-1-KO mice both show concentration defects that are not life threatening when water is available. Aqp1 has also been implicated in angiogenesis associated with tumor growth. Melanoma cells implanted in Aqp1-KO mice show less growth than those implanted in wild type mice. Further, cultured proximal tubule cells from Aqp1-KO mice show a reduced capacity for migration in vitro and a reduced capacity for healing of ischemia-reperfusion injury *in vivo*. It was suggested that cell movements associated with wound healing and morphogenesis requires osmotic water movement in conjunction with changes in the actin cytoskeleton and ionic transport along the leading edge of lammelipodia (reviewed by Belge and Devuyst 2006).

Aqp2 (12q13) was initially described as WCH-CD in collecting duct principal cells (Fushimi et al. 1993). It is translocated into the apical membrane by exocytosis from a subapical pool of vesicles during vasopressin stimulation to increase water reabsorption and urine concentration. Activation of V_2 receptors promotes phosphorylation of serine 256 by way of an adenylate cyclase-cyclic AMP-protein kinase A pathway (Fushimi et al. 1997). Aqp2-null humans display diabetes insipidus that is lethal in Aqp2-KO mice. Although a common model for ADH stimulation is from a subapical pool, Aqp2 also occurs in the basolateral membrane of collecting duct cells (reviewed by Brown 2003). Various studies with different species and cell cultures have implicated increased basolateral expression with ADH stimulation and hyperosmolality (van Balkom et al. 2003; Hasler 2009). Aqp2 is also an integrin binding protein (Chen et al. 2012). These authors note that Aqp2-KO mice show abnormalities in basolateral integrin binding and abnormalities in tubular structure. They propose Aqp2 expressed in the basolateral membrane plays a role in cell migration and epithelial morphogenesis.

Aqp3 (9p13) is an aquaglyceroporin that was cloned from rat kidney and is expressed in the basolateral membrane of principal cells in the medullary collecting duct (Echevarria et al. 1994). Permeability of Aqp3 to water and small organic molecules was inhibited by the organomercurial compound PCMBS. Ecelbarger et al. (1995) subsequently showed Aqp3 to be localized in the basolateral membrane of cells in the cortical and outer medullary collecting duct with greatest expression in the inner medulla. Water deprivation for 48 h doubled the expression of Aqp3 in the inner medulla and appears to serve an important role in transepithelial water transport in conjunction with Aqp2 in the apical membrane. Aqp3 is also present in basolateral membrane of secretory epithelia including airway, salivary glands, lacrimal glands and sweat glands. In this capacity water from the serosal fluid enters the cells and is expelled via Aqp5 located in the apical membrane, generally in response to an osmotic gradient generated by serosa-mucosa ion transport. Another phenotype of Aqp3-KO mice is dry skin due to the role of Aqp3 in transporting water and glycerol to the basal layer of keratinocytes (Hara-Chikuma and Verkman 2008). Aqp3-KO mice also show delayed proliferation of enterocytes of the colon following an episode of colitis (Thiagarajah et al. 2007).

Aqp4 (18q22) was first described as a mercurial insensitive water channel (MIWC) that was co-localized with Aqp3 (then called glycerol intrinsic protein, GLIP) in the basolateral membrane of several epithelial tissues, including principal cells of the collecting duct (Frigeri et al. 1995). Aqp4-KO mice show a mild urinary concentration defect while the review of Ishibashi et al. (2009) report that no phenotype has been reported for Aqp4-null humans. Aqp4 is expressed to a great degree in glial and ependymal cells of the brain and has opposing roles in the formation or reduction of edema depending on whether it results from cytotoxic or vasogenic sources (Papadopoulos and Verkman 2007). Aqp4 also affects cell migration and neural excitability with reduced hearing, vision and olfaction observed in Aqp4-KO mice.

Aqp5 (12q13) is expressed in the apical membrane of many exocrine glands where it serves to mediate water movement associated with glandular secretion and often functions with AQP3 in the basolateral membrane. Aqp5-KO mice show reduced sweating (mice have sweat glands in their feet) and salivary secretions. Stimulation of muscarinic (M_3) receptors causes Aqp5 to become translocated from intracellular lipid rafts into the apical membrane of cells in the interlobular ducts of rat parotid gland (Ishikawa et al. 2005). Aqp5-KO mice also show a lower level of paracellular transport than males indicating a possible connection with tight junction proteins (Kawedia et al. 2007). This was more prominent in female mice but a reason for this difference was not obvious. In the lung, Aqp5 mediates fluid secretion into the alveolar air space by type 1 alveolar pneumocytes (Nielsen et al. 1997).

Aqp6 (12q13) is of interest because it is activated by mercury. It has been identified in a number of tissues including the kidney where it is localized in the endosome of type A intercalated cells of the collecting duct. Aqp6-KO mice appear normal and no phenotypes are noted by Ishibashi et al. (2009) in humans.

Aqp7 (9p13) is an aquaglyceroporin described initially in testis and sperm although Aqp7-KO mice have normal reproductive function. It is localized in brush border epithelia of proximal tubules and Aqp7-KO mice show an increase in urinary glycerol loss. Aqp7-KO mice and Aqp7-null humans also show a reduced capacity for glycerol mobilization from adipose tissue. Dumas et al. (2007) observed humans are unique among primates in that there are five copies of Aqp7 gene in the human genome that resulted from segmental gene duplication. These authors suggest this is related to the large capacity for endurance running relative to other primates.

Aqp8 (16p12), initially described in testis and pancreatic acinar cells, is unique among class 1 aquaporins in having six exons and a sequence identity that is similar to that of tonoplast integral proteins (TIPs) of plants. Aqp8 has been identified in numerous tissues including the inner mitochondrial membrane of liver hepatocytes (Calamita et al. 2005). Aqp8 is also permeable to urea but its role in urea transport may be considered to be redundant because more specific urea transport proteins are expressed in the liver and the role of Aqp8 is not clear.

Aqp9 (15q22) is an aquaglyceroporin initially described in leucocytes and also in liver. Aqp9-KO mice have a slightly reduced capacity for glycerol uptake but the lack of Aqp9 may be compensated by other aquaglyceroporins (e.g. Aqp7). Aqp9 mRNA levels are increased in osteoclasts but no bone defects are seen in Aqp9-KO mice.

Aqp10 (1q21) is a pseudogene in mice. In humans it is an aquaglyceroporin that is expressed in a variety of tissues however its physiological function remains poorly understood.

Aqp11 (11q 13) KO mice suffer fatal polycystic kidney disease with renal cysts forming from proximal tubule cells beginning shortly after birth. Vacuoles appear to have originated from the endoplasmic reticulum and it is suggested Aqp11 serves a role in intracellular water transport.

Aqp 12 (2q37.2) in the human genome exists as two genes denoted as Aqp12A and Aqp12B that appear to have resulted from local gene duplication that is not apparent in the mouse genome. Although expressed in the pancreas its physiological role remains to be identified.

2.3.2 *Ray Finned Fish (Actinopterygii)*

The presence of a diverse genome among the vertebrates has been ascribed to two rounds of whole genome duplication (ca 600 mya), prior to that seen in the teleosts, at the very base of vertebrate evolution (Nakatain et al. 2007; Vandepoele et al. 2004). The increase in genetic diversity, in particular Hox genes and related anatomical complexity, has been ascribed to whole genome duplication (Ohno 1970). An increase in Aqp paralogs from genomic duplication and in orthologs from evolutionary changes in homologous Aqps could account for their varied functions in tissues of the diverse array of vertebrate species. Figure 2.5 provides an approximate time line for vertebrate evolution based largely on the fossil record. What stands out is the divergence of the tetrapod lineage from bony fishes in the upper Devonian, ca 350–400 million years ago (mya). Among the teleosts,

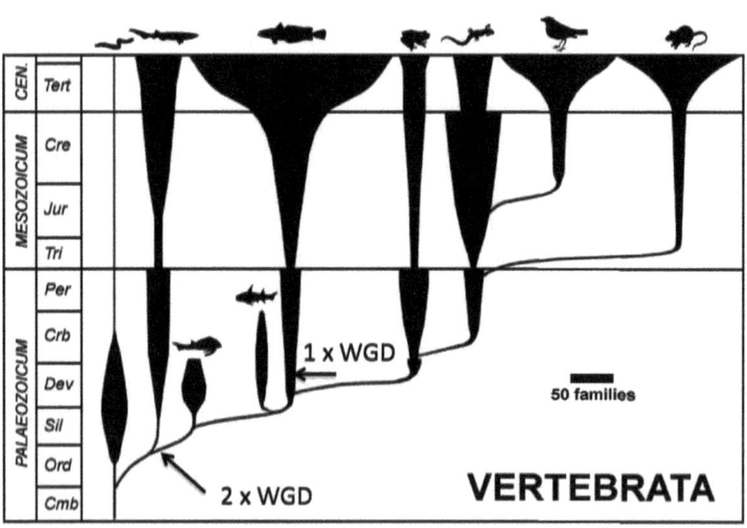

Fig. 2.5 A diagram of vertebrate evolution illustrating the time course for the divergence of the major vertebrate clades (Benton 1998). *Arrows* indicate two proposed gene duplications at the base of the vertebrate lineage and the single gene duplication at the base of the teleost crown group

examination of aquaporin transcripts from the zebrafish (*Danio rerio*) genome revealed 18 sequences related to aquaporins, aquaglyceroporins and unorthodox aquaporins that are also present in the mammalian genome, with the exception of Aqps 2, 5, and 6 (Tingaud-Sequeira et al. 2010). Of these, one is a pseudogene with one exon having similarity to the tetrapod Aqps 5 and 2 and another to tetrapod Aqp1. The deduced amino acid sequences revealed paralogs for many of the Aqps. Thus, the genes that express most of the mammalian Aqps were present in the vertebrate genome long before the divergence of land vertebrates and persist as multiple paralogs in extant fishes. Three of the duplicated Aqp genes have unlinked genetic loci, two are linked and a third, DrAqp8, is expressed in triplicate with isoforms encoded on three different linkage groups. The mechanism for multiple gene copies spread among different linkage groups has been ascribed to an additional round of whole genome duplication (WGD) at the base of the teleost crown group ca 320 mya (Vandepoele et al. 2004) with subsequent "*fusion and and/or loss of gene segments*" (reviewed by Cerdà and Finn 2010). In some cases, Aqp genes remain syntenic suggesting tandem gene duplication.

Functional expression of the different paralogs in *Xenopus* oocytes usually showed similar increases in water and glycerol permeability as seen in mammalian orthologs with the exception of DrAqp8. DrAqp8aa has a particularly high permeability to water and urea while the DrAqp8 ab and Dr.Aqp8bb have much lower permeabilities to urea and water (Fig. 2.6).

From a physiological perspective, osmotic regulation at the whole organism level requires regulated water permeability in the gills, gut and kidney. Fish in fresh water face osmodilution, and absorb salts across their gills. They drink little and form large volumes of dilute urine while fish in salt water drink copiously and absorb water across the gut while the gills eliminate excess salt and the kidneys form a small volume of isoosmotic urine (Evans and Claiborne 2009). The following is a summary of the putative roles for Aqps in these tissues taken from a recent review by Cerdà and Finn (2010).

Gills The gills of various teleost species that have been examined express Aqp3a, Aqp3b or both at levels that are generally lower in salt water adapted fish. They have been localized in the apical and basolateral membrane of chloride cells in the European eel but only the basolateral membrane of chloride cells of other species. It has been proposed that these paralogs play a role in osmoreception of chloride cells but "*the physiological roles for Aqp3a or -3b in the gill epithelium of teleosts remains speculative*". Branchial expression of Aqp1a and -b paralogs has also been reported in teleost gills. Aqp1a has been localized in epithelial cells of gill filaments of the gilthead seabream (*Sparus auratus*) and suggested to function with Aqp3 paralogs in promoting transepithelial water transport. However, no consistent expression of these paralogs has been reported for the European eel. Aqp4 exists as a single gene in fishes and has been localized in the basolateral membranes of gill pavement cells that are involved with mucous secretion. Freeze fracture electron microscopy reveals orthogonal arrays of particles that suggest Aqp4 may also function in cell adhesion and migration. Finally, various studies have detected

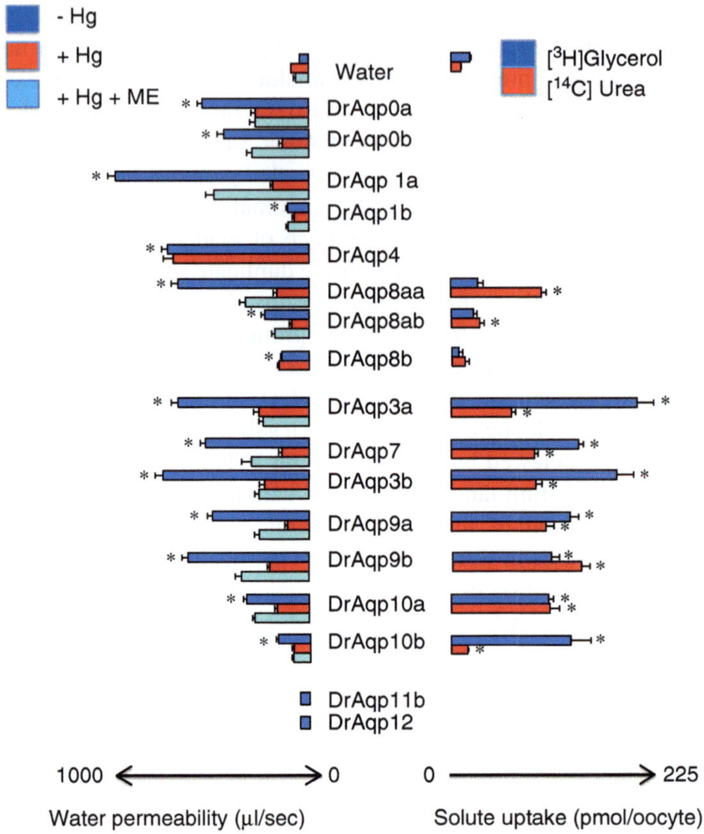

Fig. 2.6 Zebrafish express multiple paralogs of mammalian Aqps except for Aqps 2, 5 and 6. Aqp4 is expressed a single copy. When expressed in *Xenopus* oocytes, the permeabilities to water, glycerol and urea are similar to the mammalian orthologs except for DrAqp8ab and b (Tingaud-Sequeira et al. 2010)

mRNAs for virtually all of the other known piscene Aqps in the gills but their functional significance remains to be determined.

Intestine Marine and euryhaline fish in saline environments drink large amounts of water. Monovalent salts are removed by ion transport in the esophagus and along the gastrointestinal tract with water following osmotically. Transcripts for Aqp1a, 1b and Aqp3b are upregulated in the freshwater eels treated with the steroid hormone cortisol, which is known to be elevated in salt water-adapted fish. However, the esophagus is primarily a site of ion transport with low water permeability so the role of these Aqps in this region is not clear. Aqp1a and 1b are variably expressed by different species along segments of the intestine, the pyloric ceca and the rectum. This is consistent with earlier observations that isolated intestine from cortisol-treatment of freshwater eels resulted in an increase in intestinal water permeability. Cerdà and Finn (2010) caution that the presence of Aqps in the basolateral membrane of enterocytes remains to be shown so the role of Aqps in coordination of epithelial transport in these tissues is unresolved.

Kidney Although mRNAs for different paralogs of Aqp 1, 3 and 10 have been reported from the kidneys of various fish species, their cellular localization is not well described. The European eel expresses Aqp1a and -3b in the apical membrane of some renal tubules. Aqp1a is seen in the endothelia and more proximal segments while Aqp3b is seen in the more distal tubular segments. During salt water acclimation Aqp1a is upregulated while Aqp3b is downregulated. These and other observations stimulate hypotheses regarding the roles of Aqps in renal function but "*...definitive conclusions remain elusive.*" (Cerdà and Finn 2010).

Reproduction Radiation of teleosts into the oceans required the capacity of eggs to survive in hyperosmotic sea water and, for pelagic species, for eggs to be buoyant so they can disperse. This is accomplished by rapid hydration during oocyte maturation. Prior to ovulation oocytes of the sea bream (*Sparus aurata*) were found to express an Aqp1-like Aqp that was translocated to the membrane as hydration occurred during egg maturation (Fabra et al. 2005). This Aqp was termed SaAqp1o with a sequence identity of 45–54 % relative to other vertebrate orthologs while a second SaAqp1 was identified that was closer to that of the mammalian Aqp1. Further characterization resulted in reclassification of these paralogs as SaAqp1a (mammalian type) and SaAqp1b that is proposed to have resulted from gene duplication after the divergence of teleosts from the tetrapod lineage (Tingaud-Sequiera et al. 2010).

2.3.3 Chondrichthys and Primitive Chordates

Sharks and rays are descendents of a more primitive piscine ancestor and provide intriguing clues about Aqp evolution in vertebrates. The first Aqp cloned from an elasmobranch was Aqp1e from the bullshark (*Charcharhinus leucas*) that had similar amino acid sequence identities with mammalian Aqps 1, 2 and 5 (Cutler et al. 2005). A similar Aqp was cloned from a more primitive chordate, the hagfish (*Myxine glutinosa*). This was suggested to be a putative gene that duplicated to give rise to the Aqps expressed in tetrapods and may be represented as introns of the pseudogene described in the teleost genome. With improved PCR techniques Cutler (2007) found two Aqp1 paralogs in *S. acanthius*, one with a sequence identity closer to the mammalian Aqp1 and another more similar to Aqp1e. This study also identified orthologs of the human Aqp4 (78.8 % sequence identity) and the aquaglyceroprotein Aqp3 (69.9 % sequence homology). More recently, Aqp4 in *S. acanthius* was shown to be expressed in many tissues including the gills and rectal gland (Cutler et al. 2012). While acclimation to 75 % vs. 120 % seawater resulted in no difference in Aqp4 mRNA levels in the rectal gland, a significant reduction was seen in the gills. As with other fish Aqps, additional research is needed to describe their contribution to regulatory mechanisms for water transport across osmoregulatory epithelia.

2.3.4 Lobe Finned Fish (Sarcopterygii)

The evolution of terrestriality in vertebrates began with the divergence of the sarcopterigii (lobe finned) from the actinopterigii (ray finned) with the oldest sarcopterigian fossils dating from the upper Silurian, ca 420 mya. The sarcopterigii gave rise to the earliest tetrapods such as panderichthys and ichthyostega that appear as fossils in fresh water deposits in the upper Devonian ca 375 mya (Carroll 2001) and ultimately to extant amphibians and amniotes (Fig. 2.7a). That the mammalian and teleost genomes both express orthologs of Aqp 0, 1, 3, 4, 8, 10, 11 and 12 indicates these isoforms were present at the time of this divergence. With the invasion of land, new strategies for osmoregulation evolved, including the regulation of epithelial water permeability by antidiuretic hormone. This can be seen in lungfish (subclass Dipnoi) that are primitive survivors of the sarcopterigii and the closest living relatives of extant land vertebrates. Lungfish species are found in fresh water habitats in Africa, South America and Australia. They burrow in mud when water sources dry up and endure dry seasons with reduced metabolic rate. Konno et al. (2010) cloned an aquaporin (Aqp0p) from the kidney of an aestivating African species (*Protopterus annectans*) that had sequence identity similar to tetrapod Aqp0 (75–78 %), Aqp2 (74–78 %) Aqp5 (74–77 % and Aqp6 (68–70 %). An Aqp cloned from the eye had sequence identity that was 84–89 % of tetrapod Aqp0 and was a distinct ortholog of the lens Aqp0. It was further demonstrated that Aqp0p in aestivating lungfish was localized in the apical membrane of the

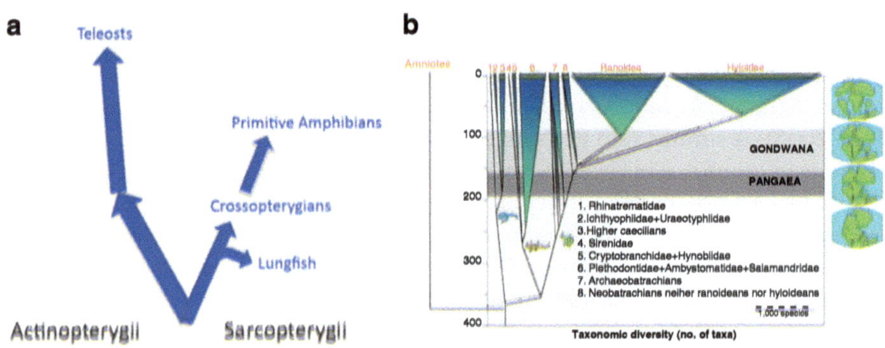

Fig. 2.7 (**a**) Divergence of actinopterigian and sarcopterigian and the emergence of tetrapods (from Romer 1962). Note the position of lungfish and the divergence of stem amphibians. (**b**) Phylogenetic tree for extant amphibians showing the estimated time course for evolution of the three orders relative to the breakup of the Pangea supercontinent. The estimated times for the divergence of anurans and urodeles (?) and amniotes (??) are rough estimates but still predate the breakup of the supercontinent (San Mauro et al. 2005)

late distal tubule and co-localized with a vasopressin type 2 (V_2) receptor in the basolateral membrane. Protein and mRNA levels for Aqp0p were reduced when animals were kept in water. The consensus peptide sequence identified a putative site (ser 263) for phosphorylation by protein kinase A and when expressed in *Xenopus* oocytes an increase in water permeability was further stimulated by cyclic AMP. Taken together, these results suggest Aqp0p to be ancestral to the Aqp that regulates reabsorption by the kidneys of terrestrial vertebrates. Duplication of Aqp0 is consistent with the observation that Aqps 0 2, 5 and 6 are syntenic in the mammalian genome. Given that amniotes diverged from amphibians in the upper carboniferous, the gene duplication that gave rise to these Aqps must have occurred in a primitive amphibian prior to this divergence. The similarity between Aqp0p and the teleost pseudogene suggests the existence of a putative Aqp0p—like Aqp that was functionally lost in teleosts but retained in species that exploited terrestrial habitats.

2.3.5 Amphibians

While amphibians have been used as an example for the adaptations required for the initial colonization of terrestrial habitats, there is a large gap in the fossil record between the earliest amphibious tetrapods and the radiation of amphibians in the Carboniferous and Permian (Carroll 2001). There is another gap between the Permian species and fossil evidence of extant species in the Mesozoic so the relationships between the physiological adaptations of the earliest tetrapods and living species is speculative. None-the-less, the expression of Aqps in living amphibians provides insight into strategies that may have been retained from these earliest species. The three orders of living amphibians, anura, caudata and gymnophiona all have species that occupy a range of habitats that range from highly aquatic to purely terrestrial (Hillman et al. 2009). A phylogeny for the three orders has been proposed by San Mauro et al. (2005) based on a combination of nuclear and mitochondrial gene sequences (Fig. 2.7b). Note the earlier divergence of the gymnophiona relative to that of the anura and caudata. The figure also shows the juxtaposition of the continental land masses of the Pangea supercontinent at the time of these divergences. The taxonomic and physiological similarities of amphibians worldwide was likely established prior to the breakup of Pangea and the subsequent tectonic events that resulted in the current distribution of amphibians in present day continents.

Despite their very different body plans, all three orders are often viewed as monophyletic and are unique among vertebrates in that they utilize their skin as a primary surface for osmotic water absorption (reviewed by Hillman et al. 2009 and Hillyard et al. 2009). Because of their greater abundance and availability anuran species are most commonly used for experiments. When water is available they are able to rehydrate quickly and form dilute urine that can be stored in a urinary bladder, which is a diverticulum off of the cloaca. When they venture on land or burrow during dry periods stored bladder water is reabsorbed to maintain osmotic balance. As noted earlier amphibian skin and bladder played an important role in the understanding of epithelial water permeability and its regulation by antidiuretic hormone. Given the similarity with the response seen in the mammalian kidney, an Aqp2-like protein would be expected.

Examination of skin from the Japanese tree frog (*Hyla japonica*) revealed two Aqps that became inserted into the apical membrane of the outer most living epithelial cell layer of dehydrated animals or when the tissue was treated with AVT (Tanii et al. 2002; Hasegawa et al. 2003). These homologs, labeled Aqp-h3 and Aqp-h2, had serine residues that were protein kinase A phosphorylation sites but were sufficiently different from the mammalian Aqp2 that they were termed anuran specific "Aqpa2" paralogs. They are suggested to have resulted from duplication of an Aqp2 precursor that remained a single gene in amniotes. Similar Aqp2as were expressed in the skin of *Bufo* species while only an Aqp-h3-like Aqp was expressed in the skin of *Rana* species. This was seen as an adaptation for the more terrestrial habitats occupied by *Hyla* and *Bufo* species that also have larger regions specialized for water absorption, extending from the ventral surface of the skin to the lower limb (Suzuki et al. 2007; Suzuki and Tanaka 2009; Ogushi et al. 2010a). The apical insertion of these Aqps and the associated increase in water permeability was also stimulated by beta-adrenergic agonists in conjunction with an increase in vascular perfusion of the skin (Ogushi et al. 2010b). This more closely resembles the response of dehydrated animals in which vascular perfusion is more prominent than observed in AVT-injected animals (Viborg and Rosenkilde 2004). Thus, the sympathetic nervous system in addition to neurohypophyseal hormones regulates water absorption by terrestrial species. As with Aqp2 in the mammalian nephron, Aqp2a paralogs are frequently observed in the basolateral membrane of the outermost living cell layer of the anuran skin in addition to the apical membrane (Ogushi et al. 2010a; Shibata et al. 2011). Anuran skin undergoes a regular cycle of cell division in the basal cell layer and constant replacement of the outermost living cell layer as cells cornify and are shed. It is possible that basolateral expression of Aqp2as is involved in cell adhesion and migration as has been suggested for Aqp2 in the mammalian kidney (Chen et al. 2012).

Surprisingly, mRNA for an Aqp-h3 homolog (Aqp-x3) was identified in the skin of the aquatic frog, *Xenopus laevis,* but was not expressed at the protein level because additional nucleotides at the C-terminal tail prevented its translation (Ogushi et al. 2010a). In its native habitat, *Xenopus* species encounter dry seasons when they seek moist substrates to avoid desiccation. At this time cutaneous water absorption could be important for survival. Regulation at the translational or post-translational level according to seasonal or husbandry conditions need to be considered in the interpretation of aquatic vs terrestrial adaptations of the amphibians.

Aqp-h2 homologs were found in the urinary bladder of *Hyla*, *Bufo*, *Rana* and even *Xenopus* species. Syntenic analysis of the genome of *X. tropicalis*, recently sequenced by Hellsten et al. (2010) showed them to be localized between the same genes (Fas apoptotic inhibitory molecule 2, FAIM2 and Rac GTPase-activating protein 1, RACGAP1) that flank Aqps2 and 5 on chromosome 12 of the human genome. Suzuki and Tanaka (2009) suggest a sequence of events whereby Aqp0 in an ancestral

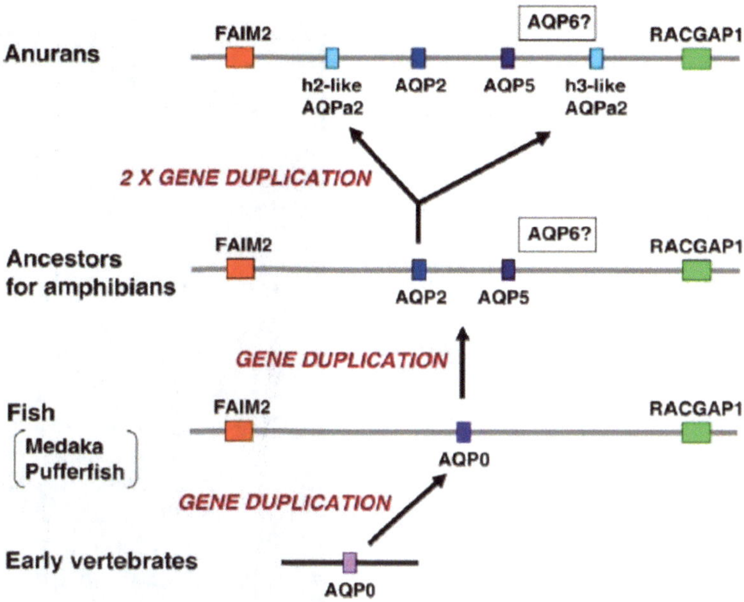

Fig. 2.8 A proposed sequence for the evolution of Aqps 2 and 5 from gene duplication in an early amphibian ancestor that persists in mammals. Two subsequent gene duplications are required to produce the Aqpa-2 paralogs (Aqp-h2 and Aqp-h3 first described in the tree frog, *Hyla japonica*) (From Suzuki and Tanaka 2009, with permission)

tetrapod duplicated to give rise to Aqps 2 and 5 that remained as single copies in amniotes but duplicated further in the course of amphibian evolution (Fig. 2.8).

Of interest, an Aqp homologous with the mammalian Aqp2 (Aqp-h2K) was localized in the apical membrane of the collecting segment of kidneys of *Hyla japonica* and a related Aqp, HC-2 has been identified in *H. chrysoscelis* (Ogushi et al. 2007; Zimmerman et al. 2007; Crane and Goldstein 2007) indicating that expression of this gene has been retained in the kidney in addition to duplication and cooptation of the Aqp2a paralogs in the skin and bladder. When amphibians become dehydrated, glomerular filtration and urine production is greatly reduced via AVT activation of V_1 receptors. The glomerular filtration rate of *H. chrysocelis* dehydrated by 20 % declined by almost 84 % relative to hydrated controls but fractional water clearance declined from 62 to 2 %. Amphibians are not able to concentrate their urine but elevated water permeability in the collecting duct could allow the tubular fluid to become iso-osmotic with the body fluids (Fig. 2.9).

Aqp5 has been immunolocalized in the apical membrane of granular glands in the skin of all anurans studied. In terrestrial species, glandular secretion has been related to thermoregulation, mucus secretion and the need to keep the skin moist as a respiratory surface (Lillywhite 2006). In this regard Aqp5 may serve a function similar to that seen in the alveoli of the mammalian lung. Recently, Shibata et al. (2014) identified two Aqp5 paralogs (Aqp-xt5a and -5b) in the genome of *X. tropicalis*. Both are syntenic

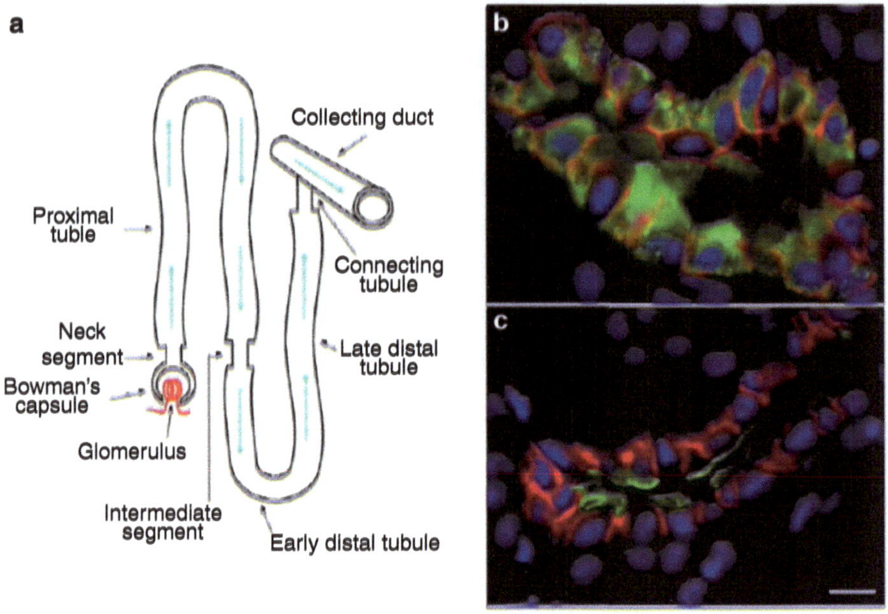

Fig. 2.9 (**a**) Schematic diagram of a nephron from an anuran kidney that is not able to concentrate urine. (**b**) Immunolabeling shows dispersal of Aqp-h2K in principal cells of the collecting duct of unstimulated kidney (*green FITC*). Aqp3 in the basolateral membrane is *labeled red*. (**c**) When stimulated with AVT Aqp2K is translocated to the apical membrane showing the pattern seen in the mammalian kidney predates the capacity to form a concentrated urine. (**a**, **b** and **c** from Suzuki and Tanaka (2009), with permission)

with Aqpa2 paralogs in the *X. tropicalis* genome. Molecular phylogenetic analysis showed Aqp-xt5a to be orthologous with mammalian Aqp5 and also to be expressed in the urinary bladder. Dehydration resulted in increased expression of Aqp-xt5a in the apical membrane of granular cells suggesting it might facilitate reabsorption of water as opposed to fluid secretion as seen in glands. AVT did not increase apical expression of Aqp-xt5a. The bladder also expresses a homolog of the urinary bladder type Aqp2a that is stimulated by AVT in more terrestrial species with larger bladder capacities. The physiological importance of this observation remains to be determined. The genomic events that resulted in the evolution of amphibian and mammalian Aqp2 and Aqp5-related Aqps, their ability to regulate epithelial water transport and their relationship to a putative ancestral Aqp0p in the early sarcopterygian lineage remain intriguing questions that can be addressed by a more extensive examination of Aqps in living amphibian species.

In order for apically expressed Aqp2 or Aqp5 related Aqps to function in epithelial water transport it is essential that a basolateral route of permeation occur as well. As with mammalian epithelia, an amphibian aquaglyceroporin (Aqp3) homolog has been identified in the basolateral membrane of absorptive and secretory epithelia (Suzuki and Tanaka 2009; Zimmerman et al. 2007). Other functions of basolateral Aqps could be to maintain the hydration status of the skin of

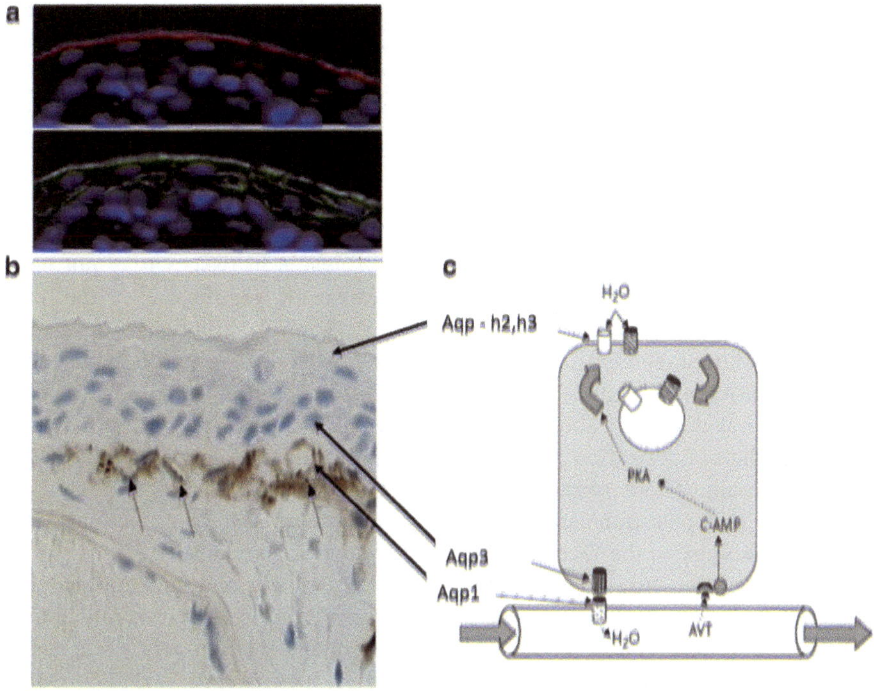

Fig. 2.10 (**a**) Immunolabeling of Aqp-h2 and Aqp-h3 in the apical membrane of the outer most living cell layer of *Hyla japonica* skin that has been stimulated by AVT. (**b**) Light micrograph of toad skin with Aqp1 immunolabeled in endothelial cells of subepithelial capillaries [**a** and **b** from Suzuki and Tanaka (2007), with permission]. (**c**) Schematic drawing of water absorption across anuran skin showing the apical entry step with Aqp-h2 and h3 stimulated by AVT, water transport across the basolateral membrane mediated by Aqp3 (not labeled) and transfer into the vasculature via Aqp1 in the subepithelial capillaries (from Hillman et al. (2009), with permission)

terrestrial species as they forage on land in a manner similar to that suggested for Aqp3 in mammalian skin (Hara et al. 2002). At the same time increased expression of Aqp3 could accommodate a greater rate of rehydration in conjunction with Aqp2a expression in the apical membrane. Exchange between the vascular and extracellular fluid compartments at the serosal face of these epithelia is required for fluid absorption and secretion. This is mediated by Aqp1 homologs that are expressed in capillary endothelia (Fig. 2.10).

A unique function for an aquaglyceroporin is seen in the freeze-tolerant frogs, *Hyla chrysoscelis* (Zimmerman et al. 2007). These anurans accumulate glycerol as a cryoprotectant when cold acclimated. Plasma glycerol concentration increased from 0.18 to 51 mM when cold acclimated while tubular reabsorption increased from 64 to 82 %. Tissue glycerol levels similarly increased from non-detectable levels in liver and muscle to 187 and 131 μM, respectively, indicating a greater synthesis, tissue distribution and renal reabsorption that would protect the tissues if freezing occurred. An aquaglyceroporin, HC-3 was cloned from these tissues and found to have high sequence identity with mammalian and amphibian AQP3. Oocyte expression of HC-3 resulted in a greater uptake of glycerol. HC-3 cDNA was used as a template to quantify HC-3 expression in a variety of tissues and showed marked increases in the liver and muscle of cold acclimated frogs. Kidney levels were relatively high in both warm and cold acclimated animals but increased in the urinary bladder of cold acclimated frogs.

These results suggest an important role for HC-3 in cryoprotection but the authors caution that correlation between mRNA levels in tissues may not be an accurate measure of protein expression because of regulation at the post-translational level. As noted above, mRNA for ventral skin type Aqp-x3 was identified in skin of *Xenopus laevis* but the protein was not expressed. In an earlier study, expression of an Aqp from toad bladder in *Xenopus* oocytes did not result in an increase in water permeability (Siner et al. 1996). It was suggested this was due to intracellular sequestration because of a YXRF sequence motif in the carboxy terminal domain. These observations suggest mechanisms for regulation at the translational and post-translational level that are important for amphibian water balance physiology in the face of their exposure to variable habitats.

Cope's grey tree frog complex includes *H. chrysoscelis* and its tetraploid sister species *H. versicolor*. Both accumulate glycerol during cold acclimation. Polyploidy is relatively common among amphibians, for example the commonly used species *Xenopus laevis* is tetraploid while its sister species, *X. tropicalis* is diploid. Mabel et al. (2011) note that among vertebrates fish and amphibians have a greater tendency to have polyploidy populations but found no evidence in the zoogeographic and conservation literature that additional genetic material in polyploid species has resulted in any selective advantage. How species level polyploidy in existing populations might affect Aqp gene distribution, in addition to that attributed to whole genome duplication events early in vertebrate evolution remains an interesting question.

2.3.6 Reptiles

Modern reptiles are descended from stem tetrapods that, based on homologies among teleost and mammalian species, could potentially express all 13 Aqp homologs identified in these taxa. In a comprehensive review of osmotic and ionic regulation in reptiles, Dantzler and Bradshaw (2009) observed that the three most abundant orders of reptiles: Crocodilia (alligators and crocodiles) Squamata (lizards, snakes and amphisbaenids) and Testunides (turtles and tortoises) have species that occupy habitats that range from terrestrial to aquatic to marine. A fourth order, Rynchocephalia contains a single terrestrial species, *Sphenodon punctatus* from New Zealand (a second species *S. guntheri* was recognized in 1989). Reported values for the relative proportions of extracellular, intracellular and vascular fluid volumes vary among taxa as does the regulation of ionic composition of the "*milieu interieur*". Physiological studies of ionic and osmotic regulation are limited to a small number of species. The genome of the lizard, *Anolis carolinensis* (Aföldi et al. 2011) contains many of the tertrapod Aqps associated with osmoregulation in mammalian and amphibian tissues but their identification in osmoregulatory tissues is limited. An overview of regulatory structures characterized in reptiles provides opportunities for future research. These include the skin, kidneys, cloaca-colon and, in some species a urinary bladder.

Skin The colonization of terrestial habitats by amniotes involved evolution of a skin that is resistant to evaporative water loss (Lillywhite 2006). Indeed, skin of desert species fulfills this requirement. It might be expected that basolateral expression of Aqp3 might be present to keep the epithelial cells hydrated, as per that seen in the skin of mice by Hara et al. (2002).

Kidneys As with the amphibia, the reptilian kidney is unable to form concentrated urine and AVT is the antidiuretic hormone. Antidiuresis is mediated to a great extent by a reduction in glomerular filtration in response to AVT binding to a V_1-type receptor. Plasma AVT concentrations correlate with plasma osmolality in the lizard, *Varanus gouldii* but levels in other reptile species have been variable. Adenylate cyclase activity in nephron segments of the lizard, *Ctenophorus ornatus* was significantly stimulated by AVT in the intermediate segment and collecting duct regions (Bradshaw and Bradshaw 1996). This would indicate the presence of an Aqp2 homolog but this remains to be demonstrated. While investigating the role of extra-renal ion regulation by salt glands, Babonis et al. (2011) cloned an Aqp3 from transcripts in the kidney of aquatic and sea snakes without salt glands. This Aqp3 was immunolocalized in the connecting segments and collecting ducts. Taken together, these observations suggest a similar function to that of amphibians for the formation of iso-osmotic urine. It should be remembered that many reptiles are uricotelic so the volume of water that must be voided to maintain nitrogen balance is less than ureotelic species.

Cloaca and Colon Urine flows in ureters to the cloaca. Muscular contraction of the cloaca forces cloacal fluid antegrade into the colon. Together they form a functional complex where solute linked fluid reabsorption is believed to occur. Dantzler and Bradshaw (2009) note, "*The transport mechanism underpinning the reabsorption of water and electrolytes from the cloacal-colonic complex are the subject of some debate.*" Babonis et al. (2012) used an antibody made against an amino acid sequence of the amphibian Aqp3 to immunolocalize this aquaporin in the basolateral membrane of the ureters and cloaca suggesting a role for fluid reabsorption or possibly mucus secretion. An apical Aqp and its regulation remain to be described.

Urinary Bladder The urinary bladder is absent in crocodilians and snakes but is found in a number of lizards, the tuatara and to varying degrees in all chelonians. Historically, Darwin (1839) noted that the Galapagos tortoises held large volumes of water in their bladders and suggested they might act as a water storage organ. The use of bladder water to offset dehydration has been demonstrated in the desert tortoise (*Gopherus agassizii*) that is able to spend months without water or even succulent vegetation (Nagy and Medica 1986). Bladder water has similarly been shown to offset dehydration in a desert lizard, the Gila monster, *Heloderma suspectum* (Davis and DeNardo 2007). The presence of Aqps in these tissues and the regulation of water permeability remain to be determined. Other reptilian species have a urinary bladder but its use as a water reservoir, in an ecological context is not well established.

Salt Glands Salt glands in reptiles are derived from a variety of embryonic sources and a variety of mechanisms have been identified that allow them to regulate the secretion of salt solutions that can become highly concentrated. The degree to which Aqps are involved in water transport associated with these secretions and interactions between transcellular and paracellular pathways is poorly understood.

2.3.7 Birds

Birds (class Aves) diverged from theropod reptiles beginning in the Jurassic and appear prominently in the Cretaceous. Current species inhabit virtually all habitats on earth from the polar regions to the tropics, oceans and the extreme deserts. As such they require a suite of osmoregulatory mechanisms that have been inherited from their reptilian heritage including the cloaca-colon complex and in many species nasal salt glands. Unlike reptiles birds have evolved a kidney capable of forming concentrated urine. The concentration of ureteral urine in the domestic fowl may be twice that of the plasma while other species may void a final urine that is 2.5 times that of the plasma which is small compared to that of desert mammals that may concentrate urine by a factor of over 20 times that of the plasma. These values and mechanisms for osmotic and ionic regulation by avian species are reviewed by Braun (2009).

Kidney The avian kidney has features similar to that of reptiles and amphibians in that glomerular circulation is fed by higher pressure arterial circulation while peritubular circulation is mediated by a lower pressure renal portal system. The avian kidney is divided into multiple lobes with an outer cortical region that has reptilian-like nephrons without loops of Henle. The inner cortical region has nephrons with highly convoluted proximal segments and loops of Henle that extend to varying degrees into a medullary cone. The distribution and function of Aqps in the avian kidney has been reviewed by Nishimura and Yang (2013).

Aqp1 A full length cDNA cloned from quail kidney has 82 % sequence identity with rat Aqp1. qAqp1 was immunolocalized in the apical brush border membrane of proximal tubule cells but not in the basolateral membrane. Unlike mammalian nephrons, qAqp1 was not localized in the descending loop of Henle. In the house sparrow, *Passer domesticus*, an Aqp1 homolog was shown to be expressed in both the proximal and distal tubules in addition to podocytes of the glomerulus (Casotti et al. 2007). Nishimura and Yang (2013) note "*that AQP1 may have a role in osmoregulation in sparrows, although the biological function of Aqp1 has not been tested.*" Further, there are no knock-out animals available to test this hypothesis.

Aqp2 A full-length Aqp2 from quail kidney has a 76 % amino acid sequence identity with rat Aqp2 and serine residues that are potential sites for phosphorylation by protein kinase A. qAqp2 mRNA was localized in branches of the collecting duct in the medulla and cortex. Using real time PCR greater levels of mRNA for qAqp2 were found in the medullary cones but water deprivation resulted in increased expression in both superficial and looped nephrons. The interpretation is an increase in Aqp2 expression resulting from antidiuretic hormone as is seen in the mammalian kidney. It is noted that the increase in expression in the cortical nephrons indicates the role of antidiuretic hormone in water retention by the kidney predates the evolution of the loop of Henle, which is consistent with what has been observed in amphibian and lungfish kidneys.

Aqp3 Aqp3 cloned from quail kidney has an 81 % sequence identity with rat and human Aqp3. mRNA for qAqp3 was localized in collecting tubules and ducts suggesting it plays a role in water exit from collecting duct cells as is seen in the mammalian kidney.

Aqp4 Two cDNAs for quail Aqp4 resulting from long and short reading frames have been cloned from medullary cones of quail kidney. Both have high degrees of homology with human Aqp4s. Orthogonal arrays of intramembranous particles are not seen in quail collecting ducts and in situ hybridization failed to show mRNA for qAqp4 in collecting ducts suggesting "*their physiological function may be different*" from that of the mammalian kidney. Aqp4 homologs have also been cloned from a variety of avian tissues including the hypothalamus and circumventricular organs where they may serve a role in osmosensation.

Cloaca-Colon Ureteral urine deposited into the cloaca is forced by muscular contraction into the distal rectum. Casotti et al. (2007) were able to clone an Aqp1 homolog in from the distal rectum and immunolocalize it in the epithelium. However, whether it is distributed at the basolateral membrane or within the cells remains to be described so a route for transepithelial water absorption is not evident.

Salt Glands Many birds utilize nasal salt glands to excrete excess salt taken in the diet and thereby conserve water. Salt glands of ducks given hypertonic salt show a down regulation of Aqp1 in the endothelial cells and Aqp5 in the epithelial cells (Müller et al. 2006). These results suggest decreased capillary ultrafiltration and epithelial water loss in the course of forming a hypertonic secretion.

In reviewing the avian literature, Nishimura (2008) observes, "*....recent evidence indicates that AQPs show functions other than water channel or glycerol transport. Comparative analysis of AQPs in primitive animals may provide useful information as to the fundamental function. The importance of AQP2 to fluid homeostasis in developing kidneys needs to be pursued. Such studies will provide insight for understanding possible mechanisms of developmental programming in adults.*" These comments mirror those of Ishibashi et al. (2011) on the different functions served by aquaporins in mammals.

2.4 Perspectives

The expression of mammalian Aqps is the result of gene duplication and co-optation for tissue-specific functions that date at least from the earliest chordates. The phylogenetic tree developed by Zardoya and Villalba (2001) using amino acid sequences from insect Aqps as outgroups provides a useful framework for interpreting comparative and evolutionary aspects of vertebrate Aqps (Fig. 2.11). The data set was analyzed with maximum parsimony (MP), neighbor joining (NJ) and maximum likelihood (ML) phylogenetic methods. In this analysis, the first branch point gave rise to Aqp4 as the most basal Aqp in eukaryotic evolution. A paralog of that duplication underwent a second duplication that resulted in Aqp1 and an ancestral Aqp for Aqps 0, 2, 5 and 6. This phylogeny is consistent with the observation that Aqp4 serves a number of functions in various vertebrate tissues, as early as the elasmobranchs (Cutler et al. 2012). Aqp1 is widely expressed in tissues of many vertebrates including elasmobranchs that express the paralog, Aqp1e. The sequence of Aqp1e is similar to mammalian Aqps 1, 2 and 5. As such, it could be representative of the putative Aqp at the branch point that gave rise to Aqps 0, 2, 5, and 6, which are syntenic in the mammalian genome. Further up the evolutionary scale, Aqp0p expressed in the collecting duct of lungfish kidney is suggested to be ancestral to Aqp2 in the tetrapod kidney. There is a branch point in the derived phylogeny in which Aqp0p, as a paralog of Aqp0, could be ancestral to Aqps 2 and 5, which are important for fluid reabsorption and secretion in the kidney and exocrine glands. Note the gene duplications proposed to have resulted amphibian-specific Aqps are not shown in this figure. Shortly after submission of this chapter, Finn et al. (2014) published an extensive analysis of aquaporins in deuterostomes and created a more complete phylogeny than depicted in Fig. 2.11. In some cases the assignment of amphibian Aqp-2a paralogs is modified from the earlier literature (Suzuki and Tanaka 2010).

On a broader evolutionary scale, it is important to note that six of the canonical Aqp genes (Aqps 1, 2, 3, 4, 6, 7) have been characterized in the nematode worm *Caenorhabditis elegans* (Huang et al. 2007) indicating these molecules were present earlier in metazoan evolution and have provided a "toolbox" of membrane proteins that serve a variety of physiological functions related to water transport, cell adhesion and cell motility. Four of the canonical Aqps identified in *C. elegans* were found to be expressed in osmoregulatory tissues. Quadruple KO animals displayed a normal phenotype under normal and hypertonic culture conditions. Recovery from hypotonic conditions was slower than controls but survivorship was the same. The authors suggested that these Aqps may confer a selective advantage to animals in their natural soil habitat where the osmotic conditions "*are likely to vary dramatically both in time and space*". A similar argument can be made that Aqp1-null mice that are viable when provided water may face conditions in their natural habitat where the inability to concentrate urine would be maladaptive.

The concept that the diversity of Aqps is related to natural selection in the face of environmental stresses [reviewed by Ishibashi et al. (2011)] is consistent with

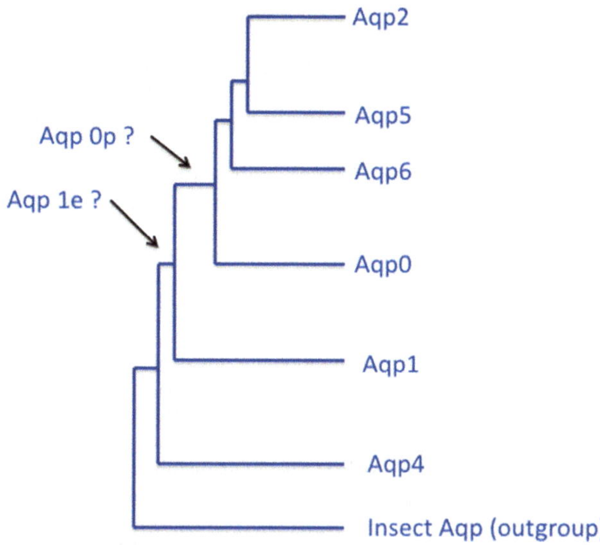

Fig. 2.11 Phylogenetic tree of metazoan Aqps. The branch points indicate gene duplications and the branch lengths are proportional to evolutionary distances. In this analysis, the first duplication gave rise to Aqp4 that remains widely distributed in metazoans. A paralog of that branch further duplicated to give Aqp1 and an ancestral gene for Aqps 0, 2, 5, and 6 that are paralogous in mammals and potentially other vertebrates. *Arrows* indicate putative roles for an Aqp1e-like and Aqp0p-like Aqps during the course of vertebrate evolution (Zardoya and Villalba 2001)

conditions associated with genome duplication in amphibians and fish (Mabel et al. 2011). For example, temperature fluctuations that interfere with spindle formation and external fertilization that may result in polyspermy. These and other factors likely occurred earlier in eukaryotic evolution and could contribute to the diversity yet the structural similarity of Aqps and aquaglyceroporins that serve a variety of functions in different tissues of living species. Ishibashi et al. (2011) conclude, "*Such interdisciplinary works as comparative physiology and endocrinology in particular should be encouraged and rewarded.*"

References

Aföldi J et al (2011) The genome of the green anole lizard and a comparative analysis with birds and mammals. Nature 477:587–591

Agre P (2006) The aquaporin water channels. Proc Am Thorac Soc 3:5–13

Agre P, Saboori AM, Asimos A, Smith BL (1987) Purification and partial characterization of the Mr 30,000 integral membrane protein associated with the erythrocyte Rh(D) antigen. J Biol Chem 262:17497–17503

Agre P, Sasaki S, Chrispeels MJ (1993) Aquaporins: a family of water channel proteins. Am J Physiol 265:F461

Babonis LS, Miller SN, Evans DH (2011) Renal responses to salinity change in snakes with and without salt glands. J Exp Biol 214:2140–2156

Babonis LS, Womack MC, Evans DH (2012) Morphology and putative function of the colon and cloaca of marine and freshwater snakes. J Morphol 273:88–102

Belge H, Devuyst O (2006) Aquaporin-1—a water channel on the move. Nephrol Dial Transplant 21:2069–2071

Benga G, Popescu O, Pop VI (1986) p-(Chloromercuri)benzenesulfonate binding by membrane proteins and the inhibition of water transport in human erythrocytes. Biochemistry 25:1535–1538

Bentley PJ (1958) The effects of neurohypophyseal extracts on water transfer across the wall of the isolated urinary bladder of the toad *Bufo marinus*. J Endocrinol 17:201–209

Benton MJ (1998) The quality of the fossil record of vertebrates. In: Donovan SK, Paul CRC (eds) The quality of the fossil record. Wiley, Hoboken, NJ, p 312

Bradshaw FJ, Bradshaw SD (1996) Arginine vasotocin: locus of action along the nephron of the ornate dragon lizard, Ctenophorus ornatus. Gen Comp Endocrinol 103:281–289

Braun EJ (2009) Osmotic and ionic regulation in birds. In: Evans DH (ed) Osmotic and Ionic regulation cells and animals. CRC, Boca Raton, FL

Broekhuyse RM, Kuhlmann ED, Stols ALH (1976) Lens membranes II. Isolation and characterization of the main intrinsic polypeptide (MIP) of bovine lens fiber membranes. Exp Eye Res 23:365–371

Brown D (2003) The ins and outs of aquaporin-2 trafficking. Am J Physiol 284:F893–F901

Brunn F (1921) Beitrag zur Kenntnis der Wirkung yon hypophysenextrakten auf den Wasserhaushalt des Frosches. Zeit f Gesam Exp Med 25:120–125

Calamita G, Bishai WR, Preston GM, Guggino WB, Agre P (1995) Molecular cloning and characterization of AqpZ, a water channel from *Escherichia coli*. J Biol Chem 270:29063–29066

Calamita G, Ferri D, Gena P, Liquori GE, Cavalier A, Thomas D, Svelto M (2005) The inner mitochondrial membrane has aquaporin-8 water channels and is highly permeable to water. J Biol Chem 280:17149–17153

Campbell EM, Ball A, Hoppler S, Bowman AS (2008) Invertebrate aquaporins: a review. J Comp Physiol B 178:35–955

Carroll RL (2001) The origin and early radiation of terrestrial vertebrates. J Paleontol 75:1202–1213

Casotti G, Waldron T, Misquith G, Powers D, Slusher L (2007) Expression and localization of an aquaporin-1 homolog in the avian kidney and lower intestinal tract. Comp Biochem Physiol A Mol Integr Physiol 147:355–362

Cerdà J, Finn RN (2010) Piscine aquaporins: an overview. J Exp Zool 313A:623–650

Chen Y, Rice W, Gu Z, Li J, Huang J, Brenner MB, Van Hoek A, Xiong J, Gunderson GG, Norman JC, Hsu VW, Fenton RA, Brown D, Lu HAJ (2012) Aquaporin 2 promotes cell migration and epithelial morphogenesis. J Am Soc Nephrol 23:1506–1517

Chevalier J, Bourget J, Hugon JS (1974) Membrane associated particles: distribution in frog urinary bladder epithelium at rest and after oxytocin treatment. Cell Tissue Res 152:129–140

Crane CM, Goldstein DL (2007) Comparative functional analysis of aquaporins/glyceroporins in mammals and anurans. Mamm Genome 18:452–462

Cutler CP (2007) Cloning and identification of four aquaporin genes in the dogfish shark (*Squalus acanthias*). Bull Mt Desert Isl Biol Lab 46:19–20

Cutler CP, Meischke L, Cramb G (2005) Evolutionary and comparative analysis of aquaporin water channels in fish. Bull Mt Desert Isl Biol Lab 44:55

Cutler CP, MacIver B, Cramb G, Zeidel M (2012) Aquaporin 4 is ubiquitously expressed in the dogfish (*Squalus acanthias*) shark. Front Physiol 2:107

Dantzler WH, Bradshaw SD (2009) Osmotic and ionic regulation in reptiles. In: Evans DH (ed) Osmotic and Ionic regulation cells and animals. CRC, Boca Raton, FL

Darwin C (1839) Journal of researches into the natural history and geography of the countries visited during the voyage of the HMS Beagle, around the world. Routledge, London

Davis JR, DeNardo DF (2007) The urinary bladder as a physiological reservoir that moderates dehydration in a large desert lizard, the Gila monster *Heloderma suspectum*. J Exp Biol 210:1472–1480

Denker BM, Smith BL, Kuhajda FP, Agre P (1988) Identification, purification, and partial characterization of a novel Mr 28,000 integral membrane protein from erythrocytes and renal tubules. J Biol Chem 263:15634–15642

Dumas L, Kim YH, Karimpour-Fard A, Cox M, Hopkins J, Pollack JR, Sikela JM (2007) Gene copy number variation spanning 60 million years of human and primate evolution. Genome Res 17:1266–1277

Dutrochet RJR (1827) Nouvelles observations sur l'endosmos et l'exosmos, et sur la cause de ce double phenomene. Anal Chim Phys 35:393–400

Ecelbarger CA, Terris J, Frindt G, Echevarria M, Marples D, Nielsen S, Knepper M (1995) Aquaporin-3 water channel localization and regulation in rat kidney. Am J Physiol 269: F663–F672

Echevarria M, Windhager EE, Tate SS, Frindt G (1994) Cloning and expression of AQP3, a water channel from the medullary collecting duct of rat kidney. Proc Natl Acad Sci U S A 91:10997–11001

Edwards WF (1824) De l'influence des agens physiques sur la vie. Crochard, Paris

Evans DH (ed) (2009) Osmotic and ionic regulation cells and animals. CRC, Boca Raton, FL

Evans DH, Claiborne JB (2009) Osmotic and ionic regulation in fishes. In: Evans DH (ed) Osmotic and Ionic regulation cells and animals. CRC, Boca Raton, FL

Fabra M, Raldua D, Power DM, Deen PM, Cerdà J (2005) Marine fish egg hydration is aquaporin-mediated. Science 307:545

Finn RN, Chauvigne F, Baldur Hlidberg J, Cutler CP, Cerda J (2014) The lineage-specific evolution of aquaporin gene clusters facilitated tetrapod terrestrial adaptation. PLoS One 9 (11):e113686

Forrester JM (1995) Malpighi's De Polypo Cordis: an annotated translation. Med Hist 39:477–492

Frigeri A, Gropper MA, Turck CW, Verkman AS (1995) Immunolocalization of the mercurial-insensitive water channel and glycerol intrinsic membrane protein in epithelial cell plasma membranes. Proc Natl Acad Sci U S A 92:4328–4331

Fujiyoshi Y, Mitsuoka A, de Groot BL, Phillippsen A, Grubmuller H, Agre P, Engel A (2002) Structure and function of water channels. Curr Opin Struct Biol 12:509–515

Fushimi K, Uchida S, Hara Y, Hirata Y, Marumo F, Sasaki S (1993) Cloning and expression of apical membrane water channel of rat collecting tubule. Nature 361:549–552

Fushimi K, Sasaki S, Marumo F (1997) Phosphorylation of serine 256 is required for cAMP-dependent exocytosis of the aquaporin 2 water channel. J Biol Chem 272:14800–14804

Gonen T, Sliz P, Kistler J, Cheng YF, Waltz T (2004) Aquaporin-0 membrane junctions reveal the structure of a closed water pore. Nature 429:193–197

Gorin WB, Yancey SB, Cline J, Revel JP, Horwitz J (1984) The major intrinsic protein (MIP) of the bovine lens fiber membrane: characterization and structure based on cDNA cloning. Cell 39:49–59

Hara M, Ma T, Verkman AS (2002) Selectively reduced glycerol in skin of aquaporin-3-deficient mice may account for impaired skin hydration, elasticity and barrier recovery. J Biol Chem 277:46616–46621

Hara-Chikuma M, Verkman AS (2008) Prevention of skin tumorigenesis and impairment of epidermal cell proliferation by targeted aquaporin-3 gene disruption. Mol Cell Biol 28:326–332

Hasegawa T, Tanii H, Suzuki M, Tanaka S (2003) Regulation of water absorption in the frog skins by two vasotocin-dependent water channel aquaporins, AQP-h2 and AQP-h3. Endocrinology 144:4087–4096

Hasler U (2009) Controlled aquaporin-2 expression in the hypertonic environment. Am J Physiol 296:C641–C653
Hays RM, Leaf A (1962) Studies on the movement of water through the isolated toad bladder and its modification by vasopressin. J Gen Physiol 45:905–919
Hellsten U et al (2010) The genome of the western clawed frog, Xenopus tropicalis. Science 328:633–636
Hevesy GV, Hofer E, Krogh A (1935) The permeability of the skin of frogs to water as determined by D2O and H2O. Skand Arch Physiol 72:199–214
Hewson R (1773) On the figure and composition of the red particles of the blood, commonly called the red globules. Philos Tran R Soc Lond 63:306–324
Heymann JB, Agre P, Engel A (1998) Progress on the structure and function of aquaporin 1. J Struct Biol 121:191–206
Hillman SS, Withers PC, Drewes RC, Hillyard SD (2009) Ecological and environmental physiology of amphibians. Oxford University Press, New York, p 469
Hillyard SD (2012) Sensing meets separation: water transport across biological membranes. In: Nielsen CH (ed) Biomimetic membranes for sensor and separation applications, biological and medical physics, biomedical engineering. Springer, Amsterdam
Hillyard SD, Møbjerg N, Tanaka S, Larsen EH (2009) Osmotic and ionic regulation in amphibians. In: Evans DH (ed) Osmotic and Ionic regulation cells and animals. CRC, Boca Raton, FL
Hooke R (ed) (1664) Micrographia some physiological descriptions of minute bodies made by magnifying glasses with observations and inquiries thereupon. Martyn and Allestry, London
Huang CG, Lamitina T, Agre P, Strange K (2007) Functional analysis of the aquaporin gene family in *Caenorhabditis elegans*. Am J Physiol 292:C1867–C1873
Ishibashi K, Sasaki S, Fushimi K, Uchida S, Kuwahara M, Saito H, Furukawa T, Nakajima K (1994) Molecular cloning and expression of a member of the aquaporin family with permeability to glycerol and urea in addition to water expressed at the basolateral membrane of kidney collecting duct cells. Proc Natl Acad Sci U S A 91:6269–6273
Ishibashi K, Hara S, Kondo S (2009) Aquaporin water channels in mammals. Clin Exp Nephrol 13:107–117
Ishibashi K, Kondo S, Hara S, Morishita Y (2011) The evolutionary aspects of aquaporin family. Am J Physiol 300:R566–R576
Ishikawa Y, Yuan Z, Inuoe N, Skowronski MT, Nakae Y, Shono M, Cho G, Yasui M, Agre P, Nielsen S (2005) Identification of AQP5 in lipid rafts and its translocation to apical membranes by activation of M3 mAChRs in interlobular ducts of rat parotid gland. Am J Physiol Cell Physiol 289:C1303–C1311
Nollet (Abbé) JA (1752) Investigations on the causes for the ebullition of liquids. English translation, J Membr Sci 100:1–3 (1995)
Jensen MØ, Dror RO, Xu H, Borhani DW, Arkin IT, Eastwood MP, Shaw DE (2008) Dynamic control of slow water transport by aquaporin 0: Implications for hydration and junction stability in the eye lens. Proc Natl Acad Sci U S A 105:14430–14435
Johnson KD, Höfte H, Chrispeels MJ (1990) An intrinsic tonoplast protein of protein storage vacuoles in seeds is structurally related to a bacterial solute transporter (GlpF). Plant Cell 2:525–532
Jørgensen CB (1997) 200 years of amphibian water economy: from Robert Townson to the present. Biol Rev 72:153–237
Jung JS, Preston GM, Smith BL, Guggino WB, Agre P (1994) Molecular structure of the water channel through aquaporin CHIP28. The hourglass model. J Biol Chem 269:14648–14654
Kachadorian WA, Wade JB, Di Scala VA (1975) Vasopressin induced structural change in toad bladder luminal membrane. Science 190:67–69
Kawedia JD, Nieman ML, Bolvin GP, Melvin JE, Kikuchi K-I, Hand AR, Lorenz JN, Menon AG (2007) Interaction between transcellular and paracellular water transport pathways through aquaporin 5 and the tight junction complex. Proc Natl Acad Sci 104:3621–3626

Kistler J, Bullivant S (1980) Lens gap junctions and orthogonal arrays are unrelated. FEBS Lett 111:73–78

Kleinzeller A (1996) William Hewson's studies of red blood corpuscles and the evolving concept of a cell membrane. Am J Physiol 271:C1–C8

Koefoed-Johnsen V, Ussing HH (1953) The contributions of diffusion and flow to the passage of D2O through living membranes. Acta Physiol Scand 28:60–76

Konno N, Hyodo S, Yamaguchi Y, Matsuda K, Uchiyama M (2010) Vasotocin/V2-type receptor/aquaporin axis exists in African lungfish kidney but is functional only in terrestrial condition. Endocrinology 151:1089–1096

Larsen EH, Deaton LE, Onken H, O'Donnell M, Grosell M, Dantzler WH, Weihrauch D (2014) Osmoregulation and excretion. Compr Physiol 4:405–473

Lillywhite H (2006) Water relations of tetrapod integument. J Exp Biol 202:222–226

Mabel BK, Alexandrou MA, Taylor MI (2011) Genome duplication in amphibians and fish: an extended synthesis. J Zool 284:151–182

Macey RI, Farmer REI (1970) Inhibition of water and solute permeability in human red cells. Biochim Biophys Acta 211:104–106

Marshall AG (1978) Biophysical chemistry principles, techniques and applications. Wiley, New York

Matteucci C, Cima A (1845) Memoire sur l'endosmose. Annal Chim et Phys 13:63–86

Maurel C, Verdoucq L, Luu D-T, Santoni V (2008) Plant aquaporins: membrane channels with multiple integrated functions. Annu Rev Plant Biol 59:595–624

Miramatsu S, Mizuno T (1989) Nucleotide sequence of the region encompassing the glpKF operon region containing a bent DNA sequence of Escherichia coli. Nucleic Acids Res 17:4378

Moon C, Preston GM, Griffin CA, Jabs EW, Agre P (1993) The human aquaporin gene. Structure, organization and chromosomal localization. J Biol Chem 21:15772–15778

Müller C, Sendler M, Hildebrandt JP (2006) Downregulation of aquaporins 1 and 5 in nasal gland by osmotic stress in ducklings, Anas platyrhynchos: implications for the production of hypertonic fluid. J Exp Biol 209:4067–4076

Murata K, Mitsuoka K, Hirai T, Walz T, Agre P, Heymann JB, Engel A, Fujiyoshi Y (2000) Structural determinants of water permeation through aquaporin-1. Nature 407:599–605

Nagy KA, Medica PA (1986) Physiological ecology of desert tortoises in southern Nevada. Herpetologica 42:73–92

Nakatain Y, Takeda H, Kohara Y (2007) Reconstruction of the vertebrae ancestral genome reveals dynamic genome reorganization in early vertebrates. Genome Res 17:1254–1265

Nielsen S, King LS, Christensen BM, Agre P (1997) Aquaporins in complex tissues. II. Subcellular distribution in respiratory and glandular tissues of rat. Am J Physiol 273:C1549–C1561

Nishimura H (2008) Urine concentration and avian aquaporin water channels. Pfleugers Arch 456:755–768

Nishimura H, Yang Y (2013) Aquaporins in avian kidneys: function and perspectives. Am J Physiol 305:R1201–R1214

Ogushi Y, Mochida H, Nakakura T, Suzuki M, Tanaka S (2007) Immunocytochemical and phylogenetic analysis of vasotocin-dependent aquaporin, AQP-h2K, specifically expressed in the kidney of the tree frog, Hyla japonica. Endocrinology 148:5891–5901

Ogushi Y, Tsuzuki A, Sato M, Mochida M, Okada R, Suzuki M, Hillyard SD, Tanaka S (2010a) The water-absorption region of ventral skin of several semiterrestrial and aquatic anuran amphibians identified by aquaporins. Am J Physiol 299:R1150–R1162

Ogushi Y, Kitagawa D, Hasegawa T, Suzuki M, Tanaka S (2010b) Correlation between aquaporin and water permeability in response to vasotocin, hydrin, and beta-adrenergic effectors in the pelvic skin of the tree frog, Hyla japonica. J Exp Biol 213:288–294

Ohno S (1970) Evolution by gene duplication. Springer, New York

Paganelli CV, Solomon AK (1957) The rate of exchange of tritiated water across the human red cell membrane. J Gen Physiol 41:259–277

Pao GM, Wu LF, Johnson KD, Hofte H, Chrispeels MJ, Sweet G, Sandal NN, Saier MH (1991) Evolution of the MIP family of integral membrane proteins. Mol Microbiol 5:33–37
Papadopoulos MC, Verkman AS (2007) Aquaporin-4 and brain edema. Pediatr Nephrol 22:778–784
Pfeffer W (ed) (1877) Osmotische Untersuchungen. Studien zur Zellmechanic. Engelmann, Leipzig
Preston GM, Agre P (1991) Isolation of the cDNA for erythrocyte integral membrane protein of 28 kilodaltons: member of an ancient channel family. Proc Natl Acad Sci U S A 88:11110–11114
Preston GM, Carroll TP, Guggino WB, Agre P (1992) Appearance of water channels in Xenopus oocytes expressing red cell CHIP28 protein. Science 256:385–387
Rao Y, Jan LY, Jan YN (1990) Similarity of the product of the Drosophila neurogenic gene big brain to transmembrane channel proteins. Nature 345:163–167
Reid EW (1890) Osmosis experiments with living and dead membranes. J Physiol 11:312–351
Romer AS (1962) The vertebrate body. WB Saunders, Philadelphia, PA, p 627
San Mauro D, Vences M, Alcobendes M, Zardoya R, Meyer A (2005) Initial diversification of living amphibians predated the breakup of Pangaea. Am Nat 165:590–599
Shibata Y, Takeuchi H-A, Hasegawa T, Suzuki M, Tanaka S, Hillyard SD, Nagai T (2011) Location of water channels in the skin of two species of desert toads, Anaxyrus (Bufo) punctatus and Incilius (Bufo) alvarius. Zool Sci 28:664–670
Shibata Y, Sano T, Tsuchiya N, Okada R, Mochida H, Tanaka S, Suzuki M (2014) Gene expression and localization of two types of AQP5 in Xenopus tropicalis under hydration and dehydration. Am J Physiol 307:R44–R56
Siner J, Paredes A, Hosselet C, Hammond T, Strange K, Harris HW (1996) Cloning of an aquaporin homolog present in water channel containing endosomes of toad urinary bladder. Am J Physiol 270:C372–C381
Smith BL, Agre P (1991) Erythrocyte 2Mr 28,000 transmembrane protein exists as a multisubunit oligomer similar to channel proteins. J Biol Chem 266:6407–6415
Stadie WC, Sunderman FW (1931) The osmotic coefficient of sodium in sodium hemoglobinate and of sodium chloride in hemoglobin solution. J Biol Chem 91:227–241
Steen WB (1929) On the permeability of the frog's bladder to water. Anat Rec 43:215–220
Stetson DL, Lewis SA, Alles W, Wade JB (1982) Evaluation by capacitance measurements of antidiuretic hormone induced membrane area changes in toad bladder. Biochim Biophys Acta 689:267–274
Sui H, Han BG, Lee JK, Wilian P, Jap BK (2001) Structural basis of water-specific transport through the AQP1 water channel. Nature 414:872–878
Suzuki M, Tanaka S (2009) Molecular and cellular regulation of water homeostasis in anuran amphibians by aquaporins. Comp Biochem Physiol A Mol Integr Physiol 153:231–241
Suzuki M, Tanaka S (2010) Molecular diversity of vasotocin-dependent aquaporins closely associated with water adaptation strategy in anuran amphibians. J Neuroendocrinol 22:407–412
Suzuki M, Hasegawa T, Ogushi Y, Tanaka S (2007) Amphibian aquaporins and adaptation to terrestrial environments: a review. Comp Biochem Physiol 148A:72–81
Takemoto LJ, Hansen JS, Nicholson BJ, Hunkapiller M, Revel JP, Horwitz J (1983) Major intrinsic polypeptide of lens membrane biochemical and immunological characterization of the major cyanogen bromide fragment. Biochim Biophys Acta 731:267–274
Tanii H, Hasegawa T, Hirakawa N, Suzuki M, Tanaka S (2002) Molecular and cellular characterization of a water channel protein AQP-h3, specifically expressed in ventral frog skin. J Mambr Biol 188:43–53
Taylor A, Mamelak M, Reaven E, Maffly R (1973) Vasopressin: possible role of microtubules and microfilaments in its action. Science 181:347–350
Thiagarajah JR, Zao D, Verkman AS (2007) Impaired enterocyte proliferation in aquaporin-3 deficiency in mouse models of colitis. Gut 56:1529–1535

Tingaud-Sequeira A, Calusinska M, Finn RN, Chauvigné F, Lozano J, Cerdà J (2010) The zebrafish genome encodes the largest vertebrate repertoire of functional aquaporins with dual paralogy and substrate specificities similar to mammals. BMC Evol Biol 10:38–56

Törnroth-Horsefield S, Hedfalk K, Fischer G, Lindkvist-Petersson K, Neutze R (2010) Structural insights into eukaryotic aquaporin regulation. FEBS Lett 584:2580–2588

Townson R (1795) Physiological observations on the amphibia. Dissertation the second. Respiration continued with a fragment upon the subject of absorption, University of Edinburgh

van Balkom BWM, van Raak M, Breton S, Pastor-Soler N, Bouley R, van der Sluijs P, Brown D, Deen PMT (2003) Hypertonicity is involved in redirecting the aquaporin-2 water channel into the basolateral, instead of the apical, plasma membrane of renal epithelial cells. J Biol Chem 278:1101–1107

van Hoek AN, Horn ML, Luthjens LH, de Jong MD, Dempstert JA, van Os CH (1991) Functional unit of 30 kDa for proximal tubule water channels as revealed by radiation inactivation. J Biol Chem 266:16633–16635

Vandepoele K, De Vos W, Taylor JS, Meyer A, Van d Peer Y (2004) Major events in the genome evolution of vertebrates: paranome age and size differ considerably between ray-finned fishes and land vertebrates. Proc Natl Acad Sci 101:1638–1643

van't Hoff JH (1887) Die Rolle des osmotischen Druckesin der Analogie zwischen Losungenund Gasen. Z Phys Chem 1:481–508

van't Hoff JH (1901) Osmotic pressure and chemical equilibrium. Nobel Lecture, 13 December 1901

Viborg A, Rosenkilde P (2004) Water potential receptors in the skin regulate blood perfusion in the ventral pelvic skin of toads. Physiol Biochem Zool 77:39–49

Wang Y, Tajkhorshid E (2007) Molecular mechanisms of conduction and selectivity in aquaporin water channels. J Nutr 137:1509S–1515S

Wang Y, Schluten K, Tajkhorshid E (2005) What makes an aquaporin a glycerol channel? A Comparative study of AqpZ and GlpF. Structure 13:1107–1118

Yamaguchi Y, Gojobori T, Marumo F (1994) Molecular cloning and expression of a member of the aquaporin family with permeability to glycerol and urea in addition to water expressed at the basolateral membrane of kidney collecting duct cells. Proc Natl Acad Sci 91:6269–6273

Yang B, Brown D, Verkman AS (1996) The mercurial insensitive water channel (AQP4) forms orthogonal arrays in stably transfected Chinese hamster ovary cells. J Biol Chem 271:4577–4580

Zardoya R (2005) Phylogeny and evolution of the major intrinsic protein family. Biol Cell 97:397–414

Zardoya R, Villalba S (2001) A phylogenetic framework for the aquaporin family in eukaryotes. J Mol Evol 52:391–404

Zeidel ML, Nielsen S, Smith BL (1994) Ultrastructure, pharmacologic inhibition, and transport selectivity of aquaporin channel-forming integral protein in proteoliposomes. Biochemistry 33:1606–1615

Zhang RB, Logee KA, Verkman AS (1990) Expression of mRNA coding for kidney and red cell water channels in Xenopus oocytes. J Biol Chem 265:15375–15378

Zimmerman SL, Frisbie J, Goldstein DL, West J, Rivera K, Krane CM (2007) Ecretion of glycerol, and expression of aquaporins and glyceroporins, during cold acclimation in Cope's gray tree frog Hyla chrysoscelis. Am J Physiol 292:R544–R555

Chapter 3
Use of Genetic Models to Study the Urinary Concentrating Mechanism

Emma T.B. Olesen, Marleen L.A. Kortenoeven, and Robert A. Fenton

Abstract Maintenance of body water homeostasis is a fundamental homeostatic mechanism in mammals. Understanding the basic mechanisms of how water balance is maintained, or dysfunctional in certain diseases is thus of clinical importance. In recent years, application of transgenic and knockout mouse technology is providing critical new information about urinary concentrating processes and thus mechanisms for maintaining body water homeostasis. In this chapter we provide a brief overview of genetic mouse model generation, and then summarize findings in transgenic and knockout mice pertinent to our understanding of the urinary concentrating mechanism, focusing predominantly on mice in which expression of specific renal transporters or receptors has been deleted.

Keywords Aquaporins • Water homeostasis

3.1 Introduction

Mammalian water homeostasis is dependent on variation in kidney water excretion over a wide range. During excessive water intake, urine is diluted to an osmolality below that of plasma, thus retaining osmolytes yet allowing water removal from the body. Conversely, during water restriction, the kidney retains water by a selective process, thereby concentrating solutes in the urine to many times the osmolality of plasma. The roles of candidate channels, transporters and regulatory hormone receptors in these processes have been studied by the use of genetic models.

In the kidney, bulk water reabsorption occurs as a constitutive process in the proximal tubule through aquaporin-1 (AQP1). However, the process of regulated water reabsorption and fine-tuning of urine concentration takes place in the hormone sensitive connecting tubule (CNT) and collecting duct (CD) epithelium,

E.T.B. Olesen (✉)
Faculty of Health and Medical Sciences, Department of Cellular and Molecular Medicine, University of Copenhagen, Blegdamsvej 3B, 12.6 Room 12.6.34, 2200 Copenhagen, Denmark
e-mail: etbo@sund.ku.dk

M.L.A. Kortenoeven • R.A. Fenton
Department of Biomedicine, Aarhus University, Aarhus, Denmark

K.A. Hyndman, T.L. Pannabecker (eds.), *Sodium and Water Homeostasis*,
Physiology in Health and Disease, DOI 10.1007/978-1-4939-3213-9_3

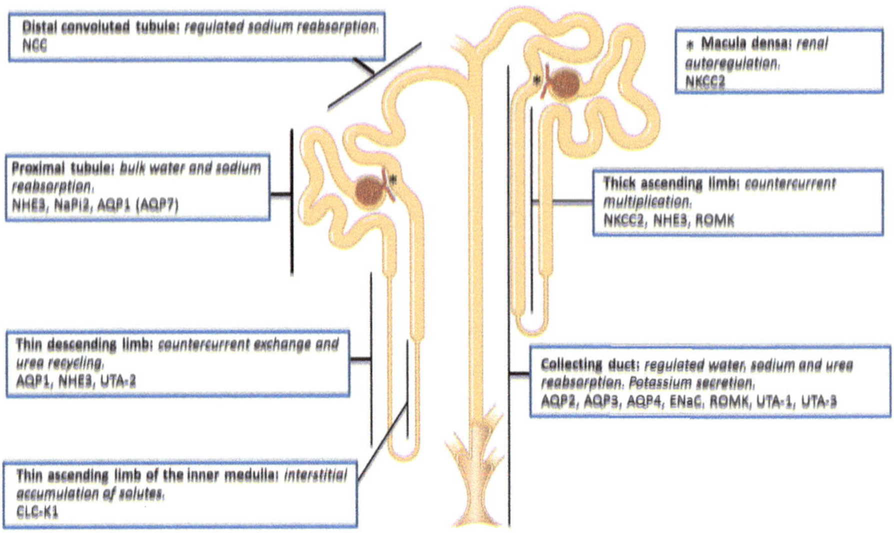

Fig. 3.1 Overview of the nephron and collecting duct with indication of important channels and transporters in each section

utilizing the osmotic gradient (set up from countercurrent multiplication) from the tubular lumen to the kidney interstitium, with water eventually re-entering the bloodstream. Entry of water into the CD and CNT epithelium occurs through the vasopressin (VP) regulated water channel Aquaporin-2 (AQP2), whereas water exits the cell through AQP3 and AQP4 which are constitutively present at the basolateral membrane. The required hypertonic medullary interstitium is obtained through deposition of salt by the urine dilution process in the thick ascending limb through the sodium-potassium-chloride co-transporter NKCC2, and through transport of urea in the inner medullary collecting duct (IMCD) through specific urea transporters UT-A1 and UT-A3 (Fig. 3.1).

The antidiuretic pituitary hormone VP plays a major role in regulating the response of the kidney to variation in water intake. It is released in response to minor increases in extracellular osmolality or to greater decreases in arterial blood pressure and binds to the Gs protein coupled vasopressin type 2 receptor (V2R). The earliest genetic model that was used for studying urinary concentration was the Brattleboro rat, which has undetectable VP levels in plasma due to a random mutation and consequently suffers a major concentrating defect. More recently, inducible V2R knockout (KO) mice were developed. Other candidates in the regulation of this process include systemic hormones such as angiotensin-2, oxytocin and secretin, as well as locally synthesized regulatory factors such as nitric oxide and PGE2; the roles of several of which have also been elucidated through use of genetic models.

During high water intake, little to no net transport of water takes place in the CD and the CD epithelium has low water permeability. High water permeability of the epithelium is achieved acutely during water restriction through insertion of AQP2 in the apical plasma membrane, a process mediated by VP. In the long term, VP increases AQP2 protein expression. In addition to modulation of CD water permeability, VP also increases osmolality in the inner medulla interstitium via its effects on facilitative urea transport and sodium reabsorption, ultimately creating an osmotic gradient for water reabsorption. Studies in mice lacking AQP2 have provided evidence that it is vital for homeostasis and thus for survival of the organism, whereas studies in conditional AQP2 knockout mice have established the relative roles of the CD and CNT. AQP2 is regulated to a large extent by posttranslational modifications that determine its intracellular trafficking processes, and likely modulate a unique conformation of the protein. Posttranslational modifications of AQP2 that are necessary to maintain body water homeostasis include phosphorylation of the C-terminal tail at ser-256. This and other properties of AQP2 have been elucidated through various knockin mouse models.

The precise roles of various intracellular signaling pathways in urinary concentration are still an area of debate. The V2R induces a range of intracellular signaling pathways in the CD principal cells, of which an adenylyl cyclase type 6 (AC6) induced cAMP increase is thought to play a major role in AQP2 phosphorylation, membrane targeting and thus urinary concentration. An animal model of mice lacking AC6 has deciphered the role of this intracellular signaling pathway for urinary concentration.

In summary, genetic models ranging from random mutations to directed genetic engineering have revealed the importance of a wide range of components in the intricate process of urinary concentration. Discussions of several of these models and what they have told us regarding urinary concentration is covered later in this chapter.

3.1.1 Overview of Mouse Models

The Nobel Prize in Physiology or Medicine was jointly awarded in 2007 to Mario R. Capecchi, Sir Martin J. Evans and Oliver Smithies for their work to develop gene knockout mice (e.g. Koller et al. 1989; Thompson et al. 1989). Advances in these techniques and the ability to engineer the mouse genome has proven useful for a variety of applications in research, medicine and biotechnology. In particular for this chapter, the use of various gene ‘knockouts’, gene ‘knockins’ and transgenic mouse models has added essential information to our understanding of the molecular mechanisms underlying urinary concentration. A variety of mouse models with defective renal water handling or urinary concentrating defects exist (currently over 100 listed in PubMed). Several of these mouse models arise from deletion of genes/proteins that are regulated by VP or play a role in VP signaling e.g. aquaporins, urea transporters, ion transporters and channels (NHE3, NKCC2, ENaC, ROMK, ClC-K1), G protein-coupled receptors (V2R, PGE2 receptors), or various signaling molecules (summarized in Fenton and Knepper 2007). Other models with body water balance defects arise 'by accident', often the result of alterations in renal architecture following gene deletion. As the phenotype of several of these models will be discussed in detail later in this chapter, a brief overview of the gene manipulation techniques utilized to generate the models is informative.

3.1.2 Basics of Gene Manipulation

A large number of genetic models are generated via the use of homologous recombination in embryonic stem (ES) cells, which has become a relatively standard methodology to modify the mouse genome at a specific position (locus). Although initially developed for altering the genome of yeast, it was adapted for use in mammalian cells 30 years ago by Oliver Smithies (Smithies et al. 1985) and is now routinely used for manipulation of mouse ES cells. The basis of genetic recombination for manipulation of the mouse genome is that a segment of DNA, identical to the region of interest in the majority of aspects, is generated via standard molecular techniques and used to replace the original piece of DNA in the genome. This engineered fragment has the desired 'mutation' (e.g. altered DNA sequence, or selection marker) flanked by several kilobases of DNA that are homologous to the mouse genome flanking the mutation—this is where the genomic recombination occurs (see Fig. 3.2a, b). As homologous recombination in ES cells occurs in less than 0.01 % of cells, positive recombination events are screened for drug resistance/sensitivity combined with screening via Southern blot or PCR. Adaptations to this technique, alongside novel gene editing tools, molecular biology techniques and advances in cell biology mean that any gene deletion, DNA point mutation, inversion or translocation can now be modeled in mice. In addition, these modifications of the genome can be designed to occur both in a spatial and temporal manner. Detailed explanations of the generation, characterization techniques, advantages and pitfalls of genetically modified mice are described in detail in various excellent reviews e.g. (Capecchi 2005; Eisener-Dorman et al. 2009; Kohan 2008; Matthaei 2007). The types of genetic manipulation used to generate mouse models in this chapter can be subdivided into the following (Fig. 3.2):

1. *Gene Knockout Mice*. Although random gene deletion can be achieved through the use of chemical mutagenesis e.g. the highly potent mutagen N-ethyl-N-nitrosourea (ENU) that induces 1 new mutation in every 700 loci, gene targeting via homologous recombination allows complete removal of large regions of coding DNA at a specific locus, resulting in the production of non-functional mutated or truncated protein (Fig. 3.2b). The region of DNA to be removed is flanked by DNA identical to the surrounding region, allowing homologous recombination to occur and target the desired region rather than allowing the DNA to incorporate itself randomly into multiple locations.

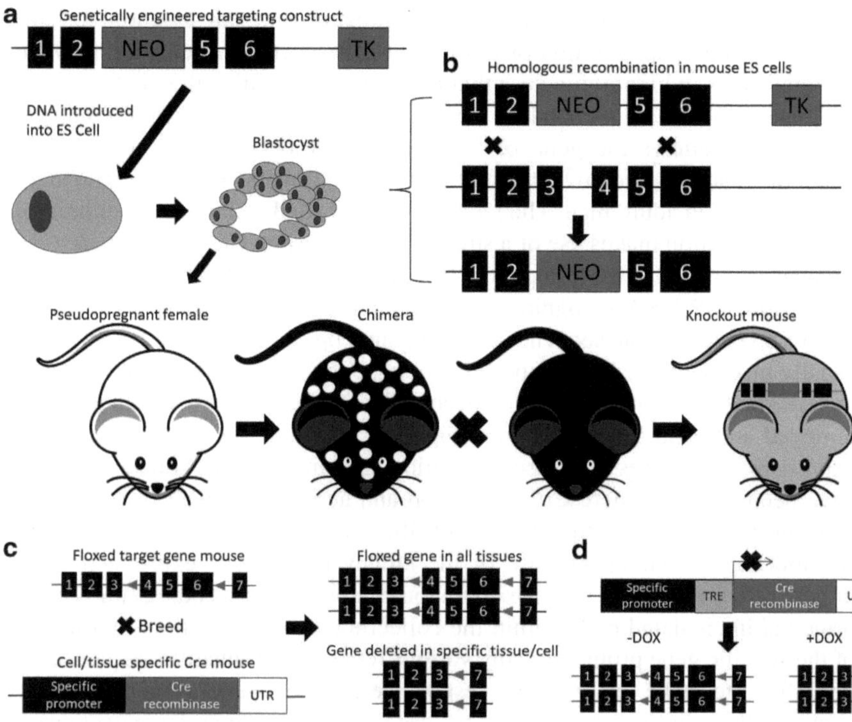

Fig. 3.2 Overview of the generation of various gene modified mice. (**a**) a targeting construct containing a neomycin selection cassette (NEO) instead of exons essential for normal target gene function and a negative selection thymidine kinase cassette (TK) is generated by standard molecular biology techniques and introduced into embryonic stem (ES) cells (usually from a SV129/J mouse background, agouti coat color) by homologous recombination. (**b**) Homologous recombination results in insertion of the modified gene sequence into a single allele of genomic DNA. Correctly targeted ES cells are microinjected into a blastocyst derived from a mouse host with an alternative genetic backgrounds (e.g., C57BL/6, black coat color) and these blastocysts are transplanted into a pseudopregnant female. The resulting chimeric offspring (mixed coat color) are backcrossed to a mouse of the same genetic background as the host blastocyst (e.g., C57BL/6) resulting in agouti offspring if they contain the modified DNA sequence. (**c**) Overview of cell/tissue specific knockout. Gene deletion occurs in selected cells due to exon removal only in cells expressing cre recombinase. See text for details. (**d**) Overview of inducible knockout mouse. In this situation the cre recombinase is only 'active' in the presence of a specific compound e.g. doxycycline (DOX), allowing time-specific control over gene deletion

2. *Gene Knockin Mice*. These are generated in a similar fashion to knockout mice, but the region of interest is modified in the targeting construct to contain a point mutation, or alterations in gene sequence to ultimately result in altered amino acid incorporations into a specific protein. The technique has often been used to generate mouse models of genetic diseases resulting from frameshift mutations or amino acid substitutions.

3. *Conditional gene modified mice*. A large number of genes are essential for either mouse development, viability or fertility, or in respect to this chapter, play such essential roles in kidney function that creation of a traditional gene knockout is not plausible due to prenatal or neonatal death. However, conditional gene modification allows the gene of interest to be knocked-out in a spatial and temporal manner, circumventing lethality and allowing researchers to define gene function in adult mice. The most commonly used technique for conditional gene modification makes use of a site-specific integrase, Cre recombinase (Cre), isolated from the P1 bacteriophage. Cre catalyzes recombination between two consensus loxP DNA recognition sites, which are 34 bp in length and contain two 13 bp palendromic sequences flanking an 8 bp central region. LoxP sites are usually placed either side of a protein coding exon or other essential region of a gene, so that if loxP recombination occurs, the DNA segment between the sites is removed and gene function is lost (Fig. 3.2c). Regulation of Cre recombinase expression determines DNA recombination, and an extensive collection of mice have been created, where Cre expression and activity is driven by a variety of gene promoters with either tissue specific, cell specific, or developmentally specific expression. In the case of the kidney tubule, Cre expression can be targeted to, for example the proximal tubule cells, thick ascending limb cells, or principal/intercalated cells within the collecting duct. In addition, modification of the specific gene promoter using various repressor or enhancer sequences can be utilized to make the promoter responsive to an exogenous agent like doxycycline, allowing Cre expression to be turned on/off in a time-dependent manner (Fig. 3.2d).

3.2 Central Diabetes Insipidus

In medical terminology, defects in the urinary concentrating mechanism leading to increased water excretion are referred to as diabetes insipidus. Central diabetes insipidus occurs when VP release from the neurohypophysis is compromised, e.g. as a side effect to intra-cranial surgery. This condition is routinely treated by substitution of the hormone. Nephrogenic diabetes insipidus (NDI) is either hereditary or acquired and is caused by a deficient renal response to VP (Moeller et al. 2013). The treatment of this condition is far more complex and is still under scrutiny in the fields of basic and clinical science. Thiazide diuretics alongside PGE2 synthesis inhibitors or amiloride improve the condition and are a widely used treatment in patients (Knoers and Monnens 1992). The mechanism of thiazide diuretics is largely unknown.

3.2.1 *Brattleboro Rats*

A litter of laboratory rats with a spontaneous genetic mutation causing familial hypothalamic diabetes insipidus was born in Brattleboro in 1962 (Valtin et al. 1962). In this original strain, the rats were characterized as having depleted levels of VP and oxytocin in the pituitary gland (Sawyer et al. 1964) and emerged as the most widely utilized and characterized model for studying the urinary concentrating mechanism. In the earliest reports, Brattleboro rats had urine osmolality as low as 114 mOsm/kg (Sawyer et al. 1964) and a slight increase was observed during 6 h dehydration up to 250 mOsm/kg (Valtin et al. 1962). In 1979, the urinary concentrating ability of rats bred from this original litter was studied in more detail. Baseline urine osmolality was 155 mOsm/kg, and during 6 h dehydration, this increased to around 464 mOsm/kg (Gellai et al. 1979) alongside an increase in papillary osmolality and urine-to-plasma osmolality. However, plasma VP levels remained undetectable during dehydration (Michimata et al. 2003). The rats showed clear signs of severe dehydration during a 24 h experiment, including 20 % weight loss, increased plasma osmolality and a decline in GFR to 29 % of baseline levels. These studies demonstrate the importance of VP for maintaining whole body water homeostasis during water restriction, but also provide evidence that VP independent mechanisms exist for increasing urinary concentration. This was later established as a V2R independent mechanism, as the same effect was seen in Brattleboro rats treated with a V2R antagonist (Michimata et al. 2003). Notably, the urinary concentrating ability of Brattleboro rats appears to increase over generations of offspring from 124 mOsm/kg in one of the first studies from 1964 (Sawyer et al. 1964) to 155 mOsm/kg in a study from 1979 (Gellai et al. 1979) and in 2003, researchers observed an average urine osmolality of 182 mOsm/kg (Michimata et al. 2003). This renders two notes of caution: (1) it is possible that rats that lack VP evolve a compensatory mechanism over generations of offspring, making them less suitable for studying the role of VP and the development of an inducible model for VP deficiency may be relevant; (2) Direct comparison of urine parameters between different models, or within the same model analyzed in two different labs has to be performed with caution, as technical differences between urine collection methods, animal housing conditions and animal dietary intake can greatly affect the data obtained.

3.2.2 *EP1 Receptor Knockout Mice*

A mild urinary concentrating defect was observed in mice lacking the E-prostanoid receptor EP1 (Kennedy et al. 2007). Stimulation of the pituitary gland with PGE2 induces VP release ex vivo and in vivo (Hashimoto et al. 1988, 1989; Negro-Vilar et al. 1985) and plasma VP increases in response to an osmotic stimulus or hemorrhage are inhibited by COX inhibitors, which decrease prostanoid synthesis

(Brooks et al. 1986; Yamaguchi et al. 1998). After 4 h of water deprivation, urine osmolality and hypothalamic VP mRNA increases were blunted in the EP1 knockout mice (Kennedy et al. 2007). Thus, it appears that PGE2, acting through the EP1 receptor, is a requirement for an acute increase in VP release in response to hyperosmolality, but not for VP release as such, resulting in mild central diabetes insipidus.

3.3 Nephrogenic Diabetes Insipidus: V2R and Adenylyl Cyclase

Although other hormones and local regulatory factors also play a role, VP is by far the most important antidiuretic hormone, as illustrated in Brattleboro rats. In the kidney, VP binds to its type 2 receptor, V2R; a Gs coupled receptor, which increases cAMP and causes Ca^{2+} oscillations in the CD (Fig. 3.3). Although the V2R induces a range of signaling pathways, Gαs mediated cAMP increases are believed to be the major pathway for increased osmotic water permeability, and several of the actions of VP are mimicked by forskolin (a non-specific activator of various adenylyl cyclases) or cAMP treatment.

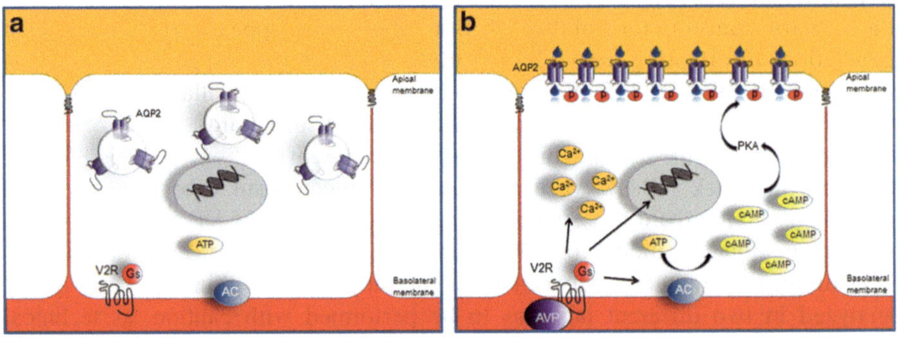

Fig. 3.3 Schematic overview of vasopressin action in the kidney collecting duct principal cells. During water loading and in Brattleboro rats (**a**), the collecting duct has low permeability to water. The water channel AQP2 is located in intracellular vesicles and the apical (luminal) plasma membrane has low water permeability and is rate limiting for water transport across the epithelium. During water restriction (**b**) AQP2 is translocated and its insertion in the apical plasma membrane increases water permeability of the collecting duct and thereby reabsorption of water from the tubular lumen back to the general circulation. This event requires phosphorylation of AQP2 at ser256 mediated predominantly by protein kinase A (PKA). This process is stimulated by the antidiuretic hormone vasopressin (VP), which binds to the G-protein coupled receptor V2R and induces a range of effects including increased AQP2 transcription and diverse signaling pathways e.g. known exocytosis mediators Ca^{2+} and, through activation of adenylyl cyclase (AC), cyclic AMP (cAMP). The precise involvement of the specific signaling pathways in AQP2 translocation events is not fully understood, but a role for cAMP has recently been elucidated using AC-VI knockout mice (Rieg et al. 2010)

3.3.1 V2R Knockout Mice

Mice harboring global knockout of the VP type-2 receptor V2R are not viable beyond 5 days after birth without treatment (Yun et al. 2000). Thiazide treatment was able to increase animal survival up to an average of 16 days. Male homozygous mice suffered a severe concentrating defect alongside hydronephrosis, while heterozygous female mice had a milder concentrating defect that responded to 24 h water deprivation or treatment with dDAVP, a V2R agonist. Interestingly, there was no significant difference in AQP2 expression in the males or females and moreover, heterozygous females displayed no decrease in AQP2 apical membrane targeting.

Mice lacking the V2R have also been developed as a tamoxifen induced knockout model (Li et al. 2009). These mice have a clear urinary concentrating defect, displaying urine osmolality of 175 ± 18 mOsm/kg at baseline, distension of the renal pelvis, decreased GFR and a marked decrease in whole kidney AQP2 and AQP3 protein expression. However, they were viable for at least 3 months after tamoxifen treatment, providing evidence that the V2R is involved in the maintenance of urinary concentrating ability during free water intake that is not due to an effect on kidney development. The ability to concentrate urine during water restriction has not been studied in mice lacking the V2R, but the model was utilized to study the effect of a selective agonist for the PGE2 receptor EP4, which increased urine osmolality and AQP2 protein expression and restored GFR in the V2R knockouts (Li et al. 2009).

3.3.2 Adenylyl Cyclase 6 Knockout Mice

Mice lacking adenylyl cyclase type 6 (AC6) are viable and display normal body weight and no obvious phenotype by gross examination (Tang et al. 2008). The mice have a low basal level of cAMP in the IMCD that does not respond significantly to VP (Rieg et al. 2010). They display a moderate urinary concentrating defect with a 2.5-fold decrease in urine osmolality and a corresponding increased fluid intake during baseline conditions (free water intake), alongside maintained AQP2 protein expression. Phosphorylation at ser-256 decreases to half the levels observed in wildtype. When water deprived, AC6 knockout mice increase urine osmolality proportionately more than wildtypes, but maintain lower urine osmolality and develop plasma hyperosmolality. In contrast, when water loaded, AC6 knockouts display a clear, but blunted increase in urine osmolality and phosphorylation of AQP2 at ser-256 in response to dDAVP injections. These observations suggest that AC6 is important in maintaining baseline urinary concentration, but is not an absolute requirement for VP mediated AQP2 regulation.

In general, mouse models lacking either the V2R or AC6 have provided evidence that the V2R is pivotal to perinatal survival and maintenance of water homeostasis, with no other receptor sufficiently activated to maintain normal urine production. However, this vital role of the V2R does not entirely depend on regulation of AQP2 abundance and trafficking or on cAMP production. Many questions still remain concerning the role of VP in kidney development and the potential alternative signaling pathways that can lead to AQP2 regulation.

3.4 Nephrogenic Diabetes Insipidus: Aquaporins

3.4.1 AQP1 Knockout Mice

AQP1 is expressed in the apical and basolateral membrane of kidney proximal tubule cells, thin descending limb cells and the endothelium of vasa recta. AQP1 is also abundant in various extrarenal epithelia and endothelia and in erythrocytes. In the kidney, AQP1 does not undergo acute hormonal regulation i.e. rapid trafficking to/from target sites, but is constitutively present at the plasma membrane. Humans with loss-of-function mutations in the gene for AQP1 have no observable phenotype during normal water intake, but fail to increase urine osmolality during water deprivation (King et al. 2001).

AQP1 knockout (AQP1-KO) mice were generated by targeted gene deletion and displayed a 2.3-fold lower urine osmolality than wildtype littermates (Ma et al. 1998; Verkman 2008) (Fig. 3.4). They were viable, appeared normal at birth and survived to adulthood, albeit with mild growth retardation. During a 36 h period of water deprivation, wildtype and heterozygote mice responded with a marked increase in urine concentration and 20–22 % weight loss, whereas AQP1-KO mice did not increase urine osmolality and as a consequence suffered from lethargy, serum hyperosmolality and 35 % weight loss. This study also demonstrated that AQP1 is the predominant water channel of the proximal tubule; apical membrane vesicles from proximal tubules of AQP1-KO mice displayed water permeability similar to what would be expected in membranes without water channels. Moreover, water permeability of similar vesicles from wildtype mice decreased to levels seen in AQP1-KO mice when treated with the non-specific water channel blocker HgCl. Vesicles from heterozygote mice displayed intermediate water permeability, although the mice displayed no NDI phenotype. This is consistent with the hypothesis that under normal circumstances, AQP1 is constitutively expressed to a level where it is not rate limiting for whole kidney water transport.

Studies of proximal tubule fluid flow rate and single nephron GFR in AQP1-KO mice confirmed an important role for AQP1 in proximal tubule water reabsorption (Schnermann et al. 1998). Interestingly, water delivery to the distal nephron was similar to wildtype mice, indicating that increased urine volume in AQP1-KO mice

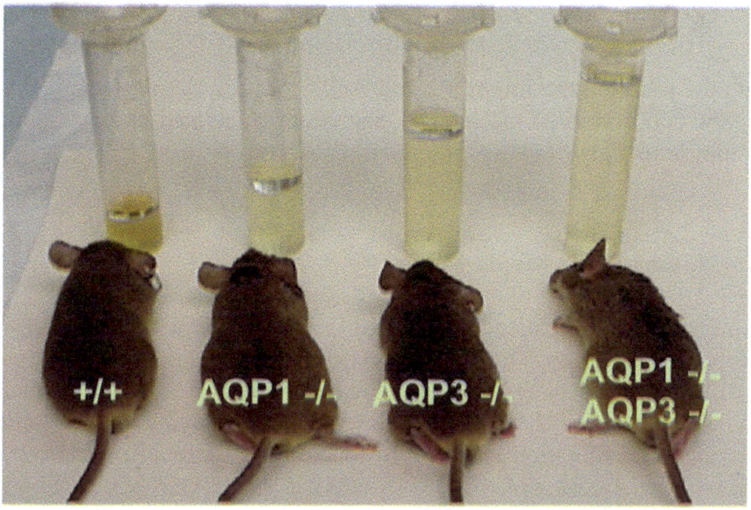

Fig. 3.4 Urinary concentrating ability in mice lacking AQP1, AQP3 or AQP1 and AQP3. Mice are viable and appear grossly normal, but suffer a urinary concentrating defect visualised by the 24 h urine output from each mouse. Image from: *Verkman A.S., Dissecting the Roles of Aquaporins in Renal Pathophysiology Using Transgenic Mice, Seminars in Nephrology, Volume 28, Issue 3, 2008, 217–226* (Verkman 2008) Used with permission

is due to decreased CD water transport and hence uncovering an important role for AQP1 expression in the descending thin limb of the loop of Henle and vasa recta for the countercurrent multiplication and exchange mechanism. Moreover, single nephron GFR was decreased by around 50 % when measured from tubular fluid in the proximal tubule, but flow rates in the distal nephron are not different between wildtype and AQP1-KO mice. This suggests a role for tubuloglomerular feedback (TGF) in retaining water in the AQP1-KO mice, which was later studied in mice lacking both AQP1 and adenosine-1 receptor (AQP1/A1AR-KO) (Hashimoto et al. 2004). Mice lacking the adenosine-1 receptor have a defective TGF response (Brown et al. 2001; Sun et al. 2001) and AQP1/A1AR-KO mice displayed reduced proximal tubular transport coupled with a normal GFR. However, they had only a mild salt-wasting phenotype, presumably due to compensation of the distal nephron. To date, the mechanism of TGF activation in AQP1-KO mice remains unknown.

In the outer medullary descending vasa recta, AQP1 is required for NaCl driven transport of water across the endothelium (Pallone et al. 2000), which may play a role in the concentration of solutes in the interstitium. Mathematical modeling suggested that water is shunted from the descending to the ascending vasa recta; thereby ‘bypassing’ the inner medulla and preventing washout of solutes. Studies in AQP1-KO mice also revealed an AQP1 independent mechanism for water transport driven by hydrophilic solutes (raffinose), but the physiological relevance of this mechanism is unclear. The mechanism for AQP1 independent water transport is not accounted for at the molecular level, except that the mechanism is insensitive to mercury.

3.4.2 AQP2 Mouse Models

The absolute requirement for AQP2 and AQP2 posttranslational modification for the urinary concentrating mechanism has been elucidated through a range of various mouse models.

3.4.2.1 Mutant AQP2 Mouse Models

Approximately 10 % of patients with hereditary NDI have a mutation in the AQP2 gene (Robben et al. 2006). Mice with loss-of-function mutation of AQP2 have been developed using various techniques: (1) Characterization of a family of mice with a spontaneous recessive trait described as congenital progressive hydronephrosis revealed a genetic mutation resulting in the serine residue at position 256 to be replaced with leucine (S256L-AQP2) (McDill et al. 2006). (2) A knock-in mouse model was generated containing the T126M point mutation of AQP2 known to cause NDI in humans (Mulders et al. 1997). (3) AQP2Δ230 mutant mice lacking a large portion of the COOH terminal of AQP2, including all known phosphorylation sites, were generated as a knock-in model (Shi et al. 2007) and 4) C57/BL6 mice were administered the potent germline mutagen ethylnitrosourea, which induces random point mutations throughout the genome of this mouse strain and loss of function alleles in 1 in 700 gametes (Miosge et al. 2002). Subsequent screening for altered whole body metabolism revealed a family of mice with an autosomal recessive NDI phenotype (Lloyd et al. 2005), and by genetic analysis it was predicted that the highly conserved phenylalanine residue at position 204 of AQP2 was substituted for valine (F204V-AQP2). Common to these mouse models is a severe urinary concentrating defect with urine osmolality of 150–500 mOsm/kg H_2O alongside hydronephrosis and either intracellular mis-routing of AQP2 or downregulation of AQP2 protein levels. Despite these similar traits, growth rates and survival were different between the models; T126M-AQP2 mice did not survive more than 8 days even with fluid supplementation (Yang et al. 2001), whereas F204V-AQP2 mice had normal lifespan and displayed no other obvious phenotype than a severe urinary concentrating defect (Lloyd et al. 2005). In contrast to the models discussed above that result in recessive NDI, mice with a frame-shift mutation in the C-terminus of AQP2 develop dominant NDI (Sohara et al. 2006). This phenotype is similar to patients with a frame-shift in the C-terminus (Kuwahara et al. 2001). The mutants form heterotetramers with the wildtype AQP2, resulting in missorting of AQP2 to the basolateral plasma membrane, explaining the dominant phenotype.

3.4.2.2 AQP2 Knock Out Mouse Models

Mice with global knockout of AQP2 (AQP2 KO) suffer severe growth retardation and do not survive beyond 13 days (Rojek et al. 2006). Their phenotype regarding urinary concentration is unknown due to their failure to thrive (Fig. 3.5). The importance of AQP2 for urinary concentration was subsequently studied in CD specific AQP2 KO mice and in inducible AQP2 KO mice. Mice with tamoxifen-induced AQP2 KO were generated by mating mice heterozygous for floxed AQP2 and cre/Esr1 in order to yield mice homozygous for floxed AQP2 with the cre/Esr1 gene (Yang et al. 2006). Mice underwent tamoxifen treatment at 5–8 weeks of age resulting in an AQP2 KO with >90 % efficiency. All mice survived the 16 day duration of the study and were reported to survive for 6 weeks, but the animals suffered a concentrating defect at baseline which peaked 7–10 days after tamoxifen treatment commenced. At this timepoint, an ~11-fold higher urine volume was observed, alongside urine osmolality of 182 mOsm/kg H_2O, which was ~12-fold lower than wildtype mice. During total deprivation of food and water for 18 h, floxed + cre-AQP2 mice did not increase urine osmolality and suffered detrimental weight loss of >30 % compared with ~16 % in wild types. Morphological changes in the kidney included papillary atrophy and dilated CDs in both cortical and medullary regions. Expression of AQP3 was markedly increased at the transcriptional and protein level, with preserved basolateral membrane localization (Yang et al. 2006).

CD specific AQP2 KO mice survive until adulthood, albeit suffering growth retardation, papillary atrophy and hydronephrosis. They display a severe concentrating defect at baseline with a tenfold increase in urine volume and a urine osmolality of 170 mOsm/kgH_2O compared to 1630 ± 135 mOsm/kgH_2O in wildtype mice. Three hours of total water deprivation resulted in moderately decreased urine volume to 87 ± 7 % of baseline levels with unchanged urine

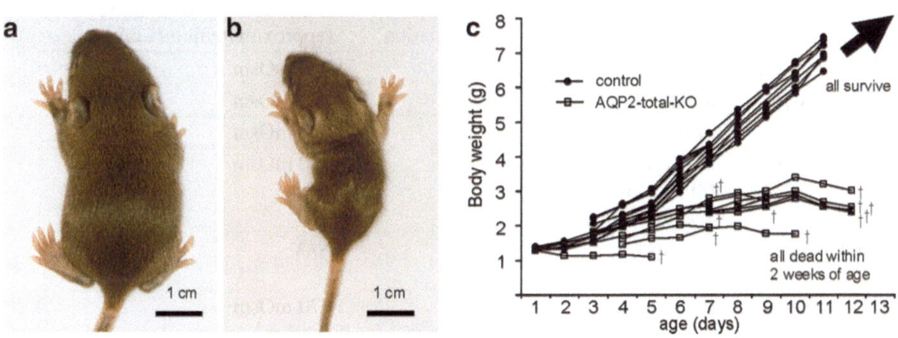

Fig. 3.5 Growth retardation of AQP2-total-KO mice. A 10-day-old wildtype mouse (**a**) and a AQP2-total-KO mouse (**b**) from the same litter. The AQP2-total-KO mouse grew markedly less. (**c**) The growth of AQP2-total-KO and wildtype mice over time. All AQP2-total-KO mice died before day 13, whereas all controls survived. Image from: *Rojek A* et al., *Severe urinary concentrating defect in renal collecting duct-selective AQP2 conditional-knockout mice PNAS 2006;103:6037–6042* (Rojek et al. 2006). Used with permission. Copyright (2006) National Academy of Sciences, USA

osmolality. Owing to the ability of CD specific AQP2 KO mice to survive to adulthood, whereas total AQP2 deletion is lethal (when occurring from birth), it was hypothesized that the connecting tubule (CNT) may compensate for the low water permeability of the CD in CD specific AQP2 KO mice. Subsequently, CNT specific AQP2 KO mice were developed (Kortenoeven et al. 2013) by mating mice harboring loxP sites around exon 3 of the AQP2 gene (Rojek et al. 2006) with mice expressing Cre recombinase driven by the V-ATPase B1-subunit promoter (Miller et al. 2009). These mice had a mild concentrating defect during free water intake with a ~1.5-fold lower urine osmolality, which was less apparent during 24 h water restriction (55 % of baseline water intake mixed with a fixed amount of food), with a ~1.2-fold lower urine osmolality. During acute VP treatment, urine osmolality was not significantly different in AQP2-CNT KOs compared to littermates that consisted of either floxed mice with no Cre recombinase expression or non-floxed mice with Cre recombinase expression, suggesting that the CD has the major role in VP-mediated antidiuresis.

In general, mice with deletion or mutation of AQP2 display a severe phenotype with a vast urinary concentrating defect, suggesting that the involvement of AQP2 in urinary concentration cannot be compensated for. Interestingly, viability in the mouse models is not directly related to the severity of the concentrating defect (see Table 3.1 for overview), which may reflect a role for AQP2 in kidney development and morphology, which could be affected differently by the mutations (Chen et al. 2012) and is independent of VP.

Another study investigated the involvement of AQP2 interacting proteins in the apical membrane accumulation of AQP2 and thus water balance. AQP2 contains a PDZ interaction motif (PDZ ligand) at its carboxyl-terminus (Noda et al. 2004),

Table 3.1 Overview of AQP2 mouse models

Model	Viable	Growth retardation	Baseline urine osmolality (approximate numbers)
T126M-AQP2	<8 days	Yes	220 mOsm
F204V-AQP2	Adulthood	No	160 mOsm
S256L-AQP2	<4 weeks	Yes	500 mOsm
AQP2Δ230	30 % viable >16 weeks 100 % viable with water loading	Yes	200 mOsm
Global AQP2-KO	<13 days	Yes	N/A
CD-AQP2-KO	Adulthood	Yes	170 mOsm
Inducible AQP2 KO	>6 weeks	N/A	180 mOsm
CNT-AQP2-KO	Adulthood	No	1200 mOsm

which can interact with PDZ domain containing proteins. PDZ domains are common to many proteins involved in organization of signaling complexes at the cellular membrane (Harris and Lim 2001). SPA1, a PDZ domain containing protein, can bind to AQP2 in vitro (Noda et al. 2004). Mice lacking SPA1 have decreased apical membrane accumulation of AQP2 during water restriction, likely resulting in a urinary concentrating defect (although not measured in the study).

3.4.3 AQP3 and AQP3/AQP4 Knockout Mice

AQP3 is in the family of aquaglyceroporins with high permeability for glycerol and moderate water transport properties. AQP3 is the main basolateral water channel of the cortical portion of the CD, where the bulk of distal nephron water transport takes place (Knepper and Burg 1983). Although the apical membrane is thought to be rate limiting for water transport and the main site for VP regulation of body water homeostasis, AQP3 abundance is upregulated by water restriction or VP in both normal and Brattleboro rats (Terris et al. 1996).

To investigate its role in body water homeostasis, AQP3 knockout mice were developed by targeted gene deletion (Ma et al. 2000). AQP3 null mice appeared grossly normal with similar behaviour, weight gain and kidney morphology as wildtype littermates. With free water intake, they displayed marginally increased plasma osmolality and a ~5-fold decrease in urine osmolality alongside a ~10-fold increase in urine volume and water intake (Fig. 3.4). Urine osmolality greatly increased in response to dDAVP or water restriction, leaving questions as to whether AQP3 is rate-limiting for CD water reabsorption. Although it might be speculated that AQP2 or the other basolateral water channel, AQP4, may be upregulated and compensate for the missing AQP3; these proteins are in fact downregulated at baseline (Kim et al. 2005) but have not been assessed during water restriction of AQP3 null mice.

3.4.4 AQP4 Knockout Mice

AQP4 is expressed in many different tissues including the kidney (Ma et al. 1996). In the kidney AQP4 is found in the basolateral membrane of the CD, with highest levels in the medullary CD principal cells (Poulsen et al. 2013). AQP4 null mice were generated by targeted gene deletion and displayed normal growth, tissue morphology and urinary concentration at baseline, but a slightly impaired ability to concentrate the urine during 36 h water restriction, which was not corrected by dDAVP administration (Ma et al. 1997). AQP4 deletion did not affect transcription levels or protein abundance of other kidney AQPs; AQP1, AQP2, AQP3 or AQP5. These findings indicate that bulk distal nephron water transport takes place in the cortical CD, whereas intact IMCD water transport is required for maximal urinary concentration. Furthermore, AQP3 expression in the IMCD is not rate limiting for water reabsorption in normal mice and together with basolateral AQP2 (Chen et al. 2012) is sufficient to compensate for a lack of AQP4 at baseline, but not during water restriction. Mice lacking AQP4 have a fourfold reduction in osmotic water permeability of isolated IMCDs during dDAVP stimulation (Chou et al. 1998), which further indicates that during baseline, the cortical portion of the CD is the main site for maintenance of kidney water transport during baseline conditions, whereas the inner medullary portion comes into play during times of insufficient water intake. AQP3/4 knockout mice have also been developed by breeding of the individual knockout models. These mice were viable and had normal growth rate, but displayed a further disruption of urinary concentrating capacity compared to the individual knockout mice, yet they still responded to dDAVP and to water restriction with increased urine osmolality. This strongly suggests that the basolateral membrane has the capacity to increase transport of water in the absence of both basolateral water channels. An explanation for this may be either constitutive basolateral targeting of AQP2 (Yui et al. 2013) enhancing the water permeability of the basolateral membrane, or a relatively water permeable basolateral plasma membrane in IMCD cells (diffusion of water directly through basolateral lipid bilayer).

3.4.5 AQP7/AQP1 Knockout Mice

AQP7 is an aquaglyceroporin, which in addition to water can also transport urea and arsenite. In the kidney, AQP7 is expressed in the apical membrane of proximal straight tubules. AQP7 knockout mice have slightly decreased water permeability of brush border membrane vesicles obtained from the outer medulla, corresponding to reduced water permeability of the apical membrane of proximal straight tubules (Sohara et al. 2005). Knockout mice have normal plasma glycerol, although urinary glycerol was increased. AQP7 knockout mice do not have a urinary concentrating defect during baseline conditions and renal urea handling was unchanged as assessed by the urea concentration of urine, plasma and inner medullary interstitium. AQP7/AQP1 double knockout mice display a slightly more severe concentrating defect than AQP1 KO mice. These studies suggest that AQP7 is involved in tubular transport of water in proximal straight tubules at low levels that can be compensated for in AQP7 knockout mice (AQP1 may compensate), whereas AQP7 reabsorbs glycerol at levels that cannot be compensated for by other transport mechanisms.

3.5 Nephrogenic Diabetes Insipidus: Urea Transporters

Urea is a product of protein metabolism that is generated in the liver through the TCA and urea cycles. The net effect of these cycles is to convert ammonium ions derived from excess amino groups, generated by transamination reactions during gluconeogenesis and anabolic processes, into the non-toxic compound urea, which is excreted in the urine. As urea is an osmolyte, its excretion in large enough amounts would be predicted to cause osmotic diuresis and severely compromise the kidney's ability to retain water during high protein intake. This effect is counteracted by the unique handling of urea in the kidney inner medulla owing to the high permeability of the IMCD for urea. Urea recycling has also been hypothesized to play a major role in the urinary concentrating mechanism, but studies in UT-A2 and UT-B knockout mice do not fully support this view. Interstitial accumulation of urea in the inner medulla constitutes a significant portion of inner medullary interstitial hypertonicity and the driving force for water reabsorption. The molecular mechanism involves facilitated transport of urea across the plasma membrane through specific transporters (Fenton and Knepper 2007). Urea transporters are derived from the genes UT-A and UT-B (Fenton et al. 2002a; Yang et al. 2002). Although several UT-A proteins exist due to alternative splicing, only 2 isoforms seem to play an essential role in urine concentration. Urea exits the IMCD by passive transport through UT-A1 on the apical membrane (Fenton et al. 2002b; Nielsen et al. 1996) and UT-A3 on the basolateral side (Stewart et al. 2004; Terris et al. 2001). UT-A2 is expressed in the thin descending limbs (Wade et al. 2000), and UT-B in the descending vasa recta (Tsukaguchi et al. 1997). Urea transporter abundance is increased during dehydration and in response to VP (Fenton and Knepper 2007).

Mice with targeted deletion of UT-A1 and UT-A3 displayed a ~3.5-fold decrease in inner medullary urea concentration alongside a decrease in baseline IMCD urea permeability that did not respond to VP stimulation (Fenton et al. 2004). IMCD osmotic water permeability was normal in the knockouts. When fed a 'normal' 20 % protein diet, UT-A1/3 knockout mice presented with a urinary concentrating defect, with ~3.5-fold increase in water intake and urine volume and ~3.5-fold decrease in urine osmolality compared with wildtype littermates, which was similar during free access to drinking water and during 24–36 h water restriction (1.7 ml water per 20 g body weight per day). The differences were more pronounced when mice were fed a 40 % protein diet (Fenton et al. 2005). When fed a 4 % protein diet, UT-A1/3 knockout mice displayed a marginal concentrating defect that was only evident during water restriction (Fenton et al. 2004, 2005). These studies are consistent with the view that CD urea transport through UT-A1/3 is a mechanism that counteracts protein metabolism induced osmotic diuresis.

UT-B knockout mice display a concentrating defect both at baseline and during water restriction alongside decreased urinary and papillary urea concentration and increased plasma urea levels (Yang et al. 2002). This is in line with the hypotheses that (1) UT-B expression in the descending vasa recta is involved in accumulation of urea in the inner medullary interstitium and (2) this accumulation of urea is necessary for the facilitated transport of urea from the IMCD. In contrast, UT-A2 knockout mice displayed no discernable urinary concentrating defect, although the papillary urea concentration was also decreased in this model (Uchida et al. 2005). Thus, papillary urea concentration does not in itself explain the concentrating defect in UT-B knockout mice. The difference in phenotype between the two mouse models may be explained by the expression pattern of UT-A2 and UT-B. Urea enters the bloodstream from the highly concentrated papillary tip interstitium into the ascending vasa recta from where it exits back into the interstitium at the less concentrated papilla base. The low concentration of urea at the papillary base is maintained by the re-entry of urea into the descending thin limb through UT-A2 and the descending vasa recta through UT-B. In UT-A2 knockout mice, a compensatory increase in UT-B expression is observed, which to some degree would maintain urea accumulation in the papillary interstitium. It may be hypothesized that the reverse situation is seen in UT-B knockout mice, with a VP mediated upregulation of UT-A2. Clearance from the papillary base would occur through increased re-entry of urea into the circulation and the tubular lumen and increase the urea load to the CD in the UT-B knockout mouse model while reducing the interstitial concentration and gradient for water absorption from the IMCD. In line with this view, double UT-A2/UT-B knockout mice display a milder urinary concentrating defect then UT-B knockout alongside normalization of urea excretion (Lei et al. 2011).

Overall, studies on urea transporter knockout mice have provided insight into the mechanisms of urea accumulation in the inner medullary interstitium and can thus partly explain interstitial hypertonicity. However, the mechanisms for accumulation of NaCl in the inner medulla, another major component of the inner medullary interstitium, are still unaccounted for in the literature [see (Fenton et al. 2006) for discussion].

3.6 Nephrogenic Diabetes Insipidus: Electrolyte Transporters

Another distinct function arising from the urinary concentrating mechanism is the preservation of sodium. As with renal water handling, a defect in kidney sodium transport can result in polyuria and polydipsia as seen for example in salt wasting nephropathies such as Bartter and Gitelman's syndrome.

Functionally, kidney sodium transport is best understood as a distinction between pre-, intra- and post- macula densa transport. In general, bulk sodium transport/absorption takes place pre-macula densa in the proximal tubule; sodium transport in the thick ascending limb plays a role in urinary dilution and increasing medullary interstitial osmolality; intra-macula densa sodium transport is part of the sensing mechanism leading to TGF and finally, post-macula densa sodium transport is the hormonally regulated fine-tuning of sodium excretion in response to changes in whole-body sodium balance. Ultimately, modulation of sodium transport in each of these segments results in vastly different effects on the urinary concentrating mechanism, predominantly as modulated mechanisms in later segments can compensate for disturbances in early nephron sodium transport, but the alternative scenario is usually not possible.

3.6.1 NHE3 and NKCC2 Knockout Mice

NHE3 is a Na^+-H^+ antiporter expressed in the apical membrane of the proximal tubule and thick ascending limb. NHE3 knockout mice grow to adulthood with grossly normal phenotype besides mild diarrhea (Schultheis et al. 1998a). Upon closer examination, mice had decreased systolic and mean arterial blood pressure (Lorenz et al. 1999; Schultheis et al. 1998a) alongside slight metabolic acidosis without respiratory compensation, coupled with increased plasma K^+ without changes in plasma Na^+ (Schultheis et al. 1998a) and increased αENaC (see below) mRNA in the kidney CD and increased βENaC and γENaC mRNA in the colon (Schultheis et al. 1998a). Proximal reabsorption rates were greatly decreased in knockout mice (Lorenz et al. 1999), but micropuncture studies revealed that distal fluid delivery was similar in NHE3 knockout and wildtype mice and that this is due to a decrease in single-nephron-GFR and activation of TGF (Lorenz et al. 1999). However, this study revealed substantially decreased urinary flow rates in NHE3 knockout mice. Taken together, this indicates CD (and colonic) compensation driven by acidosis, increased renin activity and plasma aldosterone (Schultheis et al. 1998a), but the lower blood pressure indicates volume depletion and that full compensation is not achieved. Thus, NHE3 is a major determinant of Na^+ and fluid reabsorption in the proximal tubule, but a relatively mild phenotype of NHE3 knockout is accomplished by renal autoregulation. NHE3 knockout mice display a vastly decreased protein expression of AQP2 in the inner medulla and cortex despite elevated plasma VP, whereas AQP2 expression in the outer medulla is not affected (Amlal et al. 2003). These findings indicate that AQP2 expression is regulated by local factors in the model, of which PGE2 (Olesen and Fenton 2013), nitric oxide (Morishita et al. 2005), purinergic signaling (Wildman et al. 2009) and tissue osmolality (Kortenoeven et al. 2011) are known to play a role in AQP2 regulation.

The importance of renal autoregulation is further exemplified in NKCC2 knockout mice. Although NKCC2 is expressed only in the thick ascending limb and macula densa and transports less bulk Na^+ than NHE3, NKCC2 knockout mice, although grossly normal at birth, suffer severe dehydration and growth retardation and do not survive longer that 2 weeks after birth (Takahashi et al. 2000). The knockout mice suffer metabolic acidosis alongside severe electrolyte derangement including hypernatremia and hyperkalemia, indicating that NKCC2 plays a significant role in water reabsorption via its role in the countercurrent multiplication mechanism. The metabolic acidosis may reflect a generalized thick ascending limb dysfunction. A proportion of NKCC2 knockout mice could be rescued by treatment with indomethacin and survived to adulthood with a vast urinary concentrating defect that did not respond to VP or 12 h water deprivation. Electrolyte derangement was milder in the adult mice, displaying borderline alkalosis alongside hypernatremia and lower plasma K^+ than controls. Although this indicates distal nephron and CD compensation driven by the renin-angiotensin-aldosterone system can partially prevent Na^+ wasting, it cannot counteract the urinary concentrating defect due to lack of interstitial hypertonicity. As with other models with severe urinary concentrating defects, adult mice suffered progressive, extreme hydronephrosis. The explanation offered in the models is that the ureter is unable to accommodate the vast amounts of urine that are produced, which puts constant pressure on the renal pelvis and nephrons and ultimately results in increases in the pelvic space and hydronephrosis.

NKCC2 exists as three different splice variants; NKCC2F is mainly expressed in the inner stripe of outer medulla, NKCC2B predominantly in the cortex and NKCC2A in both cortex and inner stripe of outer medulla (Oppermann et al. 2006). NKCC2F and NKCC2B have been knocked out separately, resulting in mice with far milder phenotypes than total NKCC2 deletion and allowing for studies to determine the relative involvement of NKCC2 activity in different nephron segments. NKCC2B knockout mice display a grossly normal phenotype, although a mild concentrating defect both at baseline and during 48 h water restriction was observed (Oppermann et al. 2006). With normal plasma electrolyte status, the most striking observation in the mice was the apparent right shift of the TGF response coupled with an exacerbated plasma aldosterone response to an acute intravenous salt load. Thus, macula densa localization of NKCC2 is the sensor mechanism that instigates the TGF response when tubular sodium content is high, whilst modulating the effects of plasma electrolyte status on aldosterone release from the adrenal gland. NKCC2A knockout mice displayed a very mild phenotype. Their urine osmolality was slightly lower than that of wildtypes during free water intake, but was corrected during 48 h water restriction to levels similar to wildtypes. The response of the plasma renin concentration to an acute salt load was blunted and in microperfused loops of Henle, distal Cl^- concentration was increased during high perfusion rates. These data indicate that NKCC2A effects can be compensated for and that macula densa expression of NKCC2 is tightly regulated. Taken together, studies in NKCC2 knockout mouse models provide evidence that intact macula densa function depends on NKCC2 expression, which is crucial for survival of the organism in conditions of decreased proximal Na^+ reabsorption, and that NKCC2 is essential for the development of interstitial hyperosmolality and thus urinary concentrating ability.

3.6.2 ENaC and NCC Knockout Mice

The apical epithelial sodium channel (ENaC) of the CD consists of three subunits, α-, β- and γENaC. Mice lacking the β- or γ- subunit have a severe salt wasting phenotype and do not survive beyond 48 h after birth (Barker et al. 1998; McDonald et al. 1999). αENaC knockout mice also suffer perinatal death within 48 h, but this is attributable to lung edema, due to a role for αENaC in clearance of fluid from the lung after birth (Hummler et al. 1996). CD specific αENaC knockout mice have a milder phenotype, displaying metabolic acidosis and salt wasting for the first 12 h after birth, which is subsequently corrected in the survivors (Hummler et al. 1997).

40 % of males and 10 % of females survive to adulthood and male survivors have near-normal growth rates. Moreover, adult survivors display no discernable salt wasting phenotype or urinary concentrating defect (Hummler et al. 1997; Rubera et al. 2003) besides greatly increased plasma aldosterone levels, although their responsiveness to aldosterone is uncertain. Thus, the possibility that an aldosterone-induced increase in other ENaC subunits or NCC may compensate for the lack of αENaC could be considered.

In contrast to ENaC knockout mice, NCC knockout mice have a mild phenotype, displaying a decrease in mean arterial pressure only during a sodium depleted diet, with no clearly discernable increase in sodium excretion (Schultheis et al. 1998b). One hypothesis may be that deletion of NCC is compensated for by increased CD sodium reabsorption, which would still leave questions as to the mechanism of decreased arterial blood pressure. When fed a normal diet, NCC knockout mice had normal urine volume and water intake, but when fed a low-K^+ diet, NCC knockout mice developed K^+-depletion which could be brought on by (1) elevated plasma aldosterone, (2) hypocalcuria and/or (3) increased luminal flow in the CD (Morris et al. 2006). Hypokalemia induces polydipsia and polyuria in the model (Morris et al. 2006). Elevated plasma aldosterone is inconsistently observed in the knock-outs regardless of the diet.

In summary, the general insights learned from mice lacking renal sodium transporters are that defects in pre-macula densa sodium reabsorption lead to a mild phenotype if TGF is intact and defects in intra-macula densa sodium transport lead to a mild phenotype if proximal sodium reabsorption is intact. The CD compensates for blunted sodium reabsorption in other tubular segments, but disrupted CD sodium reabsorption cannot be compensated for and leads to a severe salt wasting phenotype.

3.6.3 CLC-K1

CLC-K1 mediates transepithelial chloride transport in the thin ascending limb in the inner medulla (Uchida et al. 1995). Studies in CLC-K1 knockout mice reveal a role of countercurrent exchange mechanisms driven by this protein to increase inner medullary interstitial osmolality. CLC-K1 knockout mice had polyuria with a fivefold increase in urine volume and decreased urine osmolality, whereas urinary excretion of Na^+, K^+ and Cl^- was not different from wildtype mice (Matsumura et al. 1999). In response to 24 h water restriction, a vast increase in plasma osmolality was observed, mice became lethargic and lost 27 % body weight vs 13 % in wildtypes, and they were not able to increase urine osmolality (Matsumura et al. 1999). VP injections also failed to increase urine osmolality in the mice. In vitro microperfusion of thin ascending limbs revealed that Cl^- transport was abolished. The conclusions of this study are that CLC-K1 is the predominant Cl^- transporter in the thin ascending limbs and that this process is necessary for medullary hypertonicity, resulting in water diuresis in its absence. This was subsequently confirmed in a second study showing that inner medullary interstitial osmolality was decreased at baseline and the increase seen during water restriction in wildtypes was severely blunted in knockouts (Akizuki et al. 2001). This was coupled with an impaired ability to accumulate urea in the inner medullary interstitium during water restriction, likely due to increased fluid delivery to the distal nephron and therefore decreased driving force for urea across the CD. Of note, the mechanism(s) of Na^+ deposition in the inner medullary interstitium remain poorly understood.

3.6.4 ROMK Knockout Mice

The potassium channel ROMK is expressed in the thick ascending limb, macula densa, distal convoluted tubule, connecting tubule and CD and secretes K^+ into the tubular lumen. ROMK null mice display growth retardation and severe dehydration by 1 week of age and only 5 % survive until adulthood (Lorenz et al. 2002). One week old mice have hydronephrosis alongside uncompensated hyperchloremic metabolic acidosis. (Blood gas values also indicate impaired lung function.) They display electrolyte derangement, notably severe hypernatremia and trending hyperkalemia. Hematocrit is increased which is expected due to increased whole body K^+ content leading to relative intracellular volume expansion. Adult knockout mice display lower blood pressure and urine osmolality alongside a 2.5-fold increase in water intake consistent with a persistent urinary concentrating defect in the survivors. Hyperchloremic metabolic acidosis and hypernatremia persisted, but hematocrit and plasma K^+ had normalized in the knockouts, indicating that survivors develop an alternative route for K^+ excretion, most likely through Maxi-K channels in the CNT. Indeed, surviving adult knockouts have increased Na^+ and K^+ excretion (Lu et al. 2002). ROMK knockouts had normal single nephron GFR, but displayed blunted TGF, indicating that both NKCC2 and ROMK expression in the macula densa are necessary for intact renal autoregulation. The adult surviving ROMK knockouts were crossed with heterozygotes to develop a strain with a higher survival rate of 30 % in the initial strain (Lu et al. 2002). As with CD specific αENaC knockouts, the phenotype of ROMK deletion is more severe in females (Yan et al. 2008). Their exacerbated urinary concentrating defect, hydronephrosis and reduced GFR were paralleled with increased PGE2 excretion.

3.7 Vasopressin Independent Urinary Concentration

Mouse models have shown that AQP2 is regulated by systemic hormones other than VP as well as local factors. Angiotensin II (ANGII) increases during dehydration, and mice lacking the AT1a receptor exhibit a urinary concentrating defect (Oliverio et al. 2000), which can be partly explained by direct actions of the receptor in the

CD (Stegbauer et al. 2011). As with VP, secretin is released from the neurohypophysis and increased plasma levels are observed during water restriction (Chu et al. 2009). It increases AQP2 levels ex vivo in IMCD tubular suspensions (Chu et al. 2007). Secretin acts via a Gαs protein coupled secretin receptor in the CD (Chu et al. 2007).

Secretin receptor knockout mice displayed mild polyuria at baseline with slightly increased VP levels (Chu et al. 2007). After 48 h water deprivation, AQP2 membrane targeting and mRNA expression were severely blunted in secretin receptor knockouts. Secretin treatment of CDs ex vivo clearly enhanced AQP2 membrane targeting in wildtype mice, but not in mice lacking the secretin receptor. The knockout mice have greatly decreased AQP2 membrane targeting during water deprivation, which seems surprising, as there is still a functional V2R in these mice and VP regulatory mechanisms should be in play. However, infusion of secretin has been shown to induce VP release from the neurohypophysis (Chu et al. 2009), thus it could be speculated that secretin may be involved in physiological VP release during dehydration.

Nitric oxide is a local factor that may regulate the urinary concentrating mechanism and its role has been studied in knockout mouse models. Mice lacking the three nitric oxide synthases (NOS), nNOS, iNOS and eNOS, were generated with knockout of either one, a combination of two or all three NOS (Morishita et al. 2005). Mice with deletion of a combination of any two or all three NOS displayed a urinary concentrating defect characterized by increased urine volume, plasma osmolality and water intake combined with a decrease in urine osmolality. In mice lacking all three NOS, the concentrating defect was not corrected by infusion of exogenous VP over 3 h which may be explained by the vast decrease in AQP2 protein expression also displayed in the study. The manner in which the NOS are involved in urinary concentration is still debated.

3.8 Perspectives

The kidney tubules modulate reabsorption/secretion of large quantities of water and salt, and only minor alterations in these processes are required to maintain homeostasis over a wide range of water and salt intake. Although the majority of water and salt transport takes place constitutively in the proximal part of the renal tubule—pre-macula densa—disruption of these proximal processes are detected in the macula densa and an autoregulatory response allows these changes in homeostasis to be compensated for by decreasing GFR and increasing sodium reabsorption post macula densa. In contrast, the distal nephron and CD, the hormone responsive parts of the nephron, fine-tune water and salt reabsorption by increasing their transport capacity over a large range. Defects in these processes are usually non-compensable. Studies in knockout mice have largely corroborated this view. Mice lacking AQP1 and NHE3 are viable because of renal autoregulation and compensation of the distal nephron, and AQP1 KO mice suffer severe dehydration only during water restriction, which is due to defective countercurrent exchange rather than decreased proximal tubular water transport. In contrast, mice lacking V2R, AQP2 or β/γ ENaC from birth are not viable and suffer a detrimental phenotype due to the ensuant transport defect in the CD. This shows that distal nephron and CD water and sodium transport not only come into play during a challenge to homeostasis, but also play a vital role during baseline conditions, that cannot be compensated for. Disruption of TGF does not in itself cause reduced viability, as seen in mice lacking the adenosine-1 receptor, but NKCC2 knockout mice, in which TGF is disrupted alongside decreased proximal sodium transport, suffer detrimental salt wasting.

Apart from reabsorbing water and salt back into the general circulation, tubular and renal vascular transport of water and solutes also plays a role in the interstitial accumulation of osmolytes, providing a gradient for water transport. Knockout mouse models have shown that urea and electrolytes are transported independently, and that urea handling in the kidney ensures that excess nitrogen from protein metabolism is excreted without inducing osmotic diuresis.

In conclusion, although genetic engineering directed at single proteins has provided essential insights into whole kidney function, many questions still remain. Furthermore, a limitation regarding studies in knockout mice still prevails: the studies do not distinguish between 'necessary' and 'rate-limiting'. The protein that is missing, if necessary for a given process, will always be the rate limiting component in the model regardless of its physiological role.

Acknowledgements Work from the laboratories of the authors was carried out with the support of the Danish Medical Research Council, the Novo Nordisk Foundation and the Lundbeck Foundation. We thank our many excellent colleagues for their invaluable contributions to our research endeavours over the past several years.

References

Akizuki N, Uchida S, Sasaki S, Marumo F (2001) Impaired solute accumulation in inner medulla of Clcnk1-/- mice kidney. Am J Physiol Renal Physiol 280(1):F79–F87, available from: PM:11133517

Amlal H, Ledoussal C, Sheriff S, Shull GE, Soleimani M (2003) Downregulation of renal AQP2 water channel and NKCC2 in mice lacking the apical Na^+-H^+ exchanger NHE3. J Physiol 553 (Pt 2):511–522, available from: PM:14500765

Barker PM, Nguyen MS, Gatzy JT, Grubb B, Norman H, Hummler E, Rossier B, Boucher RC, Koller B (1998) Role of gammaENaC subunit in lung liquid clearance and electrolyte balance in newborn mice. Insights into perinatal adaptation and pseudohypoaldosteronism. J Clin Invest 102(8):1634–1640, available from: PM:9788978

Brooks DP, Share L, Crofton JT (1986) Role of brain prostaglandins in the control of vasopressin secretion in the conscious rat. Endocrinology 118(4):1716–1722, available from: PM:3456303

Brown R, Ollerstam A, Johansson B, Skott O, Gebre-Medhin S, Fredholm B, Persson AE (2001) Abolished tubuloglomerular feedback and increased plasma renin in adenosine A1 receptor-deficient mice. Am J Physiol Regul Integr Comp Physiol 281(5):R1362–R1367, available from: PM:11641103

Capecchi MR (2005) Gene targeting in mice: functional analysis of the mammalian genome for the twenty-first century. Nat Rev Genet 6(6):507–512, available from: PM:15931173

Chen Y, Rice W, Gu Z, Li J, Huang J, Brenner MB, Van HA, Xiong J, Gundersen GG, Norman JC, Hsu VW, Fenton RA, Brown D, Lu HA (2012) Aquaporin 2 promotes cell migration and epithelial morphogenesis. J Am Soc Nephrol 23(9):1506–1517, available from: PM:22859853

Chou CL, Ma T, Yang B, Knepper MA, Verkman AS (1998) Fourfold reduction of water permeability in inner medullary collecting duct of aquaporin-4 knockout mice. Am J Physiol 274(2 Pt 1):C549–C554, available from: PM:9486146

Chu JY, Chung SC, Lam AK, Tam S, Chung SK, Chow BK (2007) Phenotypes developed in secretin receptor-null mice indicated a role for secretin in regulating renal water reabsorption. Mol Cell Biol 27(7):2499–2511, available from: PM:17283064

Chu JY, Lee LT, Lai CH, Vaudry H, Chan YS, Yung WH, Chow BK (2009) Secretin as a neurohypophysial factor regulating body water homeostasis. Proc Natl Acad Sci USA 106 (37):15961–15966, available from: PM:19805236

Eisener-Dorman AF, Lawrence DA, Bolivar VJ (2009) Cautionary insights on knockout mouse studies: the gene or not the gene? Brain Behav Immun 23(3):318–324, available from: PM:18822367

Fenton RA, Knepper MA (2007) Mouse models and the urinary concentrating mechanism in the new millennium. Physiol Rev 87(4):1083–1112, available from: PM:17928581

Fenton RA, Cottingham CA, Stewart GS, Howorth A, Hewitt JA, Smith CP (2002a) Structure and characterization of the mouse UT-A gene (Slc14a2). Am J Physiol Renal Physiol 282(4):F630–F638, available from: PM:11880324

Fenton RA, Stewart GS, Carpenter B, Howorth A, Potter EA, Cooper GJ, Smith CP (2002b) Characterization of mouse urea transporters UT-A1 and UT-A2. Am J Physiol Renal Physiol 283(4):F817–F825, available from: PM:12217874

Fenton RA, Chou CL, Stewart GS, Smith CP, Knepper MA (2004) Urinary concentrating defect in mice with selective deletion of phloretin-sensitive urea transporters in the renal collecting duct. Proc Natl Acad Sci USA 101(19):7469–7474, available from: PM:15123796

Fenton RA, Flynn A, Shodeinde A, Smith CP, Schnermann J, Knepper MA (2005) Renal phenotype of UT-A urea transporter knockout mice. J Am Soc Nephrol 16(6):1583–1592, available from: PM:15829709

Fenton RA, Smith CP, Knepper MA (2006) Role of collecting duct urea transporters in the kidney—insights from mouse models. J Membr Biol 212(2):119–131, available from: PM:17264985

Gellai M, Edwards BR, Valtin H (1979) Urinary concentrating ability during dehydration in the absence of vasopressin. Am J Physiol 237(2):F100–F104, available from: PM:464096

Harris BZ, Lim WA (2001) Mechanism and role of PDZ domains in signaling complex assembly. J Cell Sci 114(Pt 18):3219–3231, available from: PM:11591811

Hashimoto H, Noto T, Nakajima T (1988) A study on the release mechanism of vasopressin and oxytocin. Neuropeptides 12(4):199–206, available from: PM:3237319

Hashimoto H, Noto T, Nakajima T (1989) Effects of prostaglandin E2 and D2 on the release of vasopressin and oxytocin. Prostaglandins Leukot Essent Fatty Acids 36(1):9–14, available from: PM:2740415

Hashimoto S, Huang Y, Mizel D, Briggs J, Schnermann J (2004) Compensation of proximal tubule malabsorption in AQP1-deficient mice without TGF-mediated reduction of GFR. Acta Physiol Scand 181(4):455–462, available from: PM:15283758

Hummler E, Barker P, Gatzy J, Beermann F, Verdumo C, Schmidt A, Boucher R, Rossier BC (1996) Early death due to defective neonatal lung liquid clearance in alpha-ENaC-deficient mice. Nat Genet 12(3):325–328, available from: PM:8589728

Hummler E, Barker P, Talbot C, Wang Q, Verdumo C, Grubb B, Gatzy J, Burnier M, Horisberger JD, Beermann F, Boucher R, Rossier BC (1997) A mouse model for the renal salt-wasting syndrome pseudohypoaldosteronism. Proc Natl Acad Sci USA 94(21):11710–11715, available from: PM:9326675

Kennedy CR, Xiong H, Rahal S, Vanderluit J, Slack RS, Zhang Y, Guan Y, Breyer MD, Hebert RL (2007) Urine concentrating defect in prostaglandin EP1-deficient mice. Am J Physiol Renal Physiol 292(2):F868–F875, available from: PM:16885154

Kim SW, Gresz V, Rojek A, Wang W, Verkman AS, Frokiaer J, Nielsen S (2005) Decreased expression of AQP2 and AQP4 water channels and Na, K-ATPase in kidney collecting duct in AQP3 null mice. Biol Cell 97(10):765–778, available from: PM:15898956

King LS, Choi M, Fernandez PC, Cartron JP, Agre P (2001) Defective urinary-concentrating ability due to a complete deficiency of aquaporin-1. N Engl J Med 345(3):175–179, available from: PM:11463012

Knepper M, Burg M (1983) Organization of nephron function. Am J Physiol 244(6):F579–F589, available from: PM:6305206

Knoers N, Monnens LA (1992) Nephrogenic diabetes insipidus: clinical symptoms, pathogenesis, genetics and treatment. Pediatr Nephrol 6(5):476–482, available from: PM:1457333

Kohan DE (2008) Progress in gene targeting: using mutant mice to study renal function and disease. Kidney Int 74(4):427–437, available from: PM:18418351

Koller BH, Hagemann LJ, Doetschman T, Hagaman JR, Huang S, Williams PJ, First NL, Maeda N, Smithies O (1989) Germ-line transmission of a planned alteration made in a hypoxanthine phosphoribosyltransferase gene by homologous recombination in embryonic stem cells. Proc Natl Acad Sci USA 86(22):8927–8931, available from: PM:2573070

Kortenoeven ML, van den Brand M, Wetzels JF, Deen PM (2011) Hypotonicity-induced reduction of aquaporin-2 transcription in mpkCCD cells is independent of the tonicity responsive element, vasopressin, and cAMP. J Biol Chem 286(15):13002–13010, available from: PM:21324903

Kortenoeven ML, Pedersen NB, Miller RL, Rojek A, Fenton RA (2013) Genetic ablation of aquaporin-2 in the mouse connecting tubules results in defective renal water handling. J Physiol 591(Pt 8):2205–2219, available from: PM:23359673

Kuwahara M, Iwai K, Ooeda T, Igarashi T, Ogawa E, Katsushima Y, Shinbo I, Uchida S, Terada Y, Arthus MF, Lonergan M, Fujiwara TM, Bichet DG, Marumo F, Sasaki S (2001) Three families with autosomal dominant nephrogenic diabetes insipidus caused by aquaporin-2 mutations in the C-terminus. Am J Hum Genet 69(4):738–748, available from: PM:11536078

Lei T, Zhou L, Layton AT, Zhou H, Zhao X, Bankir L, Yang B (2011) Role of thin descending limb urea transport in renal urea handling and the urine concentrating mechanism. Am J Physiol Renal Physiol 301(6):F1251–F1259, available from: PM:21849488

Li JH, Chou CL, Li B, Gavrilova O, Eisner C, Schnermann J, Anderson SA, Deng CX, Knepper MA, Wess J (2009) A selective EP4 PGE2 receptor agonist alleviates disease in a new mouse model of X-linked nephrogenic diabetes insipidus. J Clin Invest 119(10):3115–3126, available from: PM:19729836

Lloyd DJ, Hall FW, Tarantino LM, Gekakis N (2005) Diabetes insipidus in mice with a mutation in aquaporin-2. PLoS Genet 1(2), e20, available from: PM:16121255

Lorenz JN, Schultheis PJ, Traynor T, Shull GE, Schnermann J (1999) Micropuncture analysis of single-nephron function in NHE3-deficient mice. Am J Physiol 277(3 Pt 2):F447–F453, available from: PM:10484528

Lorenz JN, Baird NR, Judd LM, Noonan WT, Andringa A, Doetschman T, Manning PA, Liu LH, Miller ML, Shull GE (2002) Impaired renal NaCl absorption in mice lacking the ROMK potassium channel, a model for type II Bartter's syndrome. J Biol Chem 277 (40):37871–37880, available from: PM:12122007

Lu M, Wang T, Yan Q, Yang X, Dong K, Knepper MA, Wang W, Giebisch G, Shull GE, Hebert SC (2002) Absence of small conductance K+ channel (SK) activity in apical membranes of thick ascending limb and cortical collecting duct in ROMK (Bartter's) knockout mice. J Biol Chem 277(40):37881–37887, available from: PM:12130653

Ma T, Yang B, Verkman AS (1996) Gene structure, cDNA cloning, and expression of a mouse mercurial-insensitive water channel. Genomics 33(3):382–388, available from: PM:8660998

Ma T, Yang B, Gillespie A, Carlson EJ, Epstein CJ, Verkman AS (1997) Generation and phenotype of a transgenic knockout mouse lacking the mercurial-insensitive water channel aquaporin-4. J Clin Invest 100(5):957–962, available from: PM:9276712

Ma T, Yang B, Gillespie A, Carlson EJ, Epstein CJ, Verkman AS (1998) Severely impaired urinary concentrating ability in transgenic mice lacking aquaporin-1 water channels. J Biol Chem 273(8):4296–4299, available from: PM:9468475

Ma T, Song Y, Yang B, Gillespie A, Carlson EJ, Epstein CJ, Verkman AS (2000) Nephrogenic diabetes insipidus in mice lacking aquaporin-3 water channels. Proc Natl Acad Sci USA 97 (8):4386–4391, available from: PM:10737773

Matsumura Y, Uchida S, Kondo Y, Miyazaki H, Ko SB, Hayama A, Morimoto T, Liu W, Arisawa M, Sasaki S, Marumo F (1999) Overt nephrogenic diabetes insipidus in mice lacking the CLC-K1 chloride channel. Nat Genet 21(1):95–98, available from: PM:9916798

Matthaei KI (2007) Genetically manipulated mice: a powerful tool with unsuspected caveats. J Physiol 582(Pt 2):481–488, available from: PM:17495035

McDill BW, Li SZ, Kovach PA, Ding L, Chen F (2006) Congenital progressive hydronephrosis (cph) is caused by an S256L mutation in aquaporin-2 that affects its phosphorylation and apical membrane accumulation. Proc Natl Acad Sci USA 103(18):6952–6957, available from: PM:16641094

McDonald FJ, Yang B, Hrstka RF, Drummond HA, Tarr DE, McCray PB Jr, Stokes JB, Welsh MJ, Williamson RA (1999) Disruption of the beta subunit of the epithelial Na^+ channel in mice: hyperkalemia and neonatal death associated with a pseudohypoaldosteronism phenotype. Proc Natl Acad Sci USA 96(4):1727–1731, available from: PM:9990092

Michimata M, Mizukami K, Suzuki M, Kazama I, Nakamura Y, Suzuki K, Yanagisawa T, Imai Y, Sasaki S, Matsubara M (2003) Vasopressin-independent renal urinary concentration: increased rBSC1 and enhanced countercurrent multiplication. Kidney Int 64(3):933–938, available from: PM:12911543

Miller RL, Lucero OM, Riemondy KA, Baumgartner BK, Brown D, Breton S, Nelson RD (2009) The V-ATPase B1-subunit promoter drives expression of Cre recombinase in intercalated cells of the kidney. Kidney Int 75(4):435–439, available from: PM:19052537

Miosge LA, Blasioli J, Blery M, Goodnow CC (2002) Analysis of an ethylnitrosourea-generated mouse mutation defines a cell intrinsic role of nuclear factor kappaB2 in regulating circulating B cell numbers. J Exp Med 196(8):1113–1119, available from: PM:12391023

Moeller HB, Rittig S, Fenton RA (2013) Nephrogenic diabetes insipidus: essential insights into the molecular background and potential therapies for treatment. Endocr Rev 34(2):278–301, available from: PM:23360744

Morishita T, Tsutsui M, Shimokawa H, Sabanai K, Tasaki H, Suda O, Nakata S, Tanimoto A, Wang KY, Ueta Y, Sasaguri Y, Nakashima Y, Yanagihara N (2005) Nephrogenic diabetes insipidus in mice lacking all nitric oxide synthase isoforms. Proc Natl Acad Sci USA 102 (30):10616–10621, available from: PM:16024729

Morris RG, Hoorn EJ, Knepper MA (2006) Hypokalemia in a mouse model of Gitelman's syndrome. Am J Physiol Renal Physiol 290(6):F1416–F1420, available from: PM:16434571

Mulders SM, Knoers NV, Van Lieburg AF, Monnens LA, Leumann E, Wuhl E, Schober E, Rijss JP, van Os CH, Deen PM (1997) New mutations in the AQP2 gene in nephrogenic diabetes insipidus resulting in functional but misrouted water channels. J Am Soc Nephrol 8 (2):242–248, available from: PM:9048343

Negro-Vilar A, Snyder GD, Falck JR, Manna S, Chacos N, Capdevila J (1985) Involvement of eicosanoids in release of oxytocin and vasopressin from the neural lobe of the rat pituitary. Endocrinology 116(6):2663–2668, available from: PM:3922746

Nielsen S, Terris J, Smith CP, Hediger MA, Ecelbarger CA, Knepper MA (1996) Cellular and subcellular localization of the vasopressin-regulated urea transporter in rat kidney. Proc Natl Acad Sci USA 93(11):5495–5500, available from: PM:8643603

Noda Y, Horikawa S, Furukawa T, Hirai K, Katayama Y, Asai T, Kuwahara M, Katagiri K, Kinashi T, Hattori M, Minato N, Sasaki S (2004) Aquaporin-2 trafficking is regulated by

PDZ-domain containing protein SPA-1. FEBS Lett 568(1–3):139–145, available from: PM:15196935

Olesen ET, Fenton RA (2013) Is there a role for PGE2 in urinary concentration? J Am Soc Nephrol 24(2):169–178, available from: PM:23160514

Oliverio MI, Delnomdedieu M, Best CF, Li P, Morris M, Callahan MF, Johnson GA, Smithies O, Coffman TM (2000) Abnormal water metabolism in mice lacking the type 1A receptor for ANG II. Am J Physiol Renal Physiol 278(1):F75–F82, available from: PM:10644657

Oppermann M, Mizel D, Huang G, Li C, Deng C, Theilig F, Bachmann S, Briggs J, Schnermann J, Castrop H (2006) Macula densa control of renin secretion and preglomerular resistance in mice with selective deletion of the B isoform of the Na, K,2Cl co-transporter. J Am Soc Nephrol 17 (8):2143–2152, available from: PM:16807402

Pallone TL, Edwards A, Ma T, Silldorff EP, Verkman AS (2000) Requirement of aquaporin-1 for NaCl-driven water transport across descending vasa recta. J Clin Invest 105(2):215–222, available from: PM:10642600

Poulsen SB, Kim YH, Frokiaer J, Nielsen S, Christensen BM (2013) Long-term vasopressin-V2-receptor stimulation induces regulation of aquaporin 4 protein in renal inner medulla and cortex of Brattleboro rats. Nephrol Dial Transplant 28(8):2058–2065, available from: PM:23615760

Rieg T, Tang T, Murray F, Schroth J, Insel PA, Fenton RA, Hammond HK, Vallon V (2010) Adenylate cyclase 6 determines cAMP formation and aquaporin-2 phosphorylation and trafficking in inner medulla. J Am Soc Nephrol 21(12):2059–2068, available from: PM:20864687

Robben JH, Knoers NV, Deen PM (2006) Cell biological aspects of the vasopressin type-2 receptor and aquaporin 2 water channel in nephrogenic diabetes insipidus. Am J Physiol Renal Physiol 291(2):F257–F270, available from: PM:16825342

Rojek A, Fuchtbauer EM, Kwon TH, Frokiaer J, Nielsen S (2006) Severe urinary concentrating defect in renal collecting duct-selective AQP2 conditional-knockout mice. Proc Natl Acad Sci USA 103(15):6037–6042, available from: PM:16581908

Rubera I, Loffing J, Palmer LG, Frindt G, Fowler-Jaeger N, Sauter D, Carroll T, McMahon A, Hummler E, Rossier BC (2003) Collecting duct-specific gene inactivation of alphaENaC in the mouse kidney does not impair sodium and potassium balance. J Clin Invest 112(4):554–565, available from: PM:12925696

Sawyer WH, Valtin H, Sokol HW (1964) Neurohypophysial Principles in rats with familial hypothalamic diabetes insipidus (Brattleboro strain). Endocrinology 74:153–155, available from: PM:14114664

Schnermann J, Chou CL, Ma T, Traynor T, Knepper MA, Verkman AS (1998) Defective proximal tubular fluid reabsorption in transgenic aquaporin-1 null mice. Proc Natl Acad Sci USA 95 (16):9660–9664, available from: PM:9689137

Schultheis PJ, Clarke LL, Meneton P, Miller ML, Soleimani M, Gawenis LR, Riddle TM, Duffy JJ, Doetschman T, Wang T, Giebisch G, Aronson PS, Lorenz JN, Shull GE (1998a) Renal and intestinal absorptive defects in mice lacking the NHE3 Na^+/H^+ exchanger. Nat Genet 19 (3):282–285, available from: PM:9662405

Schultheis PJ, Lorenz JN, Meneton P, Nieman ML, Riddle TM, Flagella M, Duffy JJ, Doetschman T, Miller ML, Shull GE (1998b) Phenotype resembling Gitelman's syndrome in mice lacking the apical Na^+-Cl^- cotransporter of the distal convoluted tubule. J Biol Chem 273 (44):29150–29155, available from: PM:9786924

Shi PP, Cao XR, Qu J, Volk KA, Kirby P, Williamson RA, Stokes JB, Yang B (2007) Nephrogenic diabetes insipidus in mice caused by deleting COOH-terminal tail of aquaporin-2. Am J Physiol Renal Physiol 292(5):F1334–F1344, available from: PM:17229678

Smithies O, Gregg RG, Boggs SS, Koralewski MA, Kucherlapati RS (1985) Insertion of DNA sequences into the human chromosomal beta-globin locus by homologous recombination. Nature 317(6034):230–234, available from: PM:2995814

Sohara E, Rai T, Miyazaki J, Verkman AS, Sasaki S, Uchida S (2005) Defective water and glycerol transport in the proximal tubules of AQP7 knockout mice. Am J Physiol Renal Physiol 289(6):F1195–F1200, available from: PM:15998844

Sohara E, Rai T, Yang SS, Uchida K, Nitta K, Horita S, Ohno M, Harada A, Sasaki S, Uchida S (2006) Pathogenesis and treatment of autosomal-dominant nephrogenic diabetes insipidus caused by an aquaporin 2 mutation. Proc Natl Acad Sci USA 103(38):14217–14222, available from: PM:16968783

Stegbauer J, Gurley SB, Sparks MA, Woznowski M, Kohan DE, Yan M, Lehrich RW, Coffman TM (2011) AT1 receptors in the collecting duct directly modulate the concentration of urine. J Am Soc Nephrol 22(12):2237–2246, available from: PM:22052052

Stewart GS, Fenton RA, Wang W, Kwon TH, White SJ, Collins VM, Cooper G, Nielsen S, Smith CP (2004) The basolateral expression of mUT-A3 in the mouse kidney. Am J Physiol Renal Physiol 286(5):F979–F987, available from: PM:15075194

Sun D, Samuelson LC, Yang T, Huang Y, Paliege A, Saunders T, Briggs J, Schnermann J (2001) Mediation of tubuloglomerular feedback by adenosine: evidence from mice lacking adenosine 1 receptors. Proc Natl Acad Sci USA 98(17):9983–9988, available from: PM:11504952

Takahashi N, Chernavvsky DR, Gomez RA, Igarashi P, Gitelman HJ, Smithies O (2000) Uncompensated polyuria in a mouse model of Bartter's syndrome. Proc Natl Acad Sci USA 97(10):5434–5439, available from: PM:10779555

Tang T, Gao MH, Lai NC, Firth AL, Takahashi T, Guo T, Yuan JX, Roth DM, Hammond HK (2008) Adenylyl cyclase type 6 deletion decreases left ventricular function via impaired calcium handling. Circulation 117(1):61–69, available from: PM:18071070

Terris J, Ecelbarger CA, Nielsen S, Knepper MA (1996) Long-term regulation of four renal aquaporins in rats. Am J Physiol 271(2 Pt 2):F414–F422, available from: PM:8770174

Terris JM, Knepper MA, Wade JB (2001) UT-A3: localization and characterization of an additional urea transporter isoform in the IMCD. Am J Physiol Renal Physiol 280(2):F325–F332, available from: PM:11208608

Thompson S, Clarke AR, Pow AM, Hooper ML, Melton DW (1989) Germ line transmission and expression of a corrected HPRT gene produced by gene targeting in embryonic stem cells. Cell 56(2):313–321, available from: PM:2912572

Tsukaguchi H, Shayakul C, Berger UV, Tokui T, Brown D, Hediger MA (1997) Cloning and characterization of the urea transporter UT3: localization in rat kidney and testis. J Clin Invest 99(7):1506–1515, available from: PM:9119994

Uchida S, Sasaki S, Nitta K, Uchida K, Horita S, Nihei H, Marumo F (1995) Localization and functional characterization of rat kidney-specific chloride channel, ClC-K1. J Clin Invest 95 (1):104–113, available from: PM:7814604

Uchida S, Sohara E, Rai T, Ikawa M, Okabe M, Sasaki S (2005) Impaired urea accumulation in the inner medulla of mice lacking the urea transporter UT-A2. Mol Cell Biol 25(16):7357–7363, available from: PM:16055743

Valtin H, Schroeder HA, BENIRSCHKE K, Sokol HW (1962) Familial hypothalamic diabetes insipidus in rats. Nature 196:1109–1110, available from: PM:13995944

Verkman AS (2008) Dissecting the roles of aquaporins in renal pathophysiology using transgenic mice. Semin Nephrol 28(3):217–226, available from: PM:18519083

Wade JB, Lee AJ, Liu J, Ecelbarger CA, Mitchell C, Bradford AD, Terris J, Kim GH, Knepper MA (2000) UT-A2: a 55-kDa urea transporter in thin descending limb whose abundance is regulated by vasopressin. Am J Physiol Renal Physiol 278(1):F52–F62, available from: PM:10644655

Wildman SS, Boone M, Peppiatt-Wildman CM, Contreras-Sanz A, King BF, Shirley DG, Deen PM, Unwin RJ (2009) Nucleotides downregulate aquaporin 2 via activation of apical P2 receptors. J Am Soc Nephrol 20(7):1480–1490, available from: PM:19423692

Yamaguchi K, Hama H, Watanabe K (1998) Possible participation of prostaglandins generated in the anteroventral third ventricular region in the hypovolemia-induced vasopressin secretion of conscious rats. Eur J Endocrinol 138(2):206–215, available from: PM:9506868

Yan Q, Yang X, Cantone A, Giebisch G, Hebert S, Wang T (2008) Female ROMK null mice manifest more severe Bartter II phenotype on renal function and higher PGE2 production. Am J Physiol Regul Integr Comp Physiol 295(3):R997–R1004, available from: PM:18579648

Yang B, Gillespie A, Carlson EJ, Epstein CJ, Verkman AS (2001) Neonatal mortality in an aquaporin-2 knock-in mouse model of recessive nephrogenic diabetes insipidus. J Biol Chem 276(4):2775–2779, available from: PM:11035038

Yang B, Bankir L, Gillespie A, Epstein CJ, Verkman AS (2002) Urea-selective concentrating defect in transgenic mice lacking urea transporter UT-B. J Biol Chem 277(12):10633–10637, available from: PM:11792714

Yang B, Zhao D, Qian L, Verkman AS (2006) Mouse model of inducible nephrogenic diabetes insipidus produced by floxed aquaporin-2 gene deletion. Am J Physiol Renal Physiol 291(2): F465–F472, available from: PM:16434568

Yui N, Lu HA, Chen Y, Nomura N, Bouley R, Brown D (2013) Basolateral targeting and microtubule-dependent transcytosis of the aquaporin-2 water channel. Am J Physiol Cell Physiol 304(1):C38–C48, available from: PM:23015545

Yun J, Schoneberg T, Liu J, Schulz A, Ecelbarger CA, Promeneur D, Nielsen S, Sheng H, Grinberg A, Deng C, Wess J (2000) Generation and phenotype of mice harboring a nonsense mutation in the V2 vasopressin receptor gene. J Clin Invest 106(11):1361–1371, available from: PM:11104789

Chapter 4
Angiotensin II and Water Balance in Amphibians

Minoru Uchiyama

Abstract Extant amphibians inhabit diverse environments, including deserts and brackish water, but are absent from Polar Regions and oceanic islands. As most amphibians move between water and land, they must cope with the challenges of each environment. The Renin–Angiotensin–Aldosterone System (RAAS) is a peptidergic hormone system involved in the regulation of blood pressure, water balance, and electrolyte homeostasis. Angiotensin II (Ang II) has been proposed to regulate thirst and oral drink responses in vertebrates. This review focuses on Ang II and cutaneous drinking in anuran amphibians. Amphibians do not drink orally but instead absorb water through their highly vascularized ventral skin. Ang II stimulates cutaneous drinking and aldosterone secretion in terrestrial toads. Regarding plasma aldosterone concentrations, the RAAS seems to be more involved in volume regulation than in changes in sodium concentration in anurans inhabiting different environments. Ang II receptors (AT1R and AT2R) were cloned and their mRNA expression was found in brain, skin, kidney, and urinary bladder. In amphibians, various types of water channel (aquaporins) and ion channels, transporters, and pumps used in Na^+ and Cl^- transport were cloned. Little is known about osmosensing and volume sensing mechanisms in both central nervous system and peripheral organs in amphibians.

Keywords Amphibian • Angiotensin II • Angiotensin II receptor • Aquaporin • C-fos immunoreactivity • Cutaneous drinking • Epithelial sodium channel • Renin–Angiotensin–Aldosterone System • Skin

4.1 Introduction

The class Amphibia is broadly distributed around the world and represents an evolutionary novelty in vertebrate history. They have a wide range of habitats, including the permafrost of Alaska, equatorial low-land jungles, forest canopies,

M. Uchiyama (✉)
Department of Life and Environmental Science, Graduate School of Science and Engineering, University of Toyama, 3190 Gofuku, Toyama 930-8555, Japan
e-mail: uchiyama@sci.u-toyama.ac.jp

K.A. Hyndman, T.L. Pannabecker (eds.), *Sodium and Water Homeostasis*, Physiology in Health and Disease, DOI 10.1007/978-1-4939-3213-9_4

rivers, subterranean burrows, caves, deserts, mountains, and ocean beaches (Frost et al. 2006). They were the first vertebrates to make the transition from water to land but need an aquatic environment for reproduction. From a different perspective, Amphibia is an interesting class because their internal milieu is under the influence of external conditions, to a greater extent than in other classes of vertebrates, i.e., there is greater exchange between the organism and the environment. As all except a few amphibian species move between water and land, they must cope with the challenges of very different environments. Since the pioneering study of amphibian water economy by Robert Townson (1795), many studies in life sciences utilized frog's skin or toad's urinary bladder. However, little is known about integrated mechanisms in control of cutaneous drinking behavior, water permeability of the skin and urinary bladder, and urine production under various environments in amphibians (Jørgensen 1997). In this group, water content is larger (total body water comprises 70–80 % of body mass) and osmolality and ion contents (plasma Na^+ is 100–120 mmol/l and Cl^- is 70–80 mmol/l) are lower than that in other vertebrates (Larsen et al. 2014). Although these values considerably differ between dry and aquatic environments when they stay for longer periods during challenging circumstances, it seems that osmoregulation is under the control of a number of endocrine systems, and that the nervous system influences the osmoregulatory organs, i.e., the skin, kidney, urinary bladder, and the lymphatic system. Recent papers reviewing aspects of amphibian osmoregulation include Boutilier et al. (1992), Shoemaker et al. (1992), Jørgensen (1997), Uchiyama and Konno (2006), Hillyard et al. (2009), Suzuki and Tanaka (2009), and Larsen et al. (2014). The object of the present review is to describe the regulation of osmoregulation by the Renin–Angiotensin–Aldosterone System (RAAS) in the amphibians. As there are few relevant works in caecilians (order Gymnophiona) and salamanders (order Urodela), this chapter will discuss only studies in frogs and toads (order Anura).

4.2 The Renin–Angiotensin–Aldosterone System

4.2.1 General

The Renin–Angiotensin–Aldosterone System (RAAS) is a peptidergic hormone system involved in the regulation of the blood pressure, water balance, and electrolytic homeostasis. Angiotensinogen, a protein produced principally in the liver is a key effecter precursor molecule of the RAAS. Angiotensinogen is cleaved to a 10-amino-acid peptide, Angiotensin I (Ang I), by a unique aspartylprotease named renin, which is produced by the juxtaglomerular apparatus in the kidney. Angiotensin converting enzyme in turn cleaves Ang I to a smaller, highly active 8-amino-acid peptide, Angiotensin II (Ang II). The major features of Ang II octapeptide (Asp1-Arg-Val-Tyr-Ile/Val-His-Pro-Phe8) that determine its biological activity were identified in early studies on the *in vivo* and *in vitro* actions of structurally modified Ang II peptides (Khosla et al. 1974). Ang II acts on the adrenal cortex to release aldosterone. Aldosterone acts in the kidney, primarily on cells of the collecting ducts, promoting the reabsorption of sodium (Na^+) and chloride (Cl^-). Ang II increases the reabsorption of Na^+ in the kidney, acts in the brain to stimulate thirst and salt appetite and increase sympathetic tone, and acts directly on blood vessel walls (primarily arterioles) to promote vasoconstriction and increase blood pressure. Production of renin is the rate-limiting step in Ang II production. Aldosterone binds to a mineralocorticoid receptor that serves as a transcription factor to stimulate the synthesis of proteins, resulting in the formation of a large number of conducting epithelial sodium channels (ENaCs) in the apical membranes of absorbing epithelia, and increasing Na^+, K^+-ATPase activity in the basolateral membranes (Connel and Davis 2005). During the last decade it has been established, apart from these classical actions, Ang II has various other effects induced by direct action on its receptors or via local effects of the angiotensin metabolites (Haulica et al. 2005).

Over 40 years ago, Epstein et al. (1970) observed that intracerebroventricular (i.c.v) injection of Ang II caused dose-responsive water intake in rats. Ganten et al. (1971) reported the existence of a brain Renin–Angiotensin System (RAS) independent of the peripheral system. This was followed by reports that intracranial injection of Ang II reliably elicited pressor and dipsogenic responses (Phillips 1987). This system has been implicated in many functions, including the regulation of body water balance and thirst, sodium appetite, blood pressure maintenance, and vasopressin release, and influences on the autonomic system and reproduction (Paul et al. 2006).

4.2.2 The Renin–Angiotensin–Aldosterone System in Amphibians

In earlier studies, renin-like activity and Angiotensin-like pressor activity were demonstrated in all classes of vertebrates (Kobayashi and Takei 1996). The amino acid sequences of native Ang I were also determined for representative species of teleost fish, amphibians, reptiles, and birds (Kobayashi and Takei 1996). These peptides differ from mammalian Ang I (Asp-Arg-Val-Tyr-Ile-His-Pro-Phe-His-Ley) at positions 1, 5, and 9. The Ang I molecule is [Asp1, Val5, Asn9] in the bullfrog (Hasegawa et al. 1983). The functional components of the RAS are present in anurans (Grill et al. 1972), and Ang II is known to make a significant contribution to both the maintenance of mean arterial pressure in resting toads and the ability of the animals to compensate for hypotensive challenge (West et al. 1998). The principal adrenocorticosteroids that are synthesized by amphibians are aldosterone and corticosterone (Carstensen et al. 1961). The synthesis and release of corticosteroids in anuran amphibians can be increased by adrenocorticotropic hormone and the RAS (Bentley 2002). In earlier *in vivo* and *in vitro* studies, aldosterone plays a major role in the regulation of active Na^+ transport in the skin, kidney, and urinary bladder (Bentley 2002; Uchiyama and Konno 2006). Thus, the RAAS appears to be involved in osmoregulation and the control of blood circulation.

4.2.3 *Plasma Angiotensin II and Aldosterone Concentrations in Amphibians*

4.2.3.1 Plasma Angiotensin II Concentration

Few papers report plasma Ang II concentration in amphibians. Garland and Henderson (1975) found no significant difference in plasma renin concentration between toads acclimated to distilled water and those acclimated to 0.9 % NaCl (hyperosmotic) solution, although the 0.9 % saline group showed high concentrations of plasma Na^+ and high excretion of urinary Na^+. Plasma and brain immunoreactive Ang II and III (ir-Angs) concentrations did not change significantly in the terrestro-fossorial toad *Scaphiopus couchii* during intracellular and extracellular dehydration (Mayer and Propper 2000). These authors also reported a significantly increased concentration of ir-Angs in the periventricular region of brain tissues but not in the plasma of toads showing a water absorption response in the field (Johnson et al. 2010). In contrast, in terrestrial *B. marinus*, dehydration stimulated increases in plasma Ang II and aldosterone as well as arginine vasotocin (AVT) concentrations, while hypervolemic treatment decreased plasma Ang II and aldosterone concentrations (Konno et al. 2005). There were significant correlations between plasma volume and concentrations of RAAS hormones and between plasma concentrations of Ang II and aldosterone. We observed that plasma Ang II and aldosterone concentrations increased significantly in the euryhaline frog *Fejervarya cancrivora* under dehydrated conditions after dry and hyperosmotic seawater acclimation (Uchiyama et al. 2014). Plasma Ang II was significantly correlated with plasma aldosterone. In amphibians, the RAAS may respond to plasma volume depletion rather than to decrease in plasma Na^+ concentration.

4.2.3.2 Plasma Aldosterone Concentration

In *Bufo marinus* acclimated to 0.9 % NaCl (hyperosmotic) solution, plasma aldosterone concentration was significantly lower than that of toads acclimated to distilled water (Garland and Henderson 1975). A similar result was reported in the semiaquatic frog *Rana pipiens*; plasma aldosterone concentration in animals acclimated to 0.9 % NaCl solution was significantly lower than that of frogs acclimated to tap water (Jungreis and Johnston 1979). An explanation for this lower serum aldosterone was that frogs in hypersaline water reduced Na^+ uptake across the ventral skin and Na^+ reabsorption in the bladder. Plasma aldosterone concentrations of semiterrestrial *Bufo japonicus formosus* were significantly increased in summer in the natural environment (Jolivet-Jaudet et al. 1984). We measured and compared plasma aldosterone concentrations after acclimation to dry, hypersaline, and freshwater conditions in five anuran species naturally inhabiting various environments (Fig. 4.1). In *Xenopus laevis* the plasma aldosterone concentration was lower as compared with that in the other anurans. In all five species, the aldosterone concentration was significantly increased in animals acclimated to the dry environment. In anurans acclimated to tap water, the plasma aldosterone concentration was not significantly different from concentrations in control frogs and toads that move freely back and forth between terrestrial and aquatic environment. The responses obtained in terrestrial toads resembled those in arboreal frogs. After hyperosmotic treatment, hypernatremic and hypervolemic toads showed a decreased plasma aldosterone concentration. Responses in semi-aquatic frogs resembled those in aquatic frogs, such that desiccation was induced during acclimation to hyperosmotic seawater as well as to a dry condition in brackish frogs. These results suggest that plasma aldosterone concentrations are correlated with body fluid volume rather than the osmolarity of body fluids in anurans (Konno et al. 2005; Uchiyama et al. 2014).

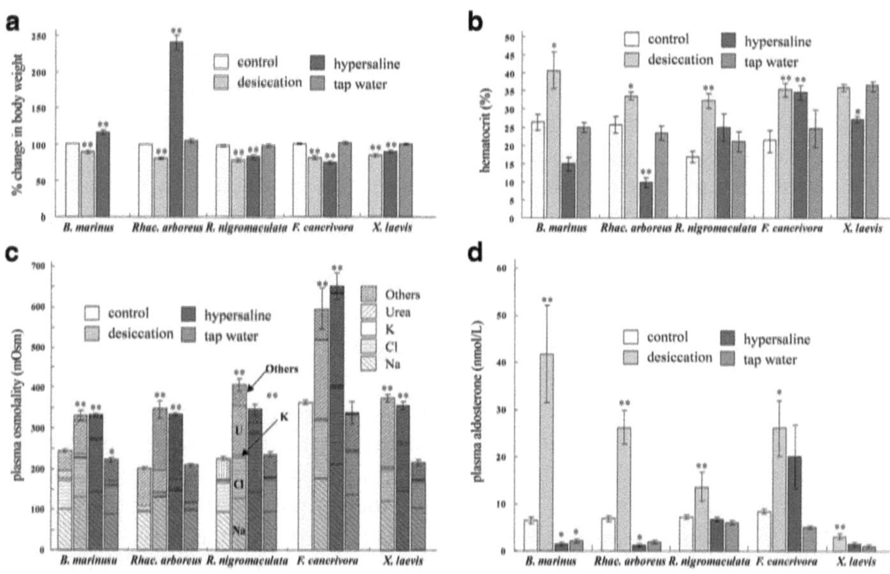

Fig. 4.1 Relative changes in body weight (**a**), hematocrit (**b**), plasma osmolality and components (**c**), and plasma aldosterone concentration (**d**) in five anuran species acclimated to various conditions. Control groups of the terrestrial species *Bufo marinus*, the arboreal *Rhacophorus arboreus*, the semi-aquatic *Rana nigromaculata*, and the brackish water species *Fejervarya cancrivora*, were kept on moist soil with free access to tap water (20 % seawater for *F. cancrivora*) *ad libitum* for 5 days. The control group of the aquatic *Xenopus laevis* was kept in tap water for 5 days. For all five species of anurans, the desiccation group was placed on dry sponges. The hypersaline group was placed in hyperosmotic NaCl solution (300 mOsm) for *B. marinus*, *Rhacophorus arboreus*, *Rana nigromaculata*, and *X. laevis*, and 80 % seawater (approximately 800 mOsm) for *F. cancrivora*. $^{*}P < 0.05$, $^{**}P < 0.01$: was significantly different from control or tap water group (for *X. laevis*) (Konno et al. 2005; Uchiyama et al. 2014)

4.2.4 The Angiotensin II Receptors

4.2.4.1 Angiotensin Receptors in Mammals

The actions of Ang II are mediated predominantly by two receptors, each with seven transmembrane domains, termed AT_1 and AT_2 based on their differential affinities for various nonpeptide antagonists. The receptors show a complex pattern of regulation and function (de Gasparo et al. 2000; Gwathmey et al. 2012). The AT_1 and AT_2 subtypes show similar Ang II binding properties but differ in genomic structure and localization as well as tissue-specific expression and regulation (de Gasparo et al. 2000; Nishimura 2001). Many classical RAS functions, such as vasoconstriction, aldosterone release, and thirst induction are mediated by AT_1. This subtype is predominantly coupled to the G protein Gq/11, and signals through phospholipases A, C, and D, inositolphosphates, calcium channels, and a variety of serine/threonine and tyrosine kinases. It also appears to modulate solute and fluid reabsorption in the proximal and distal nephron in the kidney (Gurley et al. 2011). Ang II inhibits ENaC activity of the principal cells in the cortical collecting duct through an AT_1-dependent mechanism (Mamenko et al. 2012). Unlike humans and other mammals, rats and mice express two types of AT1 receptors (AT_{1A} and AT_{1B}) (Fig. 4.2) (Burson et al. 1994). Following water deprivation or vasopressin administration, urine osmolalities were consistently lower in mice lacking AT_1 type A in collecting ducts. These results demonstrate that, in epithelial cells of the collecting duct, AT_1 type A directly modulates aquaporin-2 (AQP2) levels and contributes to the concentration of urine (Stegbauer et al. 2011). This was also shown in an *in vitro* study that Ang II via AT_1 and aldosterone increased AQP2 abundance in the principal cells of the mice collecting duct (Li et al. 2011). In studies using a selective AT_2 agonist, AT_2 were found in abundance during fetal development, but their expression generally declined after birth (de Gasparo et al. 2000). However, under pathological conditions such as congestive heart failure, renal failure or following incidents like skin lesions, vascular injury, myocardial infarction, brain ischemia or peripheral nerve transection, the AT_2 is re-expressed (de Gasparo et al. 2000). Although it has been difficult to reveal the physiological or potentially pathophysiological functions of the AT_2, anti-proliferation, neuro-regeneration, and anti-inflammation functions are suggested (de Gasparo et al. 2000; Steckelings et al. 2011). In mammals, truncated angiotensin peptides such as Ang III, Ang IV, Ang (1–7) have substantial selectivity to AT_2 over AT_1, and the AT_2 is distributed within a number of organ systems. The distribution areas of AT_2 were different from those of AT_1 (Wright and Harding 2011). See Chap. 5 for additional details of Ang receptors in mammals.

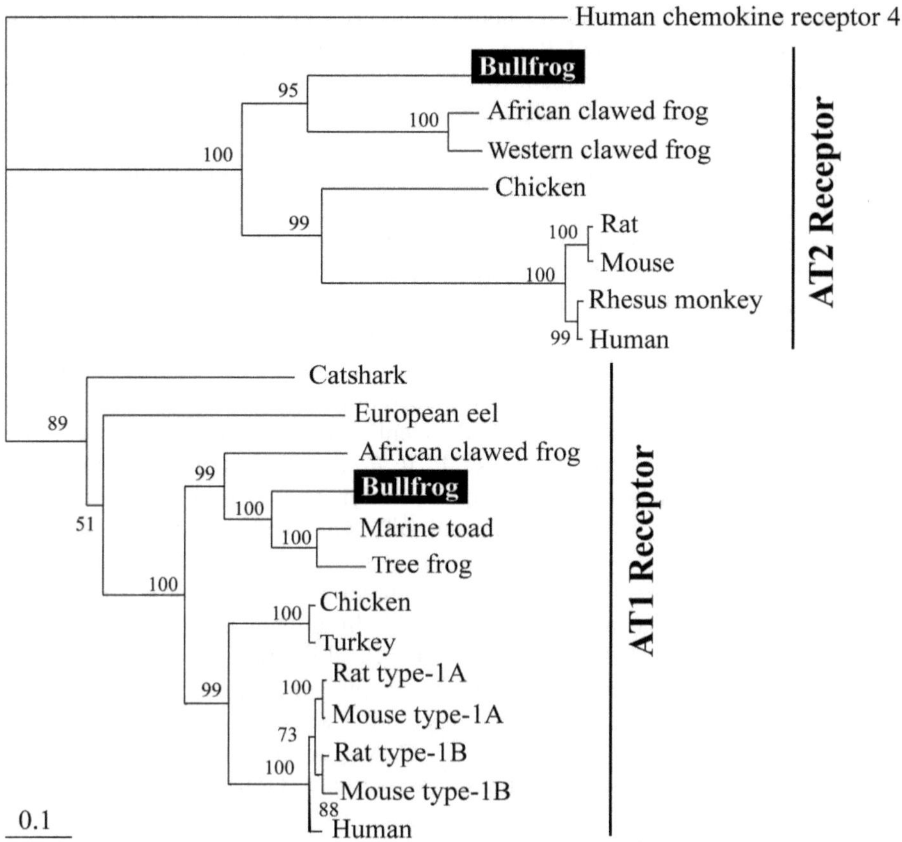

Fig. 4.2 Molecular phylogenetic tree representing vertebrate angiotensin receptor (AT1R and AT2R) generated by the neighbor-joining method. Numbers of the branches are bootstrap values. Scale bar represents 10 % amino acid substitution. The human chemokine receptor 4 was used an outogroup. GeneBank accession numbers or Ensemble gene IDs are as follows: AT2R; Bullfrog (BAJ16364), African clawed frog (NP_001085974), Western clawed frog (NP_001072452), Chicken (XP_426266), Rat (NP_036626), Mouse (NP_031455), Rhesus monkey (XP_001104306), and Human (NP_ 000677). AT1R; Catshrak (CAF02299), European eel (CAB40835), Bullfrog (BAJ16363), Marine toad (BAF48111), Tree frog (BAJ16362), Chicken (NP_990488), Turkey (XP_003209157), Rat type-1A (NP_112247), Rat type-1B (NP_112271), Mouse type-1A (NP_796296), Mouse type-1B (NP_780295), Human (NP_000676), and Human chemokine receptor 4 (AAP84352)

4.2.4.2 AT1R and AT2R in Amphibians

Two types of *Xenopus* Angiotensin receptors (*x*ATRs) share many similarities with mammalian AT_1 (*x*ATa, Ji et al. 1993; *x*ATb, Bergsma et al. 1993). We cloned ATRs from the kidney of the bullfrog *R. catesbeiana* (Uchiyama et al. 2009) and the Japanese tree frog *Hyla japonica* (Maejima et al. 2010). *Rana* AT_1 comprises 361 amino acid residues and has seven transmembrane domains. The amino acid sequence of *Rana* AT_1 is highly homologous to *x*ATR (64 %) and *Bufo* ATR (74 %). Both *Hyla* AT_1 and *Rana* AT_1 amino acid sequences are moderately homologous to AT_1 in mammals (62–64 %) and teleosts (49–51 %). RT-PCR revealed that amphibian AT_1 mRNA was present in the brain, lung, liver, large intestine, kidney, spleen, and urinary bladder (Ji et al. 1993; Uchiyama et al. 2009; Maejima et al. 2010). Although there are few studies of non-mammalian AT_2 (Nishimura 2001), in the last decade, cloning data on AT_2 have become increasingly available. Uchiyama et al. (2009) recently cloned *Rana* AT_2, a typical seven-transmembrane domain GPCR receptor in the bullfrog. *Rana* AT_2 comprises 337 amino acid residues and the amino acid sequences are highly homologous to those of *X. laevis* AT2R (60 %) and *X. tropicalis* AT2R (61 %). *Rana* AT_2 amino acid sequence is moderately homologous to that of mammalian AT_2 (45–46 %) and chicken AT_2 (50 %). The amino acid sequence of the *Rana* AT_1 has 37 % identity with the *Rana* AT_2 sequence, similarly the identity between the AT_1 and the AT_2 is 35–38 % in mammals. According to RT-PCR, *Rana* AT_2 was present in the heart, arteries, lung, stomach, and urinary bladder (Uchiyama et al. 2009). A phylogenetic tree of representative vertebrate AT_1 and AT_2 was generated by the neighbor-joining method (Fig. 4.2). Using an *in vitro* autoradiography technique, ATRs were found in the adrenal glands of several amphibians, including *R. temporaria*, but not *X. laevis* (Kloas and Hanke 1992a, b). In amphibians, it is necessary to examine the presence of ATRs in the adrenal gland because this gland comprises islets on the ventral surface of the kidney and expresses AT_1 mRNA. The distribution of brain structures possessing AT receptors is also interesting to elucidate in amphibians because the distribution areas of AT receptors are highly consistent among the mammalian species examined (Wright and Harding 2011).

4.3 Osmoregulatory Organs

4.3.1 Skin, Lymphatic System, and Circulation

In amphibians the whole of the skin is a site of exchange of water, osmolytes, energy, and respiratory gases. Water moves across the skin in three ways: evaporation, diffusion, and osmosis. Rates of evaporative water loss via the skin are very high in air (Boutilier et al. 1992; Shoemaker et al. 1992). Diffusion and osmosis are important processes for the gain or loss of ions and water in the freshwater environment. Histological observations show that the amphibian epidermis comprises about five to seven layers of epithelial heterocellular cells, such as principal cells and mitochondria-rich (MR) cells, and subepidermal glands (Hillyard et al. 2009; Larsen et al. 2014). Since the pioneering work of Hans Ussing, numerous studies of electrolytes and water transport were performed in amphibian skin (Larsen et al. 2014). Larsen et al. (2014) reviewed evidence that the ENaC and water channels (AQPs) of the principal cells and both active and passive Cl^- uptake

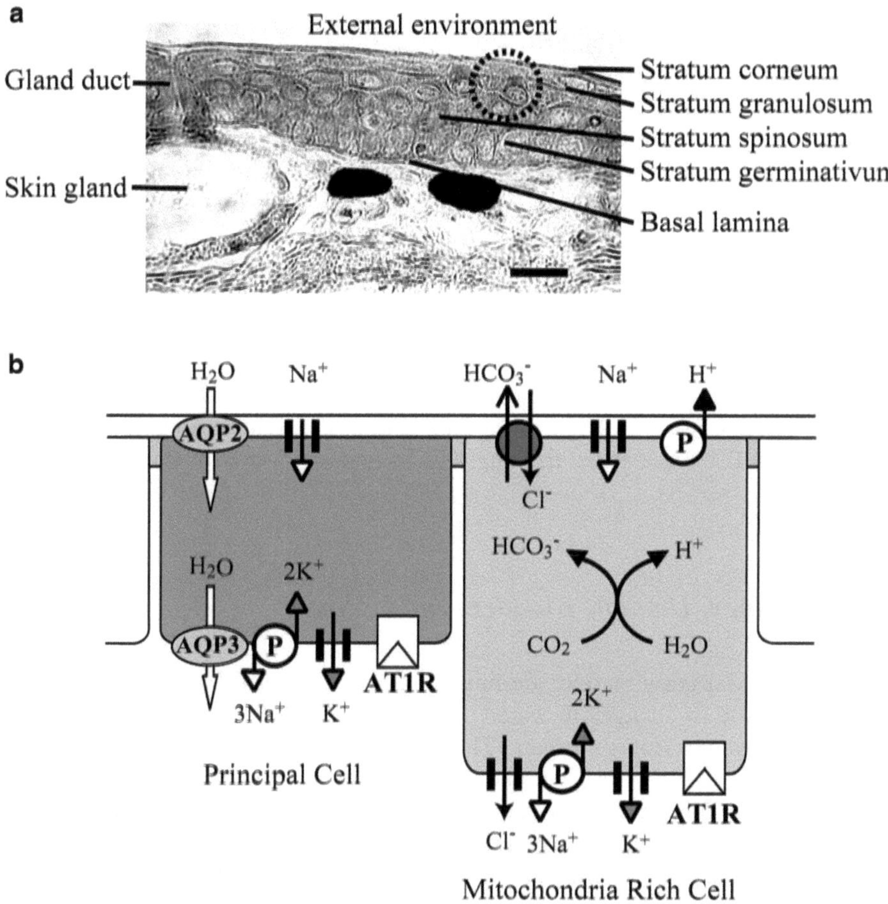

Fig. 4.3 Structure and model of ion and water transport of the epithelium in anuran skin. (**a**) A cross section of the skin from the Japanese tree frog *Hyla japonica*. A circle shows a principal cell and a mitochondria rich cell. Scale bar, 20 μm. (**b**) The current model of the functional organization of the epithelial cells. In a principal cell, uptake of Na^+ is transported through ENaC and P-type Na^+, K^+-ATPase at the apical and the basolateral membrane, respectively. In a γ-type MR cell, Cl^-/HCO_3^- antiporter is fueled by a V-type H^+-ATPase. Ang II may stimulate directly ENaC activity via AT1R. On land, the mucous gland produces and secretes cutaneous surface fluid, which covers and keeps the skin moist. The information has principally derived from studies using the mammalian kidney and the amphibian skin. For further information see the text and Larsen et al (2014). Abbreviation: *AQP2* aquaporin 2, *AQP3* aquaporin 3, *AT1R* a putative angiotensin II receptor; Na^+, ENaC; K^+, K^+-channel; $3Na^+$, $2\,K^+$ P, P-type Na^+, K^+-ATPase; Cl^-, HCO_3^-, Cl^-/HCO_3^- antiporter; Cl^-, Cl^- channel

of MR cells (of types α, β, and γ), Na^+, K^+-ATPase on the basolateral membranes in the principal and MR cells are pivotal in adaptation to life in freshwater and on land (Fig. 4.3; Hillyard et al. 2009; Larsen 2011; Larsen et al. 2014). In classical *in vivo* studies, high doses of Ang II (>10 μM) stimulated short-circuit current and water transport in frog skin (McAfee and Locke 1967; Proto et al. 1983). It has recently been suggested that Ang II directly stimulates ENaC activity via AT_1 in the distal nephron of the mammalian kidney (Mamenko et al. 2012). Thus, AT_1 may be present in the principal cells and MR cells of frog skin (Fig. 4.3). The frog on land maintains a slightly hypotonic cutaneous surface fluid by regulation of subepidermal gland secretion, which balanced water evaporation into the atmosphere and reabsorption of ion and water in the principal and MR cells of epidermis (Fig. 4.3; Larsen et al. 2014).

Anurans possess large subcutaneous lymphatic sacs that stabilize blood volume (Boutilier et al. 1992). They have a unique set of effectors (specialized skeletal muscles and lung ventilation) that are coordinated so as to mobilize lymph in the lymph hearts. There are two pairs of dorsal lymph hearts, and both pairs are innervated by spinal nerves and are under feedback control of the arterial baroreceptors as well as hormonal control by Ang II and AVT (Hedrick et al. 2013). The effect of Ang II in cutaneous drinking will be discussed later in this chapter (see Sect. 4.4.2).

4.3.2 Kidney, Urinary Bladder, and Colon

The kidney and urinary bladder are important osmoregulatory organs. The kidney responds rapidly to changes in water status (shortage or surplus) and the urinary bladder is a water storage organ. The kidney is a mesonephros in the adult amphibian and a pronephros in the larvae. The nephron usually comprises a glomerulus and Bowman's capsule, a ciliated neck segment, a proximal tubule, an intermediate segment, an early and a late distal tubule, and a collecting tubule, which opens into a collecting duct shared with other nephrons. In freshwater, amphibians exhibit high glomerular filtration rates and excrete copious dilute urine to balance cutaneous water uptake. On land, amphibians produce a small amount of concentrated urine and show anuria when subjected to a desiccated condition. Urine volume is controlled by the glomerular filtration rate and ion and water reabsorption along the segments of the nephron. In mammals, all RAAS components are present and AT_1 is localized in the renal arteries, glomerular mesangial cells, and the cell membranes of the proximal and distal tubules of the kidney. Ang II has multiple direct inter-renal effects, including renal vasoconstriction, tubular sodium reabsorption, tubuloglomerular feedback sensitivity, and modulation of pressure natriuresis (Carey and Siragy 2003; Mamenko et al. 2012). The functions of Ang II in the amphibian kidney are unknown, although AT_1 is abundant in amphibian kidney (Uchiyama et al. 2009; Maejima et al. 2010).

Amphibians on land reabsorb water stored in the urinary bladder. The bladder epithelium comprises several types of cells, such as basal cells, principal cells, MR cells, and Goblet cells. The epithelium of the bladder provides the final adjustment of the ion concentrations of excreted urine. During desiccation on land, reabsorption of water from urinary storage maintains body fluid volume.

Active Na^+ absorption is regulated by aldosterone (Snyder 2002). ENaC and Na^+, K^+-ATPase were expressed on the apical membranes and basolateral membrane, respectively, in principal cells (Konno et al. 2007).

Amphibians have a well-developed large intestine, across which Na^+ is actively transported. Expression of ENaC mRNA was reported in the large intestine of *B. marinus* (Konno et al. 2007). Such Na^+ transport can result in the production of osmotic gradients along which water can be transported.

4.4 Drinking

4.4.1 Oral Drinking

Drinking plays a pivotal role in the regulation of body fluids of most vertebrates. Regulatory or primary drinking occurs in response to thirst (Fitzsimons 1998). In mammals, birds, and reptiles, primary drinking involves the stimulation of the thirst center in the hypothalamus as a result of cellular dehydration or a decrease in the volume of extracellular fluids (Kobayashi and Takei 1996). The animals obtain information from two primary sources: the peripheral and visceral sensory systems, which transmit information through the classical sensory pathways to the central nervous system and sensory systems in the brain, which monitor the constituents of the circulation to assess the physiological status of the individual (Kobayashi and Takei 1996; Fitzsimons 1998).

4.4.1.1 Effect of Angiotensin II

It has been proposed that Ang II regulates thirst responses in vertebrates by two mechanisms. In the first of these, circulating Ang II may act on receptors in the circumventricular organs (CVOs) of the brain, and in the second mechanism Ang II produced in central regions may act as a neurotransmitter coordinating osmotic and hormonal information (Phillips 1987; Ferguson 2014). The local angiotensin synthesis is established in the brain: the hypothalamic paraventricular nucleus, supraoptic nucleus, CVOs, and neuronal cell bodies of the nucleus of the tractus solitarii (Paul et al. 2006). These proposed mechanisms are based on experiments in which Ang II, injected either by intraperitoneal (i.p.) or intracerebralventricular (i.c.v.), induced thirst-related behavior (i.e., oral drinking) in mammals, birds reptiles, and fish (Kobayashi and Takei 1996). In vertebrates, dehydration and hypovolemic treatments induced increases in serum osmolality and plasma Na^+ concentrations associated with an increased plasma Ang II concentration. Thirst-related behavior was also related to an increase in plasma Ang II concentration (Kobayashi and Takei 1996).

4.4.2 Cutaneous Drinking in Amphibians

It is generally believed that amphibians, unlike other vertebrates, do not normally drink but instead obtain water through the ventral skin, sometimes at a high rate when in freshwater (Jørgensen 1997; Bentley 2002). This cutaneous "drinking" involves a behavioral response called the water absorption response (WR) (Stille 1958), and is physiological regulation of water uptake. Cutaneous drinking is associated with increases of osmotic permeability and capillary blood flow in epithelium of pelvic region (Boutilier et al. 1992; Viborg and Hillyard 2005). Water uptake through skin is more rapid in terrestrial species than in semi-aquatic or aquatic species. Terrestrial toads have a greater area of ventral skin specialized for water absorption, which is effective for rapid rehydrating (Hillman et al. 2009). Hillyard and Willumsen (2011) hypothesized that osmotic stimuli mediate sensory processes for longer-term detection of conditions with low water potential, while ionic stimuli (an influx of Na^+ via ENaC) are more important for shorter-term analysis of rehydration sources.

4.4.2.1 Effect of Ang II in Cutaneous Drinking

In terrestrial toads and arboreal frogs, either peripheral or i.c.v. administration of Ang II induced WR behavior, in which the pelvic patch is pressed against a moist surface (Hoff and Hillyard 1991; Propper et al. 1995; Maejima et al. 2008, 2010). The effect of Ang II was blocked by saralasin (a peptide antagonist for Ang II receptors; Hoff and Hillyard 1993) and candesartan (an AT_1 antagonist; Maejima et al. 2010) in toads and tree frog, respectively. These findings may support the hypothesis that Ang II is involved in the induction of thirst-related behavior in anurans, as has been found for oral drinking in other vertebrates. An i.c.v. injection of Ang II stimulated the WR and more effective in inducing the WR than a peripheral injection in terrestrial anurans (Propper et al. 1995; Maejima et al. 2010). Expression of AT_1 mRNA in the diencephalon of the brain was increased in dehydrated frogs (Maejima et al. 2010). Furthermore, we observed immunohistochemically that expression of c-fos-like protein was increased in cells in areas adjacent to the anterior commissure, the preoptic area, and the dorsomedial hypothalamic nuclei after dehydration and hypersaline injection, and an i.c.v. microinjection of Ang II into the third ventricular cavity, where CVOs probably localize (Fig. 4.4; Duvernoy and Risold 2007). In mammals, c-fos protein expression was stimulated in response to plasma Ang II or a microinjection of Ang II into areas of CVOs (Herbert et al. 1992). Thus, amphibian WR behavior seems to be regulated via AT_1 in the central nervous system as well as by regulating oral drinking in mammals.

Changes in extracellular fluid osmolality resulting from the loss or gain of systemic water or electrolyte affect excitable properties of neurons. Transient receptor potential vaniloid types 1 and 4 (TRPV1 and TRPV4) are activated by

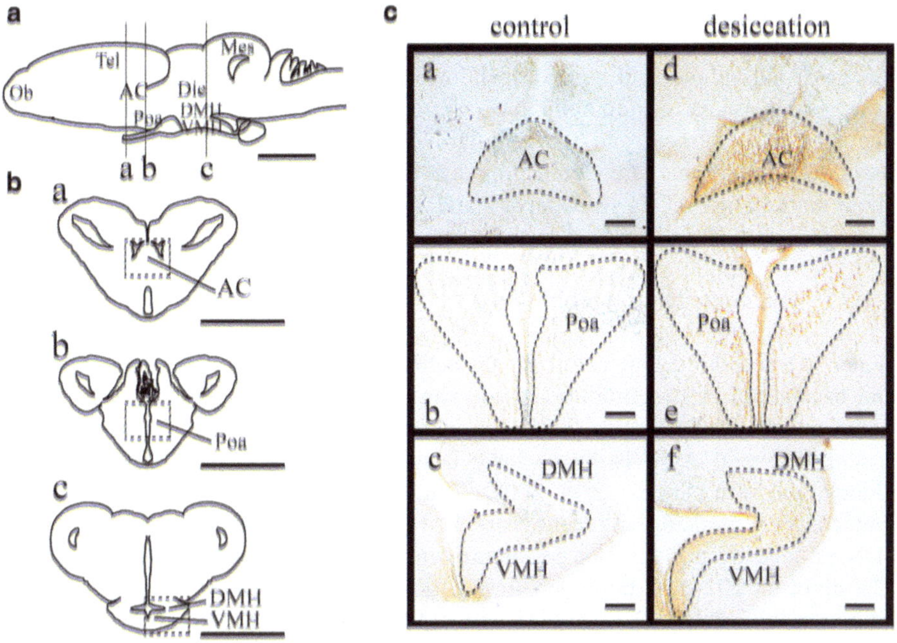

Fig. 4.4 The distribution of c-fos immunoreactivity in the brain of the Japanese tree frog *Hyla japonica*. (**a**) A medial view of the brain of a, b, and c indicate the level of sections in (**b**). (**b**) Three schematic drawings (a, b, and c) indicate the transverse sections of the brain. (**c**) The distributions of c-fos immunoreactivity in transverse sections of the brain of control frogs (a, b, and c) and desiccated frogs deprived of water for 48 h (d, e, and f). There were more c-fos-positive cells in desiccated frogs than in control frogs. Abbreviations: *AC* anterior commissure, *Die* diencephron, *DMH* dorsomedial hypothalamic nucleus, *Mes* mesencephalon, *Ob* olfactory bulb, *Poa* Preoptic area, *Tel* telencephalon, *VMH* ventromedial hypothalamic nucleus. Scale bars, 1 mm in (**a**), (**b**), and 100 μm in (**c**)

osmotic stimuli and are considered to serve as osmosensors in mammals (Liedtke et al. 2000). In amphibians, mRNAs homologous to TRPV1 and TRPV4 were detected in *X. tropicalis* (Saito and Shingai 2006) and in the skin of *H. japonica* (Maejima and Uchiyama unpublished data). The nature of TRPVs action is unknown, but it clearly merits further investigation.

4.4.3 Aquaporins

The aquaporins (AQPs) are a family of water-channel proteins. The first AQP was identified in red blood cells in 1991 and is now known as AQP1 (Agre 2004). Since then, 13 AQPs (AQP0–12) have been discovered in mammals. The aquaporins are tetramers, each monomer having its own water pore. Classical AQPs (e.g., AQP2 and AQP5) primarily transport water, and the aquaglyceroporins (AQP3, AQP7, and AQP9) also transport glycerol and various small polar molecules (such as urea and ammonia). In mammals, the major AQPs expressed in kidney include AQP1 in the proximal tubule and thin descending limb of Henle epithelia, and in the descending vasa recta endothelia; the vasopressin-regulated AQP2 in the collecting duct apical membrane and intracellular vesicles; and AQP3 and AQP4 in the basolateral membrane of the collecting duct epithelia (Agre 2004). For more details about mammalian AQPs see Chaps. 2 and 3.

4.4.3.1 Aquaporins in Amphibians

Water is transported transcellularly through water channels inserted in the apical and basolateral cell membranes, while paracellular water transport occurs in the tight junction area (Bentley 2002). In anuran amphibians (*Bufo*, *Hyla*, and *Rana* species), several AQP isoforms have been reported from the urinary bladder, kidney, and skin (Suzuki et al. 2007). Tanaka and colleagues have studied these molecules energetically in recent years. So far the full-length sequences of 30 AQP cDNAs have been elucidated in anuran amphibians in a species-specific manner (Saitoh et al. 2014). A phylogenetic analysis of AQPs indicated the existence of anuran-specific AQPs, i.e., AQPa1 and AQPa2 (the letter "a" represents anuran). AQPa2s are further subdivided into urinary bladder and ventral skin types (Suzuki and Tanaka 2009). In response to arginine vasotocin and isoproterenol (a - β-adrenergic agonist), the AQPa2s are translocated mostly into the apical plasma membrane of the principal cells of the stratum granulosum in the epidermis, a key site in the control of water transport in the ventral skin and urinary bladder (Suzuki et al. 2007; Saitoh et al. 2014). In the skin of the pelvic and femoral regions, both the ventral skin type and the urinary bladder type of AQPa2 were observed in *B. japonica* and *H. japonica*, but only the ventral skin type of AQP was observed in *Rana* species. In *Xenopus* species, the ventral skin type AQPa2 was expressed at the mRNA level but not at the protein level. The distribution of cells expressing AQPa2s in the abdominal regions seems to be correlated with the natural habitat in anuran species (Ogushi et al. 2010). Little is known about any potential interaction between Ang II and AQPs in amphibians, although it is suggested that Ang II in cooperation with vasopressin plays a significant role in regulating AQPs expression and modulating urine concentration in mammals (Li et al. 2011).

Chapter 2 gives more detailed information on amphibian aquaporins.

4.5 Perspectives

Over the past decades our understanding of the molecular mechanisms and transport proteins (such as Na^+, K^+-ATPase, aquaporins, ENaC, and Cl^-/HCO_3^-) associated with epithelial membranes that participate in amphibian osmoregulation has been greatly advanced. The mechanism of the renin/angiotensin-converting enzyme (ACE)/Ang II/(AT_1/AT_2) axis is slowly becoming clear. However, nothing is known for other angiotensin axes, such as the prorenin/prorenin receptor/Map kinase axis, the ACE2/Ang (1–7)/Mas receptor axis, and the Ang IV/AT4R/insulin-regulated aminopeptidase axis which have been recently discovered in mammals (see Chap. 5 for further descriptions). In addition, the following topics remain to be investigated for the further understanding of the RAAS in amphibians. Little is known about molecular characterizations and localization of osmoreceptors and baroreceptors, and their relation with RAAS in the central nervous system and peripheral tissues. It is also interesting to investigate whether neural networks mediating the cutaneous drinking in amphibians are similar with those mediating the oral drinking in terrestrial tetrapods. In mammals, estrogen regulates AT_1 expression, while little is known about sex differences in mechanisms of Ang II or hydromineral balance in amphibians. Furthermore, it is not known whether similar mechanisms are present in all anurans and in salamanders and caecilians.

Acknowledgements The author thanks Dr. Sho Maejima and MSc. Shohei Kitani for their technical assistance.

References

Agre P (2004) Aquaporin water channels (Nobel lecture). Angew Chem Int Ed 43(33):4278–4290
Bentley PJ (2002) Endocrines and osmoregulation: a comparative account in vertebrates (Zoophysiology Vol 39). Springer, Berlin
Bergsma DJ, Ellis C, Nuthulaganti PR et al (1993) Isolation and expression of a novel angiotensin II receptor from *Xenopus laevis* heart. Mol Pharmacol 44:277–284
Boutilier RG, Stiffler DF, Toews DP (1992) Exchange of respiratory gases, ions, and water in amphibious and aquatic amphibians. In: Feder E, Burggren WW (eds) Environmental physiology of the amphibia. The University of Chicago Press, Chicago, pp 81–124
Burson JM, Aguilera G, Gross KW, Sigmund CD (1994) Differential expression of angiotensin receptor 1A and 1B in mouse. Am J Physiol 267:E260–E267
Carey RM, Siragy HM (2003) Newly recognized components of the renin-angiotensin system: potential roles in cardiovascular and renal regulation. Endocr Rev 24:261–271
Carstensen H, Burgers AC, Li CH (1961) Demonstration of aldosterone and corticosterone as the principal steroids formed in incubates of adrenal glands o the American bullfrog (*Rana catesbeiana*) and stimulation of their production by mammalian adrenocorticotropin. Gen Comp Endocrinol 1:37–50
Connel JMC, Davis E (2005) The new biology of aldosterone. J Endocrinol 186:1–20
de Gasparo M, Catt KJ, Inagami T et al (2000) International union of pharmacology. XXIII. The angiotensin II receptors. Pharmacol Rev 52:415–472
Duvernoy HM, Risold PY (2007) The circumventricular organs: an atlas of comparative anatomy and vascularization. Brain Res Rev 56:119–147
Epstein AN, Fitzsimons JT, Rolls BJ (1970) Drinking induced by injection of angiotensin into the brain of the rat. J Physiol 210:457–474
Ferguson AV (2014) Circumventricular organs: integrators of circulating signals controlling hydration, energy balance, and immune function. In: De Luca LA, Jr MJV, Johnson AK (eds) Neurobiology of body fluid homeostasis: transduction and integration. CRC, Boca Raton
Fitzsimons JT (1998) Angiotensin, thirst, and sodium appetite. Physiol Rev 78:583–686

Frost DR, Grant T, Faivovich J et al (2006) The amphibian tree of life. Bull Am Mus Nat Hist 297:1–360
Ganten D, Marquez-Julio A, Granger P et al (1971) Renin in dog brain. Am J Physiol 221:1733–1737
Garland HO, Henderson IW (1975) Influence of environmental salinity on renal and adrenocortical function in the toad, *Bufo marinus*. Genl Comp Endocrinol 27:136–143
Grill G, Granger P, Thurau K (1972) The rennin angiotensin system of amphibians. I. Determination of the rennin content of amphibian kidneys. Pflugers Arch 331:1–12
Gurley SB, Riquier-Brison AD, Schnermann J et al (2011) AT_{1A} angiotensin receptors in the renal proximal tubule regulate blood pressure. Cell Metab 13:469–475
Gwathmey TM, Alzayadneh EM, Pendergrass KD et al (2012) Review: novel roles of nuclear angiotensin receptors and signaling mechanisms. Amer J Physiol Regul Integr Comp Physiol 302:R518–R530
Hasegawa Y, Watanabe TX, Sokabe H, Nakajima T (1983) Chemical structure of angiotensin in the bullfrog, *Rana catesbeiana*. Gen Comp Endocrinol 50:75–80
Haulica I, Bild W, Serban DN (2005) Review: angiotensin peptides and their pleiotropic actions. J Renin Angiotensin Aldosterone Syst 6:121–131
Hedrick MS, Hillman SS, Drewes RC et al (2013) Lymphatic regulation in nonmammalian vertebrates. J Appl Physiol 115:297–308
Herbert J, Forsling ML, Howes SR et al (1992) Regional expression of c-*fos* antigen in the basal forebrain following intraventricular infusions of angiotensin and its modulation by drinking either water or saline. Neuroscience 51:867–882
Hillman SS, Withers PC, Drewes RC et al (2009) Ecological and environmental physiology of amphibians. Oxford University Press, New York
Hillyard SD, Willumsen NJ (2011) Review Chemosensory function of amphibian skin: integrating epithelial transport, capillary blood flow and behavior. Acta Physiol 202:535–548
Hillyard SD, Møbjerg N, Tanaka S et al (2009) Osmotic and ion regulation in amphibians. In: Evans DH (ed) Osmotic and ionic regulation: cells and animals. CRC, Boca Raton, pp 367–441
Hoff KS, Hillyard SD (1991) Angiotensin II stimulates cutaneous drinking in the toad *Bufo punctatus*. Physiol Zool 64:1165–1172
Hoff KS, Hillyard SD (1993) Inhibition of cutaneous water absorption in dehydrated toads by saralasin is associated with changes in barometric pressure. Physiol Zool 66:89–98
Ji H, Sandberg K, Zang Y et al (1993) Molecular cloning, sequencing and functional expression of an amphibian angiotensin II receptor. Biochem Biophys Res Commun 194:756–762
Johnson WE, Hillyard SD, Propper CR (2010) Plasma and brain angiotensin concentrations associated with water response behavior in the desert anuran, *Scaphiopus couchii* under natural conditions in the field. Comp Biochem Physiol A Mol Integr Physiol 157:377–381
Jolivet-Jaudet G, Inoue M, Takada K et al (1984) Circannual changes in plasma aldosterone levels in *Bufo japonicus formosus*. Gen Comp Endocrinol 53:163–167
Jørgensen CB (1997) 200 years of amphibian water economy: from Robert Townson to the present. Biol Rev Camb Philos Soc 72:153–237
Jungreis AM, Johnston JW (1979) Urea production in *Rana pipiens*: effects of dehydration and aldosterone administration. Comp Biochem Physiol A Physiol 64:59–66
Khosla MC, Smeby RR, Bumpus FM (1974) Structure-activity relationship in angiotensin II analogs. In: Page IH, Bumpus RM (eds) Handbook of experimental pharmacology, vol XXXVII. Springer, Berlin, pp 126–161
Kloas W, Hanke W (1992a) Angiotensin II binding sites in frog kidney and adrenal. Peptides 13:349–354
Kloas W, Hanke W (1992b) Localization and quantification of angiotensin II (A II) binding sites in the kidney of *Xenopus laevis*—lack of A II receptors in the adrenal tissue. Gen Comp Endocrinol 86:173–183
Kobayashi H, Takei Y (1996) The renin-angiotensin system: comparative aspects (Zoophysiology Vol 35). Springer, Berlin

Konno N, Hyodo S, Takei Y et al (2005) Plasma aldosterone, angiotensin II, and arginine vasotocin concentrations in the toad, *Bufo marinus*, following osmotic treatments. Gen Comp Endocrinol 140:86–93

Konno N, Hyodo S, Yamada T et al (2007) Immunolocalization and mRNA expression of the epithelial Na^+ channel α-subunit in the kidney and urinary bladder of the marine toad, *Bufo marinus*, under hyperosmotic conditions. Cell Tissue Res 328:583–594

Larsen EH (2011) Reconciling the Krogh and Ussing interpretations of epithelial chloride transport—presenting a novel hypothesis for the physiological significance of the passive cellular chloride uptake. Acta Physiol 202:435–464

Larsen EH, Deaton LE, Onken H et al (2014) Osmoregulation and excretion. Compr Physiol 4:405–573

Li C, Wang W, Rivard CJ et al (2011) Molecular mechanisms of angiotensin II stimulation on aquaporin-2 expression and trafficking. Am J Physiol Renal Physiol 300:F1255–F1261

Liedtke W, Choe Y, Martí-Renom MA et al (2000) Vanilloid receptor–related osmotically activated channel (VR-OAC), a candidate vertebrate osmoreceptor. Cell 103:525–535

Maejima S, Yamada T, Hamada T et al (2008) Effects of hypertonic stimuli and arginine vasotocin (AVT) on water absorption response in Japanese treefrog, *Hyla japonica*. Gen Comp Endocrinol 157:196–202

Maejima S, Konno N, Matsuda K et al (2010) Central angiotensin II stimulates cutaneous water intake behavior via an angiotensin II type-1 receptor pathway in the Japanese tree frog *Hyla japonica*. Horm Behav 58:457–464

Mamenko M, Zaika O, Ilatovskaya DV et al (2012) Angiotensin II increases activity of the epithelial Na^+ channel (ENaC) in distal nephron additively to aldosterone. J Biol Chem 287:660–671

Mayer LP, Propper CR (2000) Intra- and extracellular dehydration has no effect on plasma levels of angiotensin II in an amphibian. J Exp Zool 286:343–349

McAfee RD, Locke W (1967) Effect of angiotensin amide on sodium isotope flux and short-circuit current of isolated frog skin. Endocrinology 81:1301–1305

Nishimura H (2001) Angiotensin receptors—evolutionary overview and perspectives. Comp Biochem Physiol A Mol Integr Physiol 128:11–30

Ogushi Y, Tsuzuki A, Sato M et al (2010) The water-absorption region of ventral skin of several semiterrestrial and aquatic anuran amphibians identified by aquaporins. Am J Physiol Regul Integr Comp Physiol 299:R1150–R1162

Paul M, Mehr AP, Kreutz R (2006) Physiology of local renin-angiotensin systems. Physiol Rev 86:747–803

Phillips MI (1987) Functions of angiotensin in the central nervous system. Annu Rev Physiol 49:416–435

Propper CR, Hillyard SD, Johnson WE (1995) Central angiotensin II induces thirst-related responses in an amphibian. Horm Behav 29:74–84

Proto MC, Coviello A, Khosla MC et al (1983) Effects of frog-skin angiotensin II in amphibians. Hypertension 5(6 Pt 3):V101–V104

Saito S, Shingai R (2006) Evolution of thermoTRP ion channel homologs in vertebrates. Physiol Genomics 27:219–230

Saitoh Y, Ogushi Y, Shibata Y et al (2014) Novel vasotocin-regulated aquaporins expressed in the ventral skin of semiaquatic anuran amphibians: evolution of cutaneous water-absorbing mechanisms. Endocrinology 155:2166–2177

Shoemaker V, Hillman SS, Hillyard SD et al (1992) Exchange of water, ions and respiratory gases in terrestrial amphibians. In: Feder ME, Burggren WW (eds) Environmental physiology of the amphibia. The University of Chicago Press, Chicago, pp 125–150

Snyder PM (2002) The epithelial Na^+ channel: cell surface insertion and retrieval in Na^+ homeostasis and hypertension. Endocr Rev 23:258–275

Steckelings UM, Larhed M, Hallberg A et al (2011) Non-peptide AT2-receptor agonists. Curr Opin Pharmacol 11:187–192

Stegbauer J, Gurley SB, Sparks MA et al (2011) AT1 receptors in the collecting duct directly modulate the concentration of urine. J Am Soc Nephrol 22:2237–2246
Stille WT (1958) The water absorption response of an anuran. Copeia 1958:217–218
Suzuki M, Tanaka S (2009) Molecular and cellular regulation of water homeostasis in anuran amphibians by aquaporins. Comp Biochem Physiol A Mol Integr Physiol 153:231–241
Suzuki M, Hasegawa T, Ogushi Y et al (2007) Amphibian aquaporins and adaptation to terrestrial environments: a review. Comp Biochem Physiol A Mol Integr Physiol 148:72–81
Townson R (1795) Physiological observations on the amphibia. Dissertation the second. Respiration continued with a fragment upon the subject of absorption, University of Göttingen
Uchiyama M, Konno N (2006) Hormonal regulation of ion and water transport in anuran amphibians. Gen Comp Endocrinol 147:54–61
Uchiyama M, Kitani S, Maejima S et al (2009) Cloning of angiotensin II receptors and their expression during osmotic stress in the bullfrog, *Rana catesbeiana*. In: Abstracts of the 16th international congress of comparative endocrinology, University of Hong Kong, Hong Kong, 22–26 June 2009. http://evolution.hku.hk/ICCE2009_AbstractBook_final.pdf
Uchiyama M, Maejima S, Wong MKS et al (2014) Changes in plasma angiotensin II, aldosterone, arginine vasotocin, corticosterone, and electrolyte concentrations during acclimation to dry condition and seawater in the crab-eating frog. Gen Comp Endocrinol 195:40–46
Viborg AL, Hillyard SD (2005) Cutaneous blood flow and water absorption by dehydrated toads. Physiol Biochem Zool 78:394–404
West NH, Kimmel P, Topor ZL et al (1998) The role of angiotensin in arterial blood pressure regulation in the toad Bufo marinus. J Exp Biol 201:2219–2224
Wright JW, Harding JW (2011) Brain renin-angiotensin–a new look at an old system. Prog Neurobiol 95:49–67

Chapter 5
Sex Differences in Angiotensin II Hypertension

Jennifer C. Sullivan

Abstract The renin angiotensin aldosterone system (RAAS) is critical for controlling blood pressure and body fluid volume and inhibitors of the RAAS are among the most widely used and effective drugs for the treatment of hypertension in both men and women. However, there is an extensive literature detailing sex differences in the RAAS, both in the expression levels of RAAS components and in physiological/pathophysiological responses to RAAS activation. Angiotensin (Ang) II is the primary vasoconstrictor peptide of the RAAS and male experimental animals have greater increases in blood pressure following chronic Ang II infusion than females. This chapter details basic sciences studies that have examined how males vs. females differentially respond to Ang II infusion with the goal of identifying molecular mechanisms that are responsible for the observed sex differences in Ang II hypertension.

Keywords Ang (1-7) • AT_1 receptor • AT_2 receptor • Mas receptor • ACE2 • Oxidative stress • Inflammation • Estrogen • Testosterone • Sex chromosomes

5.1 There Are Sex and Gender Differences in Hypertension

5.1.1 Clinical Introduction

Hypertension is well-recognized as having distinct sex differences in the prevalence, absolute blood pressure values, and molecular mechanisms contributing to the pathophysiology of the disease. Although first reported in 1947 that healthy, college-aged men have a significantly higher blood pressure than age-matched healthy women (Boynton and Todd 1947), current national guidelines still recommend the same approach for treating men and women with hypertension and the definition of hypertension is the same regardless if you are a man or a woman. This approach is likely out dated since a recent cross-sectional survey of the National

J.C. Sullivan, Ph.D. (✉)
Department of Physiology, Georgia Regents University, 1459 Laney Walker Blvd, Augusta, GA 30912, USA
e-mail: JENSULLIVAN@gru.edu

K.A. Hyndman, T.L. Pannabecker (eds.), *Sodium and Water Homeostasis*, Physiology in Health and Disease, DOI 10.1007/978-1-4939-3213-9_5

Health and Nutrition Examination Survey data reported that women with hypertension were more likely than men to be treated and take their medication, yet only 45 % of treated women achieved blood pressure control vs. 51 % of treated men (Gu et al. 2008). These statistics likely reflect that fact that women are still not included in clinical trials in numbers reflecting disease prevalence in the general population and pre-clinical studies remain focused on males. As a result, we still know far less about the mechanisms regulating blood pressure and cardiovascular health and function in females compared to males, calling into question the validity of applying the same treatment regimens regardless of sex. Inhibitors of the renin angiotensin aldosterone system (RAAS) are among the most effective and widely used drugs to treat hypertension irrespective of gender, despite well acknowledged differences between males and females in the RAAS. This is problematic since available data from clinical trials where results are reported separately for each gender indicate differences in the efficacy of treatment between men and women. Men exhibit greater decreases in ambulatory systolic blood pressure in response to the angiotensin converting enzyme (ACE) inhibitor lisinopril while women exhibit greater decreases in blood pressure and cardiovascular protection in response to angiotensin receptor blockers (ARBs) (Canzanello et al. 2008; Saunders et al. 2008; Miller et al. 2006; Trenkwalder et al. 2005; McInnes 1999). Therefore, better understanding of the impact of sex on the molecular mechanisms that are being targeted therapeutically may be critical in both improving blood pressure control rates in men and women and allow for a more informed decision to be made as to the first course of treatment based on gender.

5.1.2 The Renin-Angiotensin-Aldosterone System

Sex differences in the molecular mechanisms regulating blood pressure likely underlie the above observations. The goal of this chapter is to discuss sex differences in the RAAS, one of the critical physiological systems responsible for controlling blood pressure and body fluid volume. Under normal physiological conditions, renin is released from granular cells of the juxtaglomerular apparatus in the kidney in response to low blood volume, high salt content in the distal tubules, renal sympathetic nerve activity and reduced renal perfusion and catalyzes the conversion of angiotensinogen into Ang I. ACE then catalyzes the conversion of angiotensin (Ang) I into Ang II and Ang II activates angiotensin type 1 (AT_1) receptors and stimulates aldosterone secretion leading to renal salt and water reabsorption and arteriolar constriction, causing an elevation in blood pressure. However, under pathological conditions, over activation of the RAAS contributes to the development of hypertension and associated end-organ damage (Fig. 5.1).

Sex differences in the RAAS have been widely suggested to contribute to sex differences in blood pressure and cardiovascular disease; the activity and expression level of numerous components of the RAAS are modulated by the sex chromosome complement of the animal and gonadal hormone milieu. The

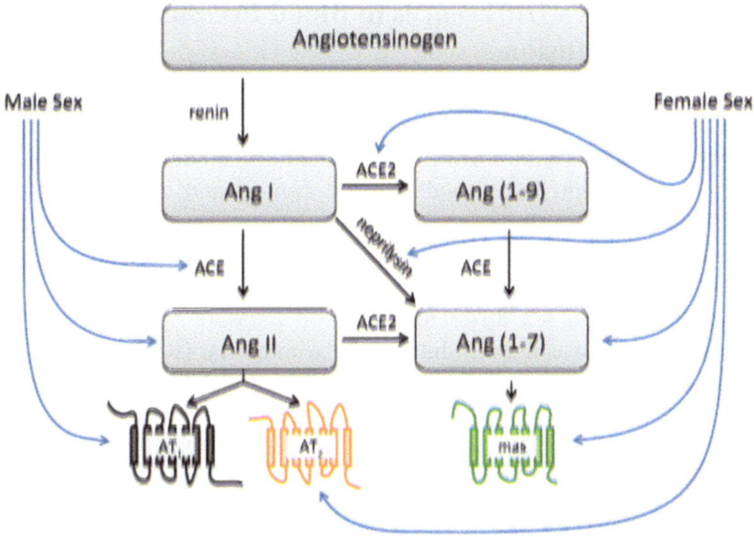

Fig. 5.1 Simplified schematic of the renin angiotensin aldosterone system (RAAS). The peptides discussed in this chapter are illustrated in *gray boxes*. The lines between the boxes represent conversion between the peptides catalyzed by the indicated enzymes. The biological effects of the RAAS are mediated by the binding of Ang II and Ang (1-7) to their respective G-protein coupled receptors AT_1, AT_2, and the mas receptor. Sex differences in the RAAS are further illustrated by the *blue lines*; the male sex favors the formation of classical RAAS components while the female sex favors non-classical RAAS components

"classical RAAS pathway" (composed of Ang II, AT_1 receptors, ACE) mediates most of the well-known biological functions of RAAS activation (Crowley and Coffman 2012; Coffman 2011). AT_1 receptors mediate the pro-hypertensive actions of Ang II, and renal AT_1 receptors in particular are critical to the development of Ang II hypertension in male mice (Crowley et al. 2006). In contrast, activation of the "non-classical RAAS pathway" (ACE2, Ang (1-7), angiotensin type 2 (AT_2), and mas receptors) opposes AT_1-mediated effects leading to vasodilation, improved blood flow, and enhanced pressure-natriuresis (Crowley and Coffman 2012; Chappell et al. 2014).

The Institute of Medicine had defined the term sex to classify human subjects as male or females according to chromosomal complement, while the term gender refers to an individual's self-representation as male or female (Committee on Understanding the Biology of Sex and Gender Differences 2001). For this reason, the term "sex" is used to refer to male/female differences in experimental animal models throughout this chapter.

5.2 There Are Sex Differences in Blood Pressure Responses to Ang II Infusion

Chronic Ang II infusion is an established model of induced hypertension and has been widely used across a number of species including rabbits, dogs, and rodents to gain mechanistic insight into blood pressure control and overall cardiovascular function in disease states. It is also acknowledged that there are sex differences in the physiological response to Ang II infusion with males having greater responses compared to females (Table 5.1).

The first report of a sex difference in blood pressure responses to Ang II in experimental animals was in dogs. Intracerebroventricular (ICV) infusion of Ang II resulted in a 30 % increase in blood pressure in male dogs with no change in

Table 5.1 Effect of Ang II on blood pressure in males and females

Species and strain	Dose	Route	Duration	Impact on blood pressure (BP)
Dogs	20 ng/min	i.c.v.	4 weeks	Males have greater increases in BP (Doursout et al. 1990)
Mice: C57/BL6J	800 ng/kg/min	Subcutaneous	7 days	Males have greater increases in BP (Xue et al. 2005)
Mice: C57Bl/6	400 ng/kg/min	Subcutaneous	14 days	Males have greater increases in BP (Ebrahimian et al. 2007)
Mice: MF1	200 ng/kg/min	Subcutaneous	12 days	Males have greater increases in BP (Ji et al. 2010)
Mice: MF1	800 ng/kg/min	Subcutaneous	6 days	Males have greater increases in BP (Ji et al. 2010)
Mice: C57Bl/6 × CBA × 129	800 ng/kg/min	Subcutaneous	7 days	Males have greater increases in BP (Tiwari et al. 2010)
Mice: FVB/N	600 ng/kg/min	Subcutaneous	10 days	Males have greater increases in BP (Brown et al. 2012)
Mice: wildtype on C57BL/6 background	490 ng/kg/min	Subcutaneous	2 weeks	Males have greater increases in BP (Ji et al. 2014)
Rats: Sprague-Dawley	0.7 mg/kg/day	Injection	10 days	Males have greater increases in BP (Tatchum-Talom et al. 2005)
Rats: Sprague-Dawley	50 ng/kg/min	Subcutaneous	13 days	Females have a decrease in BP (Sampson et al. 2008a)
Rats: Sprague-Dawley	400 ng/kg/min	Subcutaneous	13 days	Males have greater increases in BP (Sampson et al. 2008a)
Rats: Sprague-Dawley	400 ng/kg/min	Subcutaneous	14 days	Males have greater increases in BP (Sampson et al. 2008b)
Rats: SHR	200 ng/kg/min	Subcutaneous	2 weeks	Males have greater increases in BP (Sullivan et al. 2010a)
Rat: Sprague-Dawley	100 ng/kg/min	Subcutaneous	4 weeks	No sex difference in BP (Matrai et al. 2012)
Rats: Sprague-Dawley	120 ng/kg/min	Subcutaneous	2 weeks	Males have greater increases in BP (Xue et al. 2014)

icv: is Intracerebroventricular

females (Doursout et al. 1990). Since this report, numerous studies in rodents have not only confirmed this finding, but significantly expanded our understanding of how males vs. females respond to activation of the RAAS to control blood pressure. Subcutaneous infusion of Ang II results in greater increases in blood pressure in male mice compared to females (Xue et al. 2005; Ebrahimian et al. 2007; Tiwari et al. 2010; Ji et al. 2010, 2014; Brown et al. 2012). Interestingly, while there were sex differences in the blood pressure responses to both high (800 ng/kg/min) and low (200 ng/kg/min) doses of Ang II in MF1 mice, the sex difference was more pronounced at the low dose due to a more pronounced attenuation of Ang II hypertension in the female, suggesting that females require higher doses of Ang II to elicit an increase in blood pressure relative to males (Ji et al. 2010). Similarly, male Sprague-Dawley rats and spontaneously hypertensive rats (SHR) exhibit a greater increase in blood pressure in response to Ang II compared to females (Tatchum-Talom et al. 2005; Sampson et al. 2008a; Sullivan et al. 2010a; Xue et al. 2014). It was further demonstrated that the attenuation of Ang II-induced increases in blood pressure in female Sprague-Dawley rats was mediated by a smaller increase in systolic blood pressure during the night-time period in response to Ang II than in males; Ang II-induced increases in blood pressure during the day were comparable between the sexes (Sampson et al. 2008b). These data suggest that there are sex differences in the circadian response to Ang II and that during the active period of the day females have mechanisms in place to attenuate Ang II induced increases in blood pressure whereas males exacerbate Ang II hypertension.

There was not a sex difference in the blood pressure response to Ang II following 4 weeks of a much lower dose of Ang II than used in the above studies (100 ng/kg/min) (Matrai et al. 2012). However, blood pressure was only reported at the end of the study, therefore, it is unknown if the time course of the response or the overall change in blood pressure from baseline was comparable between the sexes. In contrast, 2 weeks of a similar dose of Ang II (120 ng/kg/min) resulted in a greater increase in blood pressure in male Sprague-Dawley rats compared to females (~24 mmHg change in pressure vs. ~10 mmHg, respectively) (Xue et al. 2014), further supporting the notion that the temporal effects of Ang II on blood pressure may differ by sex. Interestingly, when male and female Sprague-Dawley rats were treated with a sub-pressor dose of Ang II (10 ng/kg/min) for 1 week followed by 1 week of "rest" prior to initiating a pressor dose of Ang II for 2 weeks (120 ng/kg/min), Ang II hypertension was exacerbated exclusively in the males (Xue et al. 2014). Taken together, these data suggest that there is a "priming" effect of Ang II in males but not females, and supports the hypothesis that the mechanisms by which Ang II modulates blood pressure may be distinctly different between the sexes.

Additional studies have examined the impact of chronic Ang II infusion in male and female rats and mice when the endogenous RAAS is blocked. Based on extensive sex differences in the expression level of RAAS components, it was postulated that sex differences in blood pressure responses to Ang II were mediated by baseline differences in RAAS components upstream of Ang II receptor activation. Therefore, to eliminate this as a potentially confounding variable, male and female C57BL/6J mice and Sprague-Dawley rats were treated with the ACE inhibitor enalapril prior to Ang II infusion (800 ng/kg/min in mice and 150 ng/kg/min in rats delivered subcutaneously by osmotic mini-pump). Consistent with the above studies, male mice maintained greater increases in blood pressure following Ang II infusion compared to females (Venegas-Pont et al. 2010); however, when the RAAS was blocked by enalapril in rats, females exhibited a greater increase in blood pressure than males (Sartori-Valinotti et al. 2008). It should be noted that in the rat study, it was verified that the dose of enalapril given effectively blocked the conversion of Ang I to Ang II in both sexes; however, this same information was not provided in the mouse study. This raises the possibility that there was incomplete ACE inhibition in the mice allowing for residual RAAS activity, which could account for the differential response to Ang II in mice and rats. The mechanism responsible for the sex differences in response to Ang II was not identified in either study; however, it could be speculated that ACE is necessary for female rats to attenuate Ang II induced increases in blood pressure. Regardless, these findings suggest either there are sex differences in the endogenous RAAS that differentially impact blood pressure or that Ang II induces sex-specific signaling cascades resulting in altered blood pressure responses. It is of particular interest to identify the molecular mechanisms by which females are able to resist Ang II-induced increases in blood pressure relative to males. Understanding the signaling pathways "protecting" females can be exploited to limit Ang II induced increases in blood pressure and overall cardiovascular damage in both sexes.

5.3 Ang II Hypertension Involves Multiple Organ Systems

The robust increases in blood pressure following Ang II infusion reflects the fact that Ang II impacts all of the key organs involved in water and sodium homeostasis and blood pressure control, including the vasculature, the nervous system, and the kidney. Acute increases in blood pressure following systemic Ang II infusion are primarily mediated by AT_1-induced peripheral vasoconstriction, while increases in blood pressure following chronic Ang II infusion are attributed, in part, to centrally mediated activation of the sympathetic nervous system and alterations in renal handling of salt and water balance.

5.3.1 The Vasculature in Ang II Hypertension

Ang II induces vasoconstriction by binding to AT_1 receptors on vascular smooth muscle cells leading to an increase in calcium concentration. Abnormal calcium handling has been suggested to contribute to the pathogenesis of hypertension (Dominiczak and Bohr 1990) and there are sex differences in Ang II-induced increases in calcium. Ang II induces greater increases in intracellular calcium in isolated vascular smooth muscle cells from aorta of male SHR compared to female SHR (Loukotova et al. 2002). The authors further determined that the sex difference in calcium was dependent on calcium influx into the cell. Based on sex differences in calcium handling, it is not surprising that there are also sex differences in vascular function following the establishment of Ang II hypertension and in response to acute Ang II. Chronic Ang II infusion exacerbates vasoconstrictor responses to norepinephrine and serotonin and reduces vasodilator responses to acetylcholine in the mesenteric vascular bed of male Sprague-Dawley rats, but not female rats, suggesting a sex-specific effect of Ang II to induce vascular dysfunction (Tatchum-Talom et al. 2005). Similarly, male and female Sprague-Dawley rats exhibited a comparable increase in blood pressure in response to 100 ng/kg/day Ang II (4 weeks); however, resting vascular tone and thromboxane-induced vasoconstriction were greater in coronary arteries of males and endothelial dependent relaxation was blunted in arteries from males compared to arteries from females (Matrai et al. 2012). Consistent with this finding, treatment of male and female mice with a subpressor dose of Ang II for 1 week resulted in endothelial dysfunction in cerebral arteries of male, but not female mice (Chrissobolis and Faraci 2010). Sex differences in Ang II-induced vascular dysfunction despite comparable blood pressure between the sexes supports the hypothesis that there are innate sex differences in how arteries in males vs. females adapt and respond to Ang II-induced increases in blood pressure.

Isolated arteries from normotensive and hypertensive animal models also exhibit sex differences in vascular responses to increasing doses of Ang II. Isolated aortic rings and mesenteric arterioles from male SHR exhibit greater Ang II-induced vasoconstriction than arteries from females (Silva-Antonialli et al. 2004) and cerebral arteries isolated from female mice have a blunted constrictor response to Ang II compared to arteries from males (De Silva et al. 2009). Since impaired vascular function, and in particular decreases in endothelial function are associated with the progression of hypertension, the ability of females to "resist" Ang II-induced vascular dysfunction likely contributes to their ability to attenuate Ang II-induced increases in blood pressure relative to males.

5.3.2 Central Control of Ang II Hypertension

The central nervous system has a key role in regulating blood pressure by modulating the sympathetic and parasympathetic nervous systems, pituitary hormone release, and the baroreceptor reflex (Ramchandra et al. 2013). Ang II increases sympathetic nerve activity and modulates reflex regulation of heart rate via the sensory circumventricular organs and chronic low-dose Ang II-induced hypertension results in a slow increase in blood pressure via increased sympathetic nerve activity (DiBona 2013).

Based on the key role of the brain in mediating increases in blood pressure in response to Ang II, it is not surprising that sex differences in Ang II hypertension have been suggested to be related to the central and sympathetic control of blood pressure. Greater blood pressure responses to Ang II in male mice are associated with a blunted decrease in heart rate relative to females, reduced baroreflex sensitivity and a greater blood pressure response to ganglionic blockade with hexamethonium, suggesting that increased sympathetic nerve activity and baroreflex dysfunction contribute to the higher blood pressure response to Ang II in males (Xue et al. 2005). The authors further hypothesized that female mice maintain their ability to buffer changes in blood pressure during Ang II infusion, and additional studies extended this work to examine the role of central nitric oxide synthase (NOS) in modulating Ang II hypertension in male and female mice. NOS produces the potent vasodilator nitric oxide (NO), and females have been reported to have greater NO levels compared to males (Baylis 2012; McGuire et al. 2007; Zimmerman and Sullivan 2013; Sullivan et al. 2010b). Indeed, consistent with females being more dependent on NOS/NO to maintain cardiovascular function relative to males (Brinson et al. 2013), female mice possess a sex hormone-dependent increase in central NOS that attenuates Ang II hypertension that is not evident in males (Xue et al. 2009). More specifically, female mice exhibit an increase in NOS1 (neuronal NOS) in the subfornical organ (SFO) and paraventricular nucleus regions of the brain following Ang II infusion which increases NO production to attenuate sympathetic activity and increases in blood pressure relative to males. These studies suggest an important interaction of central Ang II and NO to differentially modulate blood pressure in a sex-dependent fashion, where enhanced NO in the female is critical in maintaining a lower blood pressure response to Ang II.

5.3.3 *The Kidney in Ang II Hypertension*

The kidney plays a central role in the long-term control of blood pressure and hypertension via alterations in the pressure-natriuresis curve (Coffman 2014) and Ang II-mediated increases in blood pressure are associated with a rightward shift in the pressure-natriuresis curve. There are sex differences in renal function and females tend to exhibit a leftward shift in the pressure-natriuresis relationship relative to males such that they excrete the same amount of sodium as males at a lower arterial pressure under physiological conditions (Khraibi et al. 2001; Hilliard et al. 2011). Although renal AT_1 receptors have been demonstrated to be essential to the ability of male mice to develop Ang II hypertension (Crowley et al. 2006), few studies have directly examined the impact of chronic Ang II infusion on renal function per se in both sexes. Our group has shown that Ang II induces greater increases in renal injury and protein excretion in male SHR compared to female SHR, consistent with males having a higher blood pressure (Sullivan et al. 2010a). Similarly, Ang II infusion increases albuminuria in male Sprague-Dawley rats treated with an ACE inhibitor, but not female rats (Sartori-Valinotti et al. 2008)

There are sex differences in blood pressure and renal hemodynamic responses to acute Ang II infusion. Male mice exhibit larger increases in blood pressure and renal vascular resistance in response to Ang II infusion than female mice, and this effect was not mediated by alterations in baroreflex function, implicating a renal mechanism (Schneider et al. 2010). Indeed, there are reported sex differences in the impact of Ang II on sodium transporters in the kidney. Ang II results in a down-regulation of NKCC2 in male, but not female mice, following Ang II treatment, while in female mice Ang II increased expression of the $\alpha-1$ subunit of Na-K-ATPase (Tiwari et al. 2010). The authors noted that these protein changes were consistent with a greater attempt to reduce sodium reabsorption in the male mice, relative to the females. Based on the central role of the kidney in modulating chronic blood pressure measurements, attenuated increases in renal injury and hemodynamic responses to Ang II infusion in the female will likely offer an additional mechanism by which Ang II induced hypertension is attenuated relative to males.

5.4 Mechanisms Mediating Sex Differences in Ang II Hypertension

5.4.1 Differential Ang II Metabolism: Ang II Versus Ang (1-7)

Our laboratory and others have noted that when Ang II levels have been assessed in males and females, be it in the plasma (Gandhi et al. 1998; Sullivan et al. 2007; Yanes et al. 2006; Pendergrass et al. 2008) or in the kidney (Yanes et al. 2006; Sullivan et al. 2007, 2010a; Pendergrass et al. 2008), levels have been reported to be comparable between the sexes, greater in females or greater in males, yet, as noted above, tissue responsiveness to exogenous Ang II is greater in males. This may be related to differential metabolism of Ang II in males and females and may reflect the finding that females have higher levels of the vasodilatory RAAS peptide Ang (1-7) (Sullivan et al. 2010a; Pendergrass et al. 2008; Bhatia et al. 2013). There is accumulating evidence supporting the hypothesis that a balance between the activation of the Ang II-mediated "classical" and Ang (1-7)-driven "non-classical" RAAS pathways is an important determinant of the overall impact of Ang II infusion and RAAS activation on blood pressure (Passos-Silva et al. 2013; Chappell et al. 2014). Ang (1-7) acts as a counterbalance to the pressor, anti-natriuretic, and proliferative actions of Ang II (van Twist et al. 2014; Zimmerman and Burns 2012) and our laboratory as well as others have demonstrated that there are sex differences in the contribution of Ang (1-7) to modulate blood pressure.

Intravenous (IV) Infusion of Ang (1-7) (24 μg/kg/h) lowers high-salt-induced increases in blood pressure in both male and female Dahl salt-sensitive rats following 1 week of treatment; however, by week 2 only female Dahl salt-sensitive rats exhibit an attenuation in salt-induced increases in blood pressure (Eatman et al. 2001), suggesting that females are more sensitive to the depressor effects of Ang (1-7) compared to males. Our laboratory has further shown that Ang (1-7) antagonizes Ang II-induced hypertension only in female SHR (Sullivan et al. 2010a). Pre-incubation with the Ang (1-7)/mas receptor blocker D-Ala^7-Ang (1-7) (A779) exacerbated Ang II hypertension only in female SHR and abolished the sex difference in Ang II hypertension. In addition, we have shown that Ang (1-7) contributes more to the blood pressure lowering effects of the ARB candesartan in females relative to males (Zimmerman et al. 2014). All of these studies support the notion that females are more dependent on Ang (1-7) to modulate blood pressure compared to males. Recent studies more specifically demonstrated that Ang (1-7) acts centrally to modulate blood pressure responses to Ang II in a sex-specific manner. As discussed previously, pretreatment with low dose Ang II exacerbates subsequent hypertension in male Sprague-Dawley rats, but not in female rats (Xue et al. 2014). ICV infusion of Ang (1-7) concurrent with low dose Ang II pretreatment abolished the increased sensitivity to Ang II hypertension in male rats, while pretreatment of female rats with the Ang (1-7)/mas receptor blocker A779 during Ang II pretreatment potentiated Ang II-induced increases in blood pressure in female rats. These elegant studies extend our understanding of sex differences in Ang II hypertension to directly implicate a central mechanism by which Ang (1-7) differentially modulates blood pressure in males and females.

While a remaining question in the field of the RAAS is why females have greater Ang (1-7) levels than males, sex differences in ACE2/neprilysin expression and activity are likely contributing factors. ACE2 is a mono-carboxypeptidase homologue of ACE discovered in 2000 that directly catalyzes the production of Ang (1-7) from Ang II, or indirectly via Ang I (Rice et al. 2004; Burrell et al. 2013; Chappell et al. 2014). As such, ACE2 is thought to act as a counterbalance to the effects of ACE through degradation of Ang II and generation of Ang-(1-7). Not surprisingly, there are reported sex differences in ACE2 expression and activity. Female Sprague-Dawley rats have greater ACE2 expression than males (Sampson et al. 2012), although paradoxically, normotensive male MF1 mice and Lewis rats as well as hypertensive mRen(2) rats have greater renal ACE2 activity than age-matched females (Pendergrass et al. 2008; Liu et al. 2010). Despite reports of greater ACE2 activity in males, ACE2 preferentially protects female mice from developing obesity-induced hypertension via increases in Ang (1-7) (Gupte et al. 2012). Increases in blood pressure following a high-fat diet in male C57BL/6 mice was accompanied by reduced renal ACE2 activity and plasma Ang (1-7) levels compared to mice on a normal fat diet. In contrast, high-fat feeding did not increase blood pressure in female mice and females exhibited increased adipose ACE2 activity with a corresponding increase in plasma Ang-(1-7) levels. Pretreatment of female mice with A779 unmasked a blood pressure response to the high fat diet in females supporting a protective role for Ang (1-7) to attenuate high fat-induced increases in blood pressure only in the female. Additional studies revealed that ACE2 deficiency increased the blood pressure response to the high-fat in both sexes, demonstrating that tissue-specific regulation of ACE2 by diet contributes to sex differences in obesity-hypertension.

Although there is no data in the literature directly assessing the relative contribution of ACE2 to Ang II hypertension in males and females, it has been reported that while there are no differences in Ang II (1000 ng/kg/min, 4 weeks) induced increases in blood pressure between male ACE2 knockout (KO) and wildtype mice, ACE2 KO mice develop severe myocardial dysfunction and greater increases in collagen deposition, oxidative stress and inflammation in response to Ang II relative to wildtype mice (Alghamri et al. 2013; Zhong et al. 2011). In addition, infusion of soluble human recombinant ACE2 prevented acute Ang II-induced increases in blood pressure (Wysocki et al. 2010). Therefore, based on these studies, while I would hypothesize that since females are more dependent on Ang (1-7) and ACE2 for blood pressure control, their blood pressure would be more sensitive to manipulation of ACE2; these studies need to be done to address this important question.

In addition to ACE2, neprilysin is a zinc metallo-endopeptidase that directly metabolizes Ang I to Ang (1-7) and neprilysin is thought to play a major role in Ang (1-7) formation in the circulation, vascular endothelium, and kidney (Chappell et al. 2014; Allred et al. 2000). Female Lewis and mRen(2) rats have greater renal neprilysin activity compared to males which corresponds with higher levels of Ang (1-7) (Pendergrass et al. 2008; Neves et al. 2006), although there are no studies to date that have examined the relative contribution of neprilysin-mediated Ang (1-7) production on blood pressure. Interestingly, our group has also reported that greater levels of Ang (1-7) in the renal cortex of female WKY rats compared to males is associated with greater ACE activity and treatment of WKY with enalapril abolished the sex difference, suggesting ACE-mediated Ang (1-7) formation in female rats (Bhatia et al. 2013). However, this increase is likely via an intermediate angiotensin peptide such as Ang (1-9). While the ACE2/Ang-(1-9)/Ang (1-7) pathway was first described in 2000, it was only recently Ang (1-9) was shown to offer cardiovascular protection and there is currently no information regarding the impact of sex on this vasodilatory peptide of the RAAS (Ocaranza et al. 2014).

5.4.2 RAS Receptors: AT_1, AT_2, Mas

Sex differences in blood pressure and overall cardiovascular responses to Ang II have been linked to sex differences in AT_1 vs. AT_2 receptor expression. The pro-hypertensive actions of the RAAS are closely linked to Ang II activation of AT_1 receptors. Activation of AT_1 receptors mediate most of the well-known biological functions of Ang II, including vasoconstriction, sodium reabsorption, mesangial cell proliferation, vascular hypertrophy, inflammation, and increases oxidative stress. Males have greater AT_1 receptor mRNA and protein expression in the kidneys and vasculature compared to females (Sullivan et al. 2007, 2010a; Silva-Antonialli et al. 2004) and specific AT_1 receptor binding is ~40 % lower in glomeruli from female Sprague-Dawley rats compared to males (Rogers et al. 2007). Therefore, less AT_1 receptor activation in females may contribute to the attenuated increase in blood pressure in response to Ang II infusion. Consistent with this hypothesis, we recently published that while hypertension in both male and female SHR is dependent on the AT_1 receptor, male SHR have a greater decrease in baseline blood pressure in response to the ARB candesartan than females (Zimmerman et al. 2014).

In contrast to the AT_1 receptor, female experimental animals have greater AT_2 expression and females have been demonstrated to have an enhanced role for the AT_2 receptor in attenuating their blood pressure compared to males (Brown et al. 2012; Hilliard et al. 2014; Sullivan et al. 2007, 2010a; Sampson et al. 2008a, 2012; Silva-Antonialli et al. 2004; Baiardi et al. 2005). Sex differences in the contribution of the AT_2 receptor to blood pressure control was first noted in response to low-dose Ang II (50 ng/kg/min, 13 days) where female Sprague-Dawley rats exhibit an AT_2 receptor-dependent decrease in blood pressure; blood pressure in male rats was not altered by low-dose Ang II (Sampson et al. 2008a). This finding was further supported by studies in AT_2 receptor knockout mice (Brown et al. 2012). Chronic Ang II infusion resulted in comparable increases in male wildtype and AT_2 receptor KO mice; however, female AT_2 KO mice have an exacerbated increase in blood pressure following Ang II infusion compared to wildtype females. In addition, the sex difference in Ang II hypertension that was present in wildtype mice was not apparent in KO mice due to an increased blood pressure response to Ang II in the females (Brown et al. 2012), these data directly support a role for AT_2 receptor activation to attenuate Ang II induced increases in blood pressure exclusively in the female.

The AT_2 receptor also contributes to sex difference in vascular and renal function. AT_2 receptor expression increases following cuff-induced vascular injury of the femoral artery in male and female mice; however females exhibit greater increases in AT_2 expression and an attenuation of injury relative to male mice (Okumura et al. 2005, 2011). Moreover, the sex differences in injury were blunted in AT_2 receptor KO mice, confirming a role for the AT_2 receptor to offer protection against cuff-induced injury in female mice. Consistent with these results, isolated renal interlobar arteries from female C57/Bl6 mice have less pronounced contraction to Ang II in comparison to males and blocking the AT_2 receptor increases Ang II-induced contraction only in females (Viegas et al. 2012). AT_2 receptor blockade also blunts pressure-natriuresis in both male and female Sprague-Dawley rats, although only in female rats did AT_2 receptor blockade also blunt auto-regulation of renal blood flow (Hilliard et al. 2011). The authors proposed that this finding reflected a role for the AT_2 receptor to blunt AT_1 receptor-mediated vasoconstriction by endogenous Ang II in response to acute hypotension in females. In support of this notion, iv infusion of an AT_2 receptor agonist, compound 21, resulted in more pronounced increases in renal blood flow and decreases in renal vascular resistance in female Sprague-Dawley rats than in males (Hilliard et al. 2012). The authors found a similar result in female SHR where compound 21 induced renal vasodilation and natriuresis, although there was no effect of compound 21 in male

SHR (Hilliard et al. 2014), suggesting the loss of AT_2 regulation of renal function may contribute to greater increases in blood pressure in hypertensive males compared to females. The ability of AT_2 receptor activation to modulate the physiological responses to Ang II in key organs responsible for blood pressure control in females is likely an important pathway by which young females are offered cardiovascular protection relative to age-matched males.

The G-protein-coupled receptor mas is the endogenous receptor for Ang (1-7). Activation of the mas receptor stimulates NO and prostaglandin production in cultured endothelial cells, transfected Chinese hamster ovary cells, and canine coronary arteries accounting for the vasodilatory, anti-proliferative, anti-thrombotic, diuretic and natriuretic properties of Ang (1-7) (van Twist et al. 2014; Zimmerman and Burns 2012). Similar to the AT_2 receptor, there is data to support the hypothesis that females are more dependent on the mas receptor to regulate blood pressure compared to males. Our group reported that female SHR exhibit an increase in renal mas receptor expression in response to chronic Ang II infusion that was not seen in males (Sullivan et al. 2010a). In addition, blockade of mas receptor activation using A779 potentiated Ang II hypertension only in female SHR abolishing the sex difference in Ang II hypertension observed in vehicle-treated SHR. These data suggest that Ang (1-7)-mas receptor activation in females acts to blunt Ang II induced hypertension. This hypothesis was supported by a recent study demonstrating that treatment with a low dose of Ang II (10 ng/kg/min) resulted in a small but significant decrease in blood pressure in female, but not male, Sprague-Dawley rats and this effect was blocked by central infusion of A779 (Xue et al. 2014). In addition, as noted above, pretreatment with central A779 potentiated subsequent Ang II hypertension in female rats verifying a critical role for Ang (1-7)/mas receptor activation in modulating Ang II hypertension in females. Therefore, we propose that the AT_2 receptor acts under basal conditions to protect females from Ang II-induced increases in blood pressure and end-organ dysfunction, while in the presence of Ang II females have greater increases in the production of Ang (1-7) resulting in additional antagonism of Ang II pathophysiology via mas receptor activation.

There is also evidence of sex differences in the interaction and cross-talk between RAAS receptors. Treatment of male and female Wistar rats with the mas receptor antagonist A779 decreased renal blood flow and increased renal vascular resistance in females with no effect in males (Safari et al. 2012). Interestingly, co-administration of an AT_2 receptor blocker abolished the effects of A779, suggesting that full expression of the renal vasoconstrictor effects of mas receptor blockade required co-activation of the AT_2 receptor. Moreover, while the response to acute Ang II infusion was not modulated by mas receptor blockade in either sex, the renal blood flow dose-response curve was attenuated by the combined blockade of mas and AT_2 receptors in female rats suggesting that, paradoxically, the combined activation of mas and AT_2 receptors may enhance Ang II-induced renal vasoconstriction. More studies are needed to better define and understand the interaction of the RAAS receptors in both sexes.

5.4.3 Oxidative Stress

Although it would be tempting to speculate that sex differences in metabolism and receptor expression can fully account for sex differences in blood pressure responses to Ang II, it does not appear to be that simple. Indeed, there is evidence to suggest that AT_1 receptor activation does not result in the same downstream signaling events in males and females. In particular, there are sex differences in Ang II-induced oxidative stress, and oxidative stress contributes to Ang II-mediated increases in blood pressure in male experimental animals (Laursen et al. 1997). Greater increases in blood pressure following Ang II infusion in male Sprague-Dawley rats compared to females is associated with the up regulation of the NAD (P)H oxidase gp67 phox subunit in the aorta of male but not female rats (Tatchum-Talom et al. 2005) and Ang II drives greater increases in NADPH oxidase activity in male mice compared to female mice (Ebrahimian et al. 2007). Consistent with greater increases in NADPH oxidase in male experimental animals, Ang II-induces greater increases in superoxide and hydrogen peroxide levels in cerebral arteries from male mice compared to female mice (De Silva et al. 2009) and our group has published that Ang II-induced oxidative stress contributes to Ang II hypertension in male but not female SHR (Bhatia et al. 2012). These results can be interpreted as (1) male rats exhibit more effective Ang II/AT_1 receptor/NAD(P)H oxidase coupling to induce increases in oxidative stress or (2) females have greater AT_2/mas receptor attenuation of Ang II-induced increases in oxidative stress, yet regardless, these data clearly indicate that not only are the physiological implications of increases in Ang II different between males and females, but the biochemical implications also differ.

The mechanism by which Ang II results in greater oxidative stress in males remains uncertain; however, multiple mechanisms have been implicated. Deletion of superoxide dismutase (SOD) 2 (SOD2 or mitochondrial SOD) resulted in endothelial dysfunction in isolated arteries from male mice following Ang II infusion, but not in arteries from female mice (Chrissobolis and Faraci 2010). Additional studies determined that reactive oxygen species signaling in the SFO, a key forebrain region of the circumventricular organ, is critical in Ang II-mediated increases in blood pressure (Zimmerman et al. 2002, 2004; Lob et al. 2010). SFO-targeted deletion of SOD3 (extracellular SOD) increases blood pressure sensitivity to systemic Ang II infusion (Lob et al. 2010) and overexpression of SOD1 (CuZn SOD) or SOD2 in the SFO attenuates Ang II hypertension in male mice (Zimmerman et al. 2002, 2004). While these same experiments have not been performed in females, central scavenging of ROS may be greater in females relative to males accounting for lower levels of Ang II-induced increases in oxidative stress. Indeed, estrogen receptor (ER) activation in the SFO inhibits Ang II-induced increases in reactive oxygen species and attenuates Ang II-induced hypertension in male mice, supporting the hypothesis that lower levels of central ROS production in females may limit Ang II induced increases in blood pressure (Xue et al. 2008). Together, these results suggest that males are more sensitive to decreases in antioxidant potential compared to females and that SOD normally offers protection during Ang II infusion in males. Since antioxidant levels have been reported to be greater in females relative to males (Lopez-Ruiz et al. 2008), females may be better able to neutralize Ang II-induced increases in oxidative stress to minimize Ang II-induced increases in blood pressure and end-organ damage.

5.4.4 *Inflammation*

Low-grade inflammation is now a recognized hallmark of hypertension and elegant studies by Guzik et al. directly implicated T cells in Ang II hypertension (Guzik et al. 2007). Rag KO mice are deficient in both B and T cells and male Rag KO mice have a blunted hypertensive response to Ang II infusion which is restored by adoptive transfer of T cells. Crowley et al. confirmed these findings using Scid mice. Scid mice are homozygous for the severe combined immune deficiency spontaneous mutation $Prkdc^{scid}$ and are characterized by an absence of functional B and T cells and male Scid mice have a blunted hypertensive response to chronic Ang II infusion (Crowley et al. 2010). Additional support for a causal role for T cells in the development of Ang II hypertension comes from studies blocking T cell activation. T cell activation requires both T cell receptor ligation and co-stimulation, often mediated by an interaction between the B7 ligands, CD80 or CD86, on antigen-presenting cells with the T cell co-receptor CD28. Either blocking B7-dependent co-stimulation of T cells pharmacologically or using mice lacking B7 ligands results in an attenuation of Ang II hypertension in male mice (Vinh et al. 2010).

There are multiple T cell subtypes, and several of them have been implicated in Ang II hypertension. Th17 cells are $CD3^{+}CD4^{+}$ T cells that secrete the pro-inflammatory cytokine IL-17 and a role for Th17 cells in hypertension has been suggested using interleukin (IL)-17 KO mice (Madhur et al. 2010). However, DOCA-salt treatment in conjunction with Ang II resulted in comparable increases in blood pressure in male wildtype, IL-17 KO, and IL-23p19 KO mice despite a decrease in Th17 cells (Krebs et al. 2014). Interestingly, the authors noted an increase in renal infiltration of γδT cells in IL17 KO and IL-23p19 KO mice raising the possibility that infiltrating γδT cells, not Th17 cells promotes Ang II hypertension. Finally, $CD4^{+}CD25^{+}$ T regulatory cells (Tregs) suppress effector T cell activation and secrete the anti-inflammatory cytokine IL-10. The adoptive transfer of Tregs from normotensive control male animals attenuates Ang II-induced increases in blood pressure and end-organ damage (Barhoumi et al. 2011; Kvakan et al. 2009). Furthermore, infusion of Ang II results in greater increases in blood pressure in male wildtype mice compared to IL-10 KO mice, and infusion of IL-10 for 2 weeks into control mice attenuates Ang II-Induced increases in blood pressure (Kassan et al. 2011).

There are sex differences in the T cell profile in kidneys of SHR, and hypertension in SHR is RAAS dependent. We published that male SHR have more renal

Th17 cells while females have a compensatory increase in Tregs in response to increases in blood pressure that is not apparent in males (Tipton et al. 2012, 2014). While the vast majority of the studies examining the role of T cells in Ang II hypertension have exclusively studied males, two recent studies have greatly expanded our understanding of the impact of sex on T cells and blood pressure control using male and female Rag KO mice (Ji et al. 2014; Pollow et al. 2014). Of particular interest with regards to the focus of this chapter, male and female Rag KO mice exhibit comparable increases in blood pressure to Ang II infusion (490 ng/kg/min; 14 days), suggesting that the sex difference in response to Ang II infusion is mediated by T cells. This is further supported by the findings that (1) adoptive transfer of T cells from wildtype male mice to male Rag KO mice restores the hypertensive response to Ang II, (2) adoptive transfer of T cells from female mice into male Rag KO mice abolishes Ang II-induced increase in blood pressure, and (3) adoptive transfer of T cells from wildtype males to female Rag KO mice does not increase blood pressure responses to Ang II. Interestingly, male Rag KO mice receiving T cells from a female donor exhibited more $CD4^+$ and $CD8^+$ T cells in the perivascular adipose tissue than when the T cells were from a male donor, although sex of the donor did not impact renal T cell infiltration. In contrast, male Rag KO mice exhibited greater increases in $CD3^+$ T cell infiltration in the SFO than females following adoptive transfer of T cells from wildtype male mice (Pollow et al. 2014). Since only the greater increase in central T cells accompanied the greater increase in blood pressure to Ang II in male mice, this further suggests a critical role for central T cell infiltration not only in Ang II hypertension, but also in sex differences in Ang II hypertension. Indeed, anteroventral third cerebral ventricle ablation of the forebrain abolishes Ang II-induced hypertension in male mice and prevents Ang II-induced increases in oxidative stress and T cell activation- directly implicating central increases in T cells as a critical component of Ang II-induced increases in blood pressure (Marvar et al. 2010). Taken together, these studies suggest that females are able to limit Ang II-induced increases in central T cell infiltration thereby attenuating increases in oxidative stress and blood pressure such that the overall increase in blood pressure following Ang II infusion is significantly less in females relative to males. A key question remaining is the mechanism by which females accomplish this; however, evidence suggests that it may be related to innate differences in T cells between the sexes. T cells from females have a less pro-inflammatory and pro-hypertensive phenotype than T cells from males such that "sex" of the T cell is a contributing factor to the impact of Ang II on blood pressure.

5.5 Is It All Just Sex Chromosomes and Hormones?

The two key parameters that drive the phenotypic differences between males and females are the sex chromosome complement (XX or XY) and levels of sex hormones (estrogen vs. testosterone) and there is evidence to support a role for both chromosomes and hormones in sex differences in Ang II hypertension. Sex hormones, particularly testosterone and estrogen, have also been well documented to impact not only blood pressure, renal, central, and vascular function, but also numerous pathways linked to blood pressure control, including the RAAS (Fig. 5.1).

5.5.1 Chromosomes

The four core genotype (FCG) mouse model is a unique animal model that allows manipulation of sex chromosome effects independent of sex hormones. In the FCG model, the Sry gene, which is the testis-determining gene, was deleted from the Y chromosome through a natural mutation resulting in XY-female mice (Arnold and Chen 2009). The Sry gene was also inserted onto an autosome, creating XY-Sry and XX-Sry transgenic mice that are both male. Sry genes (Sry1, Sry2 and Sry3) have been reported to differentially upregulate promoter activity of the angiotensinogen, renin and ACE genes and downregulate ACE2 promoter activity (Milsted et al. 2010). As a result, Sry regulation of multiple genes in the RAAS or alterations in expression of a single gene could favor increased levels of Ang II and decreased levels of Ang-(1-7) in male experimental animals. Gonadectomized FCG mice were used to determine the impact of the sex chromosomal complement on Ang II hypertension (Ji et al. 2010). Interestingly, Ang II-induced (200 ng/kg/min; 14 days) (Ji et al. 2010) increases in blood pressure were greater in XX mice than XY mice regardless of whether the mice were male or female. Moreover, plasma Ang II levels are higher in gonadectomized FCG females than males. This study verified a sex chromosome effect on Ang II hypertension, although whether this is mediated by a potentiation of Ang II hypertension by XX or inhibition of Ang II hypertension with XY remains unknown. In addition, these results were rather paradoxical based on the preponderance of evidence supporting that notion that females are "protected" from Ang II hypertension. If this is so, how can an XX chromosomal complement exacerbate Ang II hypertension? As noted by the authors, this result underscores how little we know regarding the interaction between hormonal and sex chromosomal factors. Maybe sex hormones are able to "over-ride" the signals from the chromosomes such that they are "more important" in determining the physiological responses to Ang II. While numerous questions remain, it is clear that additional studies will be needed to fully define the complexity of the interactions between hormones and chromosomes to determine the overall impact of Ang II, as well as other factors, on overall blood pressure control.

5.5.2 Female Sex Hormones

The cardiovascular protection considered to be afforded pre-menopausal women is lost with the onset of menopause, which led to a commonly held belief that female sex hormones were cardio-protective and estrogen regulates numerous components of the RAAS. Consistent with this idea, ovariectomy of female mice increases Ang II-induced hypertension, supporting the hypothesis that female sex hormones attenuate Ang II mediated increases in blood pressure (Xue et al. 2005, 2007b; Ebrahimian et al. 2007). Similarly, increased blood pressure sensitivity to Ang II in ovariectomized female Sprague-Dawley rats following pretreatment with a low dose of Ang II is blocked by ICV infusion of estrogen (Xue et al. 2014), suggesting central actions of estrogen as the female sex hormone responsible for modulating Ang II hypertension. Additional studies demonstrated that the ability of estrogen to attenuate Ang II hypertension was mediated via ER activation. ER antagonists block the estrogen attenuation of Ang II-induced increases in blood pressure in female mice and Ang II induces greater increases in blood pressure in ERα KO female mice compared with intact wildtype females, further implicating ERα as the receptor subtype responsible (Xue et al. 2007b).

The ability of estrogen to shift the balance of the classical and non-classical RAAS to favor the non-classical pathway most likely contributes to the attenuation of Ang II hypertension in estrogen-replete experimental models. This was first suggested in studies using ovariectomized transgenic rats expressing the mouse Ren2 (mRen2) gene. These mRen2 transgenic rats develop severe RAAS-dependent hypertension with age (Brosnihan et al. 1997). 17β-estradiol replacement shifted the balance of the vasoconstrictor-vasodilator components of the RAAS in these rats to favor the formation of Ang (1-7) to potentiate decreases in blood pressure to acute Ang (1-7) injections compared to vehicle-treated females. In a separate study, ovariectomy significantly increased markers of renal injury in response to renal wrap hypertension compared to estrogen-replete females and infusion of Ang (1-7) prevented the exacerbating effects of ovariectomy on renal damage (Ji et al. 2008), further supporting a role for estrogen to increase Ang (1-7) production to oppose Ang II-mediated effects.

Increases in Ang (1-7) following estrogen supplementation likely reflect the finding that female gonadal hormones positively regulate the enzymes responsible for the production of Ang (1-7). In experimental animals during pregnancy, ACE2 immunostaining and enzymatic activity are enhanced compared to non-pregnant controls (Brosnihan et al. 2003; Joyner et al. 2007). Estrogen replacement also abolishes observed decreases in ACE2 activity and expression in kidneys of ovariectomized rats in response to renal wrap hypertension (Ji et al. 2008). Neprilysin activity is also increased in the uterus, although not in the kidney, in mid-pregnancy and following 17β-estradiol treatment (Neves et al. 2006; Ottlecz et al. 1991) and ovariectomy decreases neprilysin activity in the brain while estrogen replacement restores neprilysin activity levels back to control values (Huang et al. 2004). It should be noted however, that there are no studies to date that have directly compared the relative contribution of ACE2 and neprilysin to Ang (1-7) in gonad-intact, ovariectomized, and ovariectomized females given estrogen replacement.

Female sex hormones also impact RAAS receptor expression. Greater renal AT_2 receptor expression in female mice is dependent on female sex hormones; the lower AT_1/AT_2 receptor ratio in females compared to males is due to the presence of 17β-estradiol (Baiardi et al. 2005; Silva-Antonialli et al. 2004), which could account for the greater role of the AT_2 receptors to modulate cardiovascular health in females relative to males. Indeed, ovariectomy attenuates injury-induced increases in vascular AT_2 expression in female mice, supporting the hypothesis that the vascular protective effects of estrogen are mediated, at least in part, by AT_2 receptor stimulation (Okumura et al. 2005). Female sex hormones also contribute to sex differences in AT_1 receptor expression. Ovariectomy of female Sprague-Dawley rats increases glomerular AT_1 receptor binding by ~50 % and estrogen replacement prevents ovariectomy-induced increase in glomerular AT_1 receptor binding (Rogers et al. 2007). Consistent with these findings, increases in ACE activity and AT_1 receptor binding densities in the heart, kidney, lung, aorta, adrenal and brain regions of female Wistar rats following ovariectomy are prevented by estrogen replacement (Dean et al. 2005). Moreover, Ang II induced constriction of isolated aorta is greater in arteries isolated from ovariectomized female WKY and SHR rats compared to gonad-intact rats (Silva-Antonialli et al. 2004; Nickenig et al. 1998) and this was accompanied by an increase in AT_1 receptor mRNA density; estrogen replacement blocked OVX-induced alterations in AT_1 levels (Nickenig et al. 1998). Therefore, there is strong evidence to support the hypothesis that beneficial effects of estrogen contribute to the attenuation of Ang II-induced increases in blood pressure in young females. It should be noted however, that while studies have focused on estrogen/estradiol, this is not the only female sex hormone. Less is known regarding the effect of other female sex hormones, although progesterone has been suggested to impact the RAAS (Amaral et al. 2014). Studies fully defining how the individual female sex hormones impact Ang II pathophysiology, or how fluctuations in female sex hormones differentially impact Ang II hypertension, still need to be performed.

5.5.3 Male Sex Hormones

Androgens have been suggested to be pro-hypertensive; surgical castration decreases blood pressure in multiple animal models of hypertension and testosterone supplementation increases blood pressure in females (Maranon and Reckelhoff 2013). Less is known regarding the role of testosterone in Ang II hypertension compared to the extensive literature examining the protective role of estrogen. However, available data suggests that testosterone exacerbates Ang II-induced increases in blood pressure. Testosterone mediates increased blood pressure sensitivity to Ang II in growth-restricted rats (Ojeda et al. 2010) and castration of male mice and New Zealand genetically hypertensive rats attenuates the development of Ang II-induced increases in blood pressure compared to gonad-intact controls (Song et al. 2006a). Castration of male transgenic mRen2 rats and SHR also attenuates age-related increases in blood pressure and hypertension in both of these rat strains is RAAS-dependent (Vaneckova et al. 2011; Sullivan et al. 2007; Reckelhoff et al. 1998). The testosterone receptor has been implicated in mediating increased blood pressure sensitivity to Ang II in males. ICV infusion of the androgen receptor antagonist flutamide attenuates the development of Ang II-induced hypertension in male mice suggesting that central androgen receptors contribute to Ang II-induced increases in blood pressure (Xue et al. 2007a). It has also been shown that increased testosterone levels during gestation induces hypertension in pregnant rats via heightened AT_1-mediated signaling, providing a molecular mechanism linking elevated maternal testosterone levels with gestational hypertension (Chinnathambi et al. 2014).

The mechanism(s) by which testosterone exacerbates Ang II induced increases in blood pressure have not been extensively examined. Neither castration nor testosterone replacement alters AT1 receptor expression in male SHR or New Zealand genetically hypertensive rats (Song et al. 2006a; Sullivan et al. 2007). However, testosterone potentiation of acute increases in blood pressure and renal vascular responses to Ang II are mediated, at least in part, via the amplification of Rho kinase signaling male New Zealand genetically hypertensive rats (Song et al. 2006b). Consistent with this report, Rho kinase also contributes to the enhanced pressor response to acute angiotensin II in intact male growth-restricted rats (Ojeda et al. 2013). More studies are needed to better define the molecular mechanism(s) by which testosterone exacerbates Ang II hypertension.

5.6 Perspectives

It is well established that in response to Ang II males exhibit a greater increase in blood pressure than females. A number of molecular mechanisms that have been suggested to protect females from Ang II hypertension, include (1) greater central NO and Ang (1-7) attenuate oxidative stress, T cell infiltration and sympathetic outflow to limit increases in blood pressure, (2) greater vascular AT_2 receptor expression allows for the maintenance of vascular endothelial function in response to chronic Ang II, and (3) greater AT_2 expression and Ang (1-7) production antagonizes Ang II-induced renal injury and changes in renal function relative to males (Fig. 5.2). Despite all of these advances in our understanding, a number of questions remain regarding why the RAAS of males and females differ so dramatically. Based on the central role of the RAAS in regulating cardiovascular function under both physiological and pathological conditions, a better understanding of how females increase non-classical RAAS components to limit Ang II-induced increases in blood pressure and the associated end-organ damage may be exploited to improve cardiovascular outcomes for both men and women.

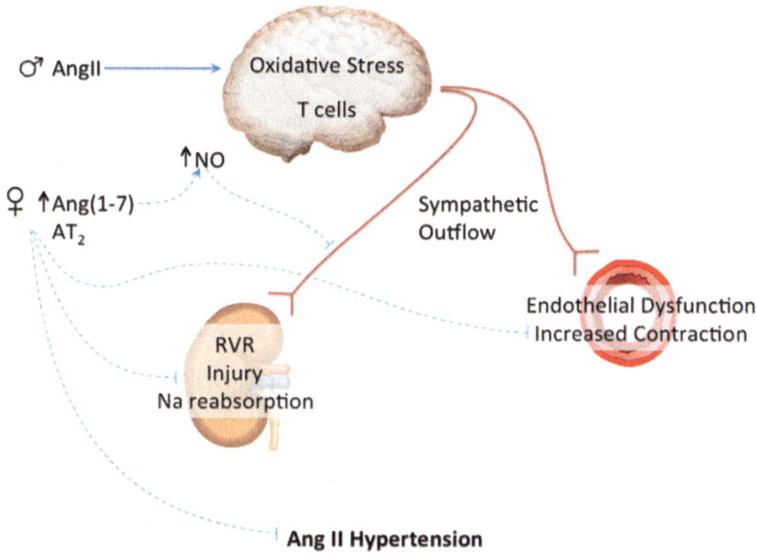

Fig. 5.2 Proposed mechanisms by which females mitigate Ang II induced increases in blood pressure relative to males. This schematic depicts a number of molecular mechanisms by which females have lower increases in blood pressure to Ang II relative to males. These include (1) greater central NO and Ang (1-7) attenuation of oxidative stress, T cell infiltration and sympathetic outflow to limit increases in blood pressure, (2) greater vascular AT_2 receptor expression allowing for the maintenance of vascular endothelial function, and (3) greater AT_2 expression and Ang (1-7) antagonism of Ang II-induced renal injury

References

Alghamri MS, Weir NM, Anstadt MP, Elased KM, Gurley SB, Morris M (2013) Enhanced angiotensin II-induced cardiac and aortic remodeling in ACE2 knockout mice. J Cardiovasc Pharmacol Ther 18(2):138–151

Allred AJ, Diz DI, Ferrario CM, Chappell MC (2000) Pathways for angiotensin-(1-7) metabolism in pulmonary and renal tissues. Am J Physiol Renal Physiol 279(5):F841–F850

Amaral LM, Kiprono L, Cornelius DC, Shoemaker C, Wallace K, Moseley J, Wallukat G, Martin JN Jr, Dechend R, Lamarca B (2014) Progesterone supplementation attenuates hypertension and the autoantibody to the angiotensin II type I receptor in response to elevated interleukin-6 during pregnancy. Am J Obstet Gynecol 211(2):158.e1–158.e2

Arnold AP, Chen X (2009) What does the "four core genotypes" mouse model tell us about sex differences in the brain and other tissues? Front Neuroendocrinol 30(1):1–9

Baiardi G, Macova M, Armando I, Ando H, Tyurmin D, Saavedra JM (2005) Estrogen upregulates renal angiotensin II AT_1 and AT_2 receptors in the rat. Regul Pept 124(1–3):7–17

Barhoumi T, Kasal DA, Li MW, Shbat L, Laurant P, Neves MF, Paradis P, Schiffrin EL (2011) T regulatory lymphocytes prevent angiotensin II-induced hypertension and vascular injury. Hypertension 57(3):469–476

Baylis C (2012) Nitric oxide synthase derangements and hypertension in kidney disease. Curr Opin Nephrol Hypertens 21(1):1–6

Bhatia K, Elmarakby AA, El-Remessy AB, Sullivan JC (2012) Oxidative stress contributes to sex differences in angiotensin II-mediated hypertension in spontaneously hypertensive rats. Am J Physiol Regul Integr Comp Physiol 302(2):R274–R282

Bhatia K, Zimmerman MA, Sullivan JC (2013) Sex differences in angiotensin-converting enzyme modulation of Ang (1-7) levels in normotensive WKY rats. Am J Hypertens 26(5):591–598

Boynton RE, Todd RL (1947) Blood pressure readings of 75,258 university students. Arch Intern Med 80(4):454–462

Brinson KN, Elmarakby AA, Tipton AJ, Crislip GR, Yamamoto T, Baban B, Sullivan JC (2013) Female SHR have greater blood pressure sensitivity and renal T cell infiltration following chronic NOS inhibition than males. Am J Physiol Regul Integr Comp Physiol 305(7):R701–R710

Brosnihan KB, Li P, Ganten D, Ferrario CM (1997) Estrogen protects transgenic hypertensive rats by shifting the vasoconstrictor-vasodilator balance of RAS. Am J Physiol 273(6 Pt 2):R1908–R1915

Brosnihan KB, Neves LA, Joyner J, Averill DB, Chappell MC, Sarao R, Penninger J, Ferrario CM (2003) Enhanced renal immunocytochemical expression of Ang-(1-7) and ACE2 during pregnancy. Hypertension 42(4):749–753

Brown RD, Hilliard LM, Head GA, Jones ES, Widdop RE, Denton KM (2012) Sex differences in the pressor and tubuloglomerular feedback response to angiotensin II. Hypertension 59 (1):129–135

Burrell LM, Harrap SB, Velkoska E, Patel SK (2013) The ACE2 gene: its potential as a functional candidate for cardiovascular disease. Clin Sci 124(2):65–76

Canzanello VJ, Baranco-Pryor E, Rahbari-Oskoui F, Schwartz GL, Boerwinkle E, Turner ST, Chapman AB (2008) Predictors of blood pressure response to the angiotensin receptor blocker candesartan in essential hypertension. Am J Hypertens 21(1):61–66

Chappell MC, Marshall AC, Alzayadneh EM, Shaltout HA, Diz DI (2014) Update on the Angiotensin converting enzyme 2-Angiotensin (1-7)-MAS receptor axis: fetal programing, sex differences, and intracellular pathways. Front Endocrinol 4:201

Chinnathambi V, More AS, Hankins GD, Yallampalli C, Sathishkumar K (2014) Gestational exposure to elevated testosterone levels induces hypertension via heightened vascular angiotensin II type 1 receptor signaling in rats. Biol Reprod 91(1):6

Chrissobolis S, Faraci FM (2010) Sex differences in protection against angiotensin II-induced endothelial dysfunction by manganese superoxide dismutase in the cerebral circulation. Hypertension 55(4):905–910

Coffman TM (2011) Under pressure: the search for the essential mechanisms of hypertension. Nat Med 17(11):1402–1409

Coffman TM (2014) The inextricable role of the kidney in hypertension. J Clin Invest 124 (6):2341–2347

Committee on Understanding the Biology of Sex and Gender Differences, Board on Health Sciences Policy Institute of Medicine (2001) Exploring the biological contributions to human health: does sex matter? Executive Summary of the Institute of Medicine Report (Wizeman TM and Pardue ML). National Academy Press, Washington

Crowley SD, Coffman TM (2012) Recent advances involving the renin-angiotensin system. Exp Cell Res 318(9):1049–1056

Crowley SD, Gurley SB, Herrera MJ, Ruiz P, Griffiths R, Kumar AP, Kim HS, Smithies O, Le TH, Coffman TM (2006) Angiotensin II causes hypertension and cardiac hypertrophy through its receptors in the kidney. Proc Natl Acad Sci U S A 103(47):17985–17990

Crowley SD, Song YS, Lin EE, Griffiths R, Kim HS, Ruiz P (2010) Lymphocyte responses exacerbate angiotensin II-dependent hypertension. Am J Physiol Regul Integr Comp Physiol 298(4):R1089–R1097

De Silva TM, Broughton BR, Drummond GR, Sobey CG, Miller AA (2009) Gender influences cerebral vascular responses to angiotensin II through Nox2-derived reactive oxygen species. Stroke 40(4):1091–1097

Dean SA, Tan J, O'Brien ER, Leenen FH (2005) 17β-estradiol downregulates tissue angiotensin-converting enzyme and Ang II type 1 receptor in female rats. Am J Physiol Regul Integr Comp Physiol 288(3):R759–R766

DiBona GF (2013) Sympathetic nervous system and hypertension. Hypertension 61(3):556–560
Dominiczak AF, Bohr DF (1990) Cell membrane abnormalities and the regulation of intracellular calcium concentration in hypertension. Clin Sci 79(5):415–423
Doursout MF, Chelly JE, Wouters P, Lawrence C, Liang YY, Buckley JP (1990) Effect of gender in centrally induced angiotensin II hypertension in dogs. Hypertension 15(2 Suppl):I117–I120
Eatman D, Wang M, Socci RR, Thierry-Palmer M, Emmett N, Bayorh MA (2001) Gender differences in the attenuation of salt-induced hypertension by angiotensin (1-7). Peptides 22 (6):927–933
Ebrahimian T, He Y, Schiffrin EL, Touyz RM (2007) Differential regulation of thioredoxin and NAD(P)H oxidase by angiotensin II in male and female mice. J Hypertens 25(6):1263–1271
Gandhi SK, Gainer J, King D, Brown NJ (1998) Gender affects renal vasoconstrictor response to Ang I and Ang II. Hypertension 31(1):90–96
Gu Q, Burt VL, Paulose-Ram R, Dillon CF (2008) Gender differences in hypertension treatment, drug utilization patterns, and blood pressure control among US adults with hypertension: data from the National Health and Nutrition Examination Survey 1999–2004. Am J Hypertens 21 (7):789–798
Gupte M, Thatcher SE, Boustany-Kari CM, Shoemaker R, Yiannikouris F, Zhang X, Karounos M, Cassis LA (2012) Angiotensin converting enzyme 2 contributes to sex differences in the development of obesity hypertension in C57BL/6 mice. Arterioscler Thromb Vasc Biol 32 (6):1392–1399
Guzik TJ, Hoch NE, Brown KA, McCann LA, Rahman A, Dikalov S, Goronzy J, Weyand C, Harrison DG (2007) Role of the T cell in the genesis of angiotensin II induced hypertension and vascular dysfunction. J Exp Med 204(10):2449–2460
Hilliard LM, Nematbakhsh M, Kett MM, Teichman E, Sampson AK, Widdop RE, Evans RG, Denton KM (2011) Gender differences in pressure-natriuresis and renal autoregulation: role of the angiotensin type 2 receptor. Hypertension 57(2):275–282
Hilliard LM, Jones ES, Steckelings UM, Unger T, Widdop RE, Denton KM (2012) Sex-specific influence of angiotensin type 2 receptor stimulation on renal function: a novel therapeutic target for hypertension. Hypertension 59(2):409–414
Hilliard LM, Chow CL, Mirabito KM, Steckelings UM, Unger T, Widdop RE, Denton KM (2014) Angiotensin type 2 receptor stimulation increases renal function in female, but not male, spontaneously hypertensive rats. Hypertension 64(2):378–383
Huang J, Guan H, Booze RM, Eckman CB, Hersh LB (2004) Estrogen regulates neprilysin activity in rat brain. Neurosci Lett 367(1):85–87
Ji H, Menini S, Zheng W, Pesce C, Wu X, Sandberg K (2008) Role of angiotensin-converting enzyme 2 and angiotensin(1-7) in 17β-oestradiol regulation of renal pathology in renal wrap hypertension in rats. Exp Physiol 93(5):648–657
Ji H, Zheng W, Wu X, Liu J, Ecelbarger CM, Watkins R, Arnold AP, Sandberg K (2010) Sex chromosome effects unmasked in angiotensin II-induced hypertension. Hypertension 55 (5):1275–1282
Ji H, Zheng W, Li X, Liu J, Wu X, Zhang MA, Umans JG, Hay M, Speth RC, Dunn SE, Sandberg K (2014) Sex-specific T-cell regulation of angiotensin II-dependent hypertension. Hypertension 64(3):573–582
Joyner J, Neves LA, Granger JP, Alexander BT, Merrill DC, Chappell MC, Ferrario CM, Davis WP, Brosnihan KB (2007) Temporal-spatial expression of Ang-(1-7) and angiotensin-converting enzyme 2 in the kidney of normal and hypertensive pregnant rats. Am J Physiol Regul Integr Comp Physiol 293(1):R169–R177
Kassan M, Galan M, Partyka M, Trebak M, Matrougui K (2011) Interleukin-10 released by $CD4^+CD25^+$ natural regulatory T cells improves microvascular endothelial function through inhibition of NADPH oxidase activity in hypertensive mice. Arterioscler Thromb Vasc Biol 31 (11):2534–2542
Khraibi AA, Liang M, Berndt TJ (2001) Role of gender on renal interstitial hydrostatic pressure and sodium excretion in rats. Am J Hypertens 14(9 Pt 1):893–896

Krebs CF, Lange S, Niemann G, Rosendahl A, Lehners A, Meyer-Schwesinger C, Stahl RA, Benndorf RA, Velden J, Paust HJ, Panzer U, Ehmke H, Wenzel UO (2014) Deficiency of the interleukin 17/23 axis accelerates renal injury in mice with deoxycorticosterone acetate+angiotensin II-induced hypertension. Hypertension 63(3):565–571

Kvakan H, Kleinewietfeld M, Qadri F, Park JK, Fischer R, Schwarz I, Rahn HP, Plehm R, Wellner M, Elitok S, Gratze P, Dechend R, Luft FC, Muller DN (2009) Regulatory T cells ameliorate angiotensin II-induced cardiac damage. Circulation 119(22):2904–2912

Laursen JB, Rajagopalan S, Galis Z, Tarpey M, Freeman BA, Harrison DG (1997) Role of superoxide in angiotensin II-induced but not catecholamine-induced hypertension. Circulation 95(3):588–593

Liu J, Ji H, Zheng W, Wu X, Zhu JJ, Arnold AP, Sandberg K (2010) Sex differences in renal angiotensin converting enzyme 2 (ACE2) activity are 17β-oestradiol-dependent and sex chromosome-independent. Biol Sex Differ 1(1):6

Lob HE, Marvar PJ, Guzik TJ, Sharma S, McCann LA, Weyand C, Gordon FJ, Harrison DG (2010) Induction of hypertension and peripheral inflammation by reduction of extracellular superoxide dismutase in the central nervous system. Hypertension 55(2):277–283

Lopez-Ruiz A, Sartori-Valinotti J, Yanes LL, Iliescu R, Reckelhoff JF (2008) Sex differences in control of blood pressure: role of oxidative stress in hypertension in females. Am J Physiol Heart Circ Physiol 295(2):H466–H474

Loukotova J, Kunes J, Zicha J (2002) Gender-dependent difference in cell calcium handling in VSMC isolated from SHR: the effect of angiotensin II. J Hypertens 20(11):2213–2219

Madhur MS, Lob HE, McCann LA, Iwakura Y, Blinder Y, Guzik TJ, Harrison DG (2010) Interleukin 17 promotes angiotensin II-induced hypertension and vascular dysfunction. Hypertension 55(2):500–507

Maranon R, Reckelhoff JF (2013) Sex and gender differences in control of blood pressure. Clin Sci 125(7):311–318

Marvar PJ, Thabet SR, Guzik TJ, Lob HE, McCann LA, Weyand C, Gordon FJ, Harrison DG (2010) Central and peripheral mechanisms of T-lymphocyte activation and vascular inflammation produced by angiotensin II-induced hypertension. Circ Res 107(2):263–270

Matrai M, Hetthessy J, Nadasy GL, Monos E, Szekacs B, Varbiro S (2012) Sex differences in the biomechanics and contractility of intramural coronary arteries in angiotensin II-induced hypertension. Gend Med 9(6):548–556

McGuire BB, Watson RW, Perez-Barriocanal F, Fitzpatrick JM, Docherty NG (2007) Gender differences in the renin-angiotensin and nitric oxide systems: relevance in the normal and diseased kidney. Kidney Blood Press Res 30(2):67–80

McInnes GT (1999) Angiotensin II antagonism in clinical practice: experience with valsartan. J Cardiovasc Pharmacol 33(Suppl 1):S29–S32

Miller JA, Cherney DZ, Duncan JA, Lai V, Burns KD, Kennedy CR, Zimpelmann J, Gao W, Cattran DC, Scholey JW (2006) Gender differences in the renal response to renin-angiotensin system blockade. J Am Soc Nephrol 17(9):2554–2560

Milsted A, Underwood AC, Dunmire J, DelPuerto HL, Martins AS, Ely DL, Turner ME (2010) Regulation of multiple renin-angiotensin system genes by Sry. J Hypertens 28(1):59–64

Neves LA, Chappell MC, Ferrario CM, Gallagher PE, Ganten D, Brosnihan KB (2006) Effect of estrogen on neprilysin expression in uterus and kidney of Sprague-Dawley normotensive and heterozygous (mRen2)27-transgenic hypertensive rats. Peptides 27(11):2912–2918

Nickenig G, Baumer AT, Grohe C, Kahlert S, Strehlow K, Rosenkranz S, Stablein A, Beckers F, Smits JF, Daemen MJ, Vetter H, Bohm M (1998) Estrogen modulates AT_1 receptor gene expression *in vitro* and *in vivo*. Circulation 97(22):2197–2201

Ocaranza MP, Michea L, Chiong M, Lagos CF, Lavandero S, Jalil JE (2014) Recent insights and therapeutic perspectives of angiotensin-(1-9) in the cardiovascular system. Clin Sci 127 (9):549–557

Ojeda NB, Royals TP, Black JT, Dasinger JH, Johnson JM, Alexander BT (2010) Enhanced sensitivity to acute angiotensin II is testosterone dependent in adult male growth-restricted offspring. Am J Physiol Regul Integr Comp Physiol 298(5):R1421–R1427

Ojeda NB, Royals TP, Alexander BT (2013) Sex differences in the enhanced responsiveness to acute angiotensin II in growth-restricted rats: role of fasudil, a Rho kinase inhibitor. Am J Physiol Renal Physiol 304(7):F900–F907

Okumura M, Iwai M, Ide A, Mogi M, Ito M, Horiuchi M (2005) Sex difference in vascular injury and the vasoprotective effect of valsartan are related to differential AT_2 receptor expression. Hypertension 46(3):577–583

Okumura M, Iwai M, Nakaoka H, Sone H, Kanno H, Senba I, Ito M, Horiuchi M (2011) Possible involvement of AT_2 receptor dysfunction in age-related gender difference in vascular remodeling. J Am Soc Hypertens 5(2):76–84

Ottlecz A, Walker S, Conrad M, Starcher B (1991) Neutral metalloendopeptidase associated with the smooth muscle cells of pregnant rat uterus. J Cell Biochem 45(4):401–411

Passos-Silva DG, Verano-Braga T, Santos RA (2013) Angiotensin-(1-7): beyond the cardio-renal actions. Clin Sci 124(7):443–456

Pendergrass KD, Pirro NT, Westwood BM, Ferrario CM, Brosnihan KB, Chappell MC (2008) Sex differences in circulating and renal angiotensins of hypertensive mRen(2). Lewis but not normotensive Lewis rats. Am J Physiol Heart Circ Physiol 295(1):H10–H20

Pollow DP, Uhrlaub J, Romero-Aleshire MJ, Sandberg K, Nikolich-Zugich J, Brooks HL, Hay M (2014) Sex differences in T-lymphocyte tissue infiltration and development of angiotensin II hypertension. Hypertension 64(2):384–390

Ramchandra R, Yao ST, May CN (2013) Organ selective regulation of sympathetic outflow by the brain Angiotensin system. Curr Hypertens Rep 15(4):401–408

Reckelhoff JF, Zhang H, Granger JP (1998) Testosterone exacerbates hypertension and reduces pressure-natriuresis in male spontaneously hypertensive rats. Hypertension 31(1 Pt 2):435–439

Rice GI, Thomas DA, Grant PJ, Turner AJ, Hooper NM (2004) Evaluation of angiotensin-converting enzyme (ACE), its homologue ACE2 and neprilysin in angiotensin peptide metabolism. Biochem J 383(Pt 1):45–51

Rogers JL, Mitchell AR, Maric C, Sandberg K, Myers A, Mulroney SE (2007) Effect of sex hormones on renal estrogen and angiotensin type 1 receptors in female and male rats. Am J Physiol Regul Integr Comp Physiol 292(2):R794–R799

Safari T, Nematbakhsh M, Hilliard LM, Evans RG, Denton KM (2012) Sex differences in the renal vascular response to angiotensin II involves the Mas receptor. Acta Physiol 206(2):150–156

Sampson AK, Moritz KM, Jones ES, Flower RL, Widdop RE, Denton KM (2008a) Enhanced angiotensin II type 2 receptor mechanisms mediate decreases in arterial pressure attributable to chronic low-dose angiotensin II in female rats. Hypertension 52(4):666–671

Sampson AK, Widdop RE, Denton KM (2008b) Sex-differences in circadian blood pressure variations in response to chronic angiotensin II infusion in rats. Clin Exp Pharmacol Physiol 35(4):391–395

Sampson AK, Moritz KM, Denton KM (2012) Postnatal ontogeny of angiotensin receptors and ACE2 in male and female rats. Gend Med 9(1):21–32

Sartori-Valinotti JC, Iliescu R, Yanes LL, Dorsett-Martin W, Reckelhoff JF (2008) Sex differences in the pressor response to angiotensin II when the endogenous renin-angiotensin system is blocked. Hypertension 51(4):1170–1176

Saunders E, Cable G, Neutel J (2008) Predictors of blood pressure response to angiotensin receptor blocker/diuretic combination therapy: a secondary analysis of the irbesartan/hydrochlorothiazide blood pressure reductions in diverse patient populations (INCLUSIVE) study. J Clin Hypertens 10(1):27–33

Schneider MP, Wach PF, Durley MK, Pollock JS, Pollock DM (2010) Sex differences in acute Ang II-mediated hemodynamic responses in mice. Am J Physiol Regul Integr Comp Physiol 299(3):R899–R906

Silva-Antonialli MM, Tostes RC, Fernandes L, Fior-Chadi DR, Akamine EH, Carvalho MH, Fortes ZB, Nigro D (2004) A lower ratio of AT_1/AT_2 receptors of angiotensin II is found in female than in male spontaneously hypertensive rats. Cardiovasc Res 62(3):587–593

Song J, Kost CK Jr, Martin DS (2006a) Androgens augment renal vascular responses to Ang II in New Zealand genetically hypertensive rats. Am J Physiol Regul Integr Comp Physiol 290(6): R1608–R1615

Song J, Kost CK Jr, Martin DS (2006b) Androgens potentiate renal vascular responses to angiotensin II via amplification of the Rho kinase signaling pathway. Cardiovasc Res 72 (3):456–463

Sullivan JC, Semprun-Prieto L, Boesen EI, Pollock DM, Pollock JS (2007) Sex and sex hormones influence the development of albuminuria and renal macrophage infiltration in spontaneously hypertensive rats. Am J Physiol Regul Integr Comp Physiol 293(4):R1573–R1579

Sullivan JC, Bhatia K, Yamamoto T, Elmarakby AA (2010a) Angiotensin (1-7) receptor antagonism equalizes angiotensin II-induced hypertension in male and female spontaneously hypertensive rats. Hypertension 56(4):658–666

Sullivan JC, Pardieck JL, Hyndman KA, Pollock JS (2010b) Renal NOS activity, expression, and localization in male and female spontaneously hypertensive rats. Am J Physiol Regul Integr Comp Physiol 298(1):R61–R69

Tatchum-Talom R, Eyster KM, Martin DS (2005) Sexual dimorphism in angiotensin II-induced hypertension and vascular alterations. Can J Physiol Pharmacol 83(5):413–422

Tipton AJ, Baban B, Sullivan JC (2012) Female spontaneously hypertensive rats have greater renal anti-inflammatory T lymphocyte infiltration than males. Am J Physiol Regul Integr Comp Physiol 303(4):R359–R367

Tipton AJ, Baban B, Sullivan JC (2014) Female spontaneously hypertensive rats have a compensatory increase in renal regulatory T cells in response to elevations in blood pressure. Hypertension 64(3):557–564

Tiwari S, Li L, Riazi S, Halagappa VK, Ecelbarger CM (2010) Sex differences in adaptive downregulation of pre-macula densa sodium transporters with Ang II infusion in mice. Am J Physiol Renal Physiol 298(1):F187–F195

Trenkwalder P, Elmfeldt D, Hofman A, Lithell H, Olofsson B, Papademetriou V, Skoog I, Zanchetti A (2005) The Study on COgnition and Prognosis in the Elderly (SCOPE)—major CV events and stroke in subgroups of patients. Blood Press 14(1):31–37

van Twist DJ, Kroon AA, de Leeuw PW (2014) Angiotensin-(1-7) as a strategy in the treatment of hypertension? Curr Opin Nephrol Hypertens 23(5):480–486

Vaneckova I, Huskova Z, Vanourkova Z, Cervenka L (2011) Castration has antihypertensive and organoprotective effects in male but not in female heterozygous Ren-2 rats. Kidney Blood Press Res 34(1):46–52

Venegas-Pont M, Sartori-Valinotti JC, Glover PH, Reckelhoff JF, Ryan MJ (2010) Sexual dimorphism in the blood pressure response to angiotensin II in mice after angiotensin-converting enzyme blockade. Am J Hypertens 23(1):92–96

Viegas VU, Liu ZZ, Nikitina T, Perlewitz A, Zavaritskaya O, Schlichting J, Persson PB, Regitz-Zagrosek V, Patzak A, Sendeski MM (2012) Angiotensin II type 2 receptor mediates sex differences in mice renal interlobar arteries response to angiotensin II. J Hypertens 30 (9):1791–1798

Vinh A, Chen W, Blinder Y, Weiss D, Taylor WR, Goronzy JJ, Weyand CM, Harrison DG, Guzik TJ (2010) Inhibition and genetic ablation of the B7/CD28 T-cell costimulation axis prevents experimental hypertension. Circulation 122(24):2529–2537

Wysocki J, Ye M, Rodriguez E, Gonzalez-Pacheco FR, Barrios C, Evora K, Schuster M, Loibner H, Brosnihan KB, Ferrario CM, Penninger JM, Batlle D (2010) Targeting the degradation of angiotensin II with recombinant angiotensin-converting enzyme 2: prevention of angiotensin II-dependent hypertension. Hypertension 55(1):90–98

Xue B, Pamidimukkala J, Hay M (2005) Sex differences in the development of angiotensin II-induced hypertension in conscious mice. Am J Physiol Heart Circ Physiol 288(5):H2177–H2184

Xue B, Johnson AK, Hay M (2007a) Sex differences in angiotensin II-induced hypertension. Braz J Med Biol Res 40(5):727–734

Xue B, Pamidimukkala J, Lubahn DB, Hay M (2007b) Estrogen receptor-alpha mediates estrogen protection from angiotensin II-induced hypertension in conscious female mice. Am J Physiol Heart Circ Physiol 292(4):H1770–H1776

Xue B, Zhao Y, Johnson AK, Hay M (2008) Central estrogen inhibition of angiotensin II-induced hypertension in male mice and the role of reactive oxygen species. Am J Physiol Heart Circ Physiol 295(3):H1025–H1032

Xue B, Singh M, Guo F, Hay M, Johnson AK (2009) Protective actions of estrogen on angiotensin II-induced hypertension: role of central nitric oxide. Am J Physiol Heart Circ Physiol 297(5): H1638–H1646

Xue B, Zhang Z, Beltz TG, Guo F, Hay M, Johnson AK (2014) Estrogen regulation of the brain renin-angiotensin system in protection against angiotensin II-induced sensitization of hypertension. Am J Physiol Heart Circ Physiol 307(2):H191–H198

Yanes LL, Romero DG, Iles JW, Iliescu R, Gomez-Sanchez C, Reckelhoff JF (2006) Sexual dimorphism in the renin-angiotensin system in aging spontaneously hypertensive rats. Am J Physiol Regul Integr Comp Physiol 291(2):R383–R390

Zhong J, Guo D, Chen CB, Wang W, Schuster M, Loibner H, Penninger JM, Scholey JW, Kassiri Z, Oudit GY (2011) Prevention of angiotensin II-mediated renal oxidative stress, inflammation, and fibrosis by angiotensin-converting enzyme 2. Hypertension 57(2):314–322

Zimmerman D, Burns KD (2012) Angiotensin-(1-7) in kidney disease: a review of the controversies. Clin Sci 123(6):333–346

Zimmerman MA, Sullivan JC (2013) Hypertension: what's sex got to do with it? Physiology 28 (4):234–244

Zimmerman MC, Lazartigues E, Lang JA, Sinnayah P, Ahmad IM, Spitz DR, Davisson RL (2002) Superoxide mediates the actions of angiotensin II in the central nervous system. Circ Res 91 (11):1038–1045

Zimmerman MC, Lazartigues E, Sharma RV, Davisson RL (2004) Hypertension caused by angiotensin II infusion involves increased superoxide production in the central nervous system. Circ Res 95(2):210–216

Zimmerman MA, Harris RA, Sullivan JC (2014) Female spontaneously hypertensive rats are more dependent on Ang (1-7) to mediate effects of low-dose AT_1 receptor blockade than males. Am J Physiol Renal Physiol 306(10):F1136–F1142

Chapter 6
The Evolution and Comparative Physiology of Endothelin Regulation of Sodium Transport

Kelly Anne Hyndman

Abstract The endothelins are a family of peptide paracrine/autocrine signaling molecules that function in a vareity of physiological processes including regulation of vascular tone and in the mammalian kidney they function to inhibit sodium reabsorption. Much of our understanding of endothelin-1 (ET1) signaling comes from studies in rodents; however, the endothelins and the endothelin receptors have an interesting evolutionary history, having evolved in the common ancestor to the vertebrates. The purpose of this chapter is to summarize our current understanding of the evolution and physiology of the endothelin signaling system in a comparative physiological context. Unlike therian mammals, which express two endothelin receptors, non-therian vertebrates express at least three endothelin receptors. Yet, similarly to the mammals, ET1 has hemodynamic effects leading to acute changes in blood pressure in a variety of non-mammalian vertebrates. Moreover, in teleost fishes, where the gills are the main site for sodium excretion, ET1 can inhibit ion transport and may act to fine-tune adjust ion excretion to maintain homeostasis. Although the mechanisms of how ET1 inhibits ion transport in non-mammalian vertebrates is not clear, evidence suggests that ET1 stimulates nitric oxide and prostaglandin production. This signaling cascade is conserved from fishes to man. Although many questions still remain about ET1 signaling in non-mammalian vertebrates, it is clear that endothelins evolved 400–500 million years ago highlighting their importance as paracrine/autocrine signaling peptides in all vertebrates.

Keywords Endothelin • Endothelin receptor • Evolution • Phylogenetics • Sodium homeostasis • Ion transport • Fish gill • Non-mammalian vertebrates • Killifish

---

K.A. Hyndman, Ph.D. (✉)
Division Nephrology, Department of Medicine, The University of Alabama at Birmingham, 1720 2nd Avenue South, KAUL 830, Birmingham, AL 35294-0024, USA
e-mail: hyndmank@uab.edu

K.A. Hyndman, T.L. Pannabecker (eds.), *Sodium and Water Homeostasis*, Physiology in Health and Disease, DOI 10.1007/978-1-4939-3213-9_6

6.1 Introduction

The excretion of sodium and water is important for the regulation of extracellular volume. This is true for all vertebrates, in which multiple organs and mechanisms have evolved in order to maintain fluid-electrolyte balance and avoid large fluctuations in extracellular volume. In teleost fishes, it is the gills that actively transport sodium to maintain homeostasis, while in mammals the kidneys complete this function. In addition to a number of humoral factors that regulate fluid-electrolyte balance, there are also local factors (paracrine and autocrine factors) that are critical for maintaining homeostasis. Endothelins are paracrine/autocrine peptides that function in physiological processes such as regulation of vascular tone and inhibit sodium reabsorption in the mammalian kidney. From fishes to man, the endothelin system is expressed, and evidence suggests that endothelin regulation of vascular tone and inhibition of sodium transport processes are conserved throughout the evolution and diversification of the vertebrates. In this chapter we will summarize our knowledge of endothelins in non-mammalian vertebrates, and present their evolutionary histories. As well, we will present the putative physiological role of endothelins in regulating hemodynamics and sodium homeostasis from studies using non-mammalian models. The majority of the studies focusing on endothelin in non-mammalian vertebrates have been conducted in teleost fishes, so we will briefly summarize the mechanisms of sodium homeostasis in teleosts and present the studies on the role of endothelins in the regulation of sodium homeostasis, as well as blood flow.

6.2 Endothelins

Endothelins (ETs) are a family of peptides with three isoforms, ET-1, ET-2 and ET-3 (Yanagisawa et al. 1988c; Inoue et al. 1989), and the genes encoding these proteins (they will be referred to as *EDN*s) are located on different chromosomes. They are regarded as the most potent vasoconstrictors yet identified. Endothelins are translated as ~200 amino acid (aa) preproendothelins (preproET) that are initially cleaved by a furin-like enzyme (Yanagisawa et al. 1988c) to form the relatively inactive 38 aa proendothelin (proET, also known as Big-ET) (Kimura et al. 1989). Proendothelin is further cleaved to form the active 21 aa ET by the endothelin converting enzyme (ECE-1 and/or ECE-2) (Shimada et al. 1994; Xu et al. 1994) (Fig. 6.1). Endothelin-1 release is stimulated by various factors including hypoxia, cortisol, growth factors, shear stress, calcium ionophores, cytokines and endotoxins (Kohan et al. 2011).

The first endothelin-like peptides sequenced from non-mammalian vertebrates were from the rainbow trout (Wang et al. 1999), frog (*Rana ridibunda*) (Wang et al. 2000) and the American alligator (*Alligator mississippiensis*) (Platzack et al. 2002). These were peptide sequences determined by mass spectrometry

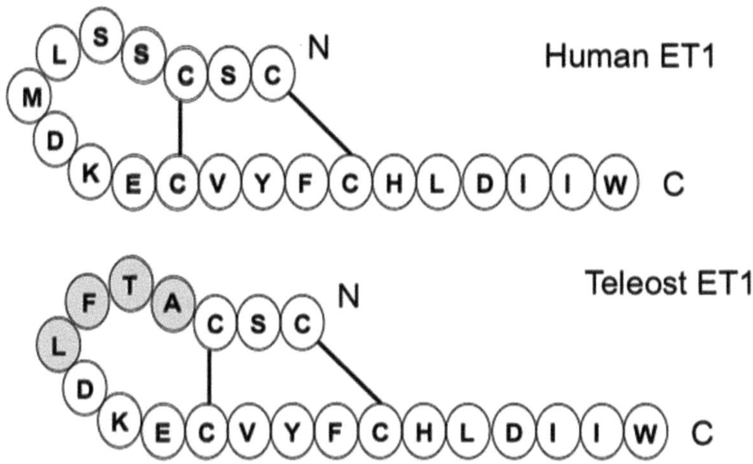

Fig. 6.1 A diagram of the 21 amino acid endothelin-1 from humans and teleost fishes. *Grey shaded* amino acids are different between human and teleosts. Modified from Hyndman and Evans (2007)

because the coding sequences for these *EDN*s had not been determined. In addition, there is a family of cardiotoxic peptides, the sarafotoxins, that shares 70 % primary sequence identity and structural identity with mammalian ET-1 (Lee and Chiappinelli 1989). The sarafotoxins are found in the venom of the Israeli burrowing asp (*Atractaspis engaddensis*), and there is little evidence to suggest they are homologous to the ETs. The most parsimonious hypothesis is that they occurred through convergent evolution (Froy and Gurevitz 1998).

6.3 Endothelin Receptors

Endothelins bind to a specific group of seven transmembrane domain, G-protein-coupled receptors (GPCRs) that are part of the rhodopsin/β-adrenergic GPCR family. The early classification of the ET GPCRs was quite confusing as it was based upon species specific or pharmacological profiles. For example, most researchers acknowledge that there are two ET-specific receptors termed, ET_A (Arai et al. 1990), and ET_B (Sakurai et al. 1990); however, a third receptor, an amphibian specific ET_C, was cloned and characterized from melanophores (Karne et al. 1993). In 1998, a fourth receptor was cloned and characterized from the chicken (*Gallus gallus*) and quail (*Coturnix japonica*), and termed the avian-specific ET_{B2}. Traditionally, the receptors were classified based upon their pharmacological profiles. For example, mammalian ET_A preferentially binds ET-1 and ET-2 over ET-3 (Arai et al. 1990). In contrast, mammalian ET_B binds all three ETs with equal affinity and binds the sarafotoxins, which are ET_B specific agonists (Sakurai et al. 1990; Lecoin et al. 1998). In addition to these receptors, pharmacological and physiological studies in mammals suggest that there are multiple ET_B-type receptors.

For example, ET-1 binding to vascular smooth muscle ET_B results in muscle contractions (Yanagisawa et al. 1988a, b). On the contrary, ET-1 binding of endothelial ET_B stimulates nitric oxide and prostacyclin production (De Nucci et al. 1988a, b), and subsequent vasodilation of smooth muscle cells. Yet, there is no molecular evidence for two ET_B-type genes in the mammals (Pollock and Highsmith 1998). Most likely these differing responses are due to splice variants of the mammalian ET_B or coupling to different second messengers (Elshourbagy et al. 1996).

The avian-specific ET_{B2} receptor binds all three ETs with equal affinity (like the mammalian ET_B), but it has a very low affinity for the sarafotoxins (like the mammalian ET_A) (Lecoin et al. 1998). In addition, the primary sequence of this receptor is more similar to mammalian ET_B than ET_A thus, Lecoin et al. (1998) termed it the ET_{B2}. Finally, the amphibian-specific ET_C preferentially binds ET-3 over ET-1 (Karne et al. 1993).

Some of the first studies to determine ET receptor expression in non-mammalian organisms used pharmacological and physiological evidence, as it wasn't until 2002 that a fish genome was sequenced (Aparicio et al. 2002). For example, physiological studies have suggested that the aortic vascular smooth muscle of the dogfish shark (*Squalus acanthias*) has ET_B-like receptors (Evans et al. 1996), but that hagfish (*Myxine glutinosa*), sea lamprey (*Petromyzon marinus*), and eel (*Anguilla rostrata*) aortic vascular smooth muscles contain ET_A-like receptors (Evans and Harrie 2001). In addition, pharmacological studies using receptor binding assays demonstrated ET_B-like receptors in the dogfish gill (Evans and Gunderson 1999), but autoradiographic studies showed ET_A-like receptors in the trout (*Oncorhynchus mykiss*) gill (Lodhi et al. 1995). The trout ET_A-like receptors were specifically localized in the gill lamellae to a region termed the lamellar sinusoid by the authors (Lodhi et al. 1995). ET_A and ET_B receptors were immunolocalized in the gills of the cod (*Gadus morhua*) (using heterologous antibodies) (Stenslokken et al. 2006), and ET_A in the gill of the tiger pufferfish (*Takifugu rubripes*) (using a homologous antibody) (Sultana et al. 2007). With the advancement of high throughput sequencing and genome sequencing, ET sequences can be found in all vertebrates and their evolutionary history is discussed below.

6.4 Evolution of the Endothelins and the Endothelin Converting Enzymes

In 2002, the first fish genome project released was that of the pufferfish, *Fugu rubripes* (Aparicio et al. 2002). As of 2014, there are more than 11 fish genomes sequenced or assembled including the elephant shark, *Callorhinchus milii* (Venkatesh et al. 2014), thus giving access to genomes that have evolved over the last 400–500 million years. The endothelins play a critical role in a variety of developmental and physiological processes. One of the most studied endothelin-dependent developmental processes is that of the development of the jaw [see

Schilling (2003) and Medeiros and Crump (2012)]. The gnathostomes (jawed vertebrates) evolved over 400 million years ago (Schilling 2003) and work in endothelin or endothelin receptor knockout mice and zebrafish have determined that endothelin signaling is critical for jaw development (Clouthier and Schilling 2004). This led to studies to determine the evolutionary and phylogenetic relationship among the endothelin gene paralogues (isoforms), which determined that *EDN1*, *EDN2* and *EDN3* mRNA were expressed in various vertebrates from fishes to man, but not in the cephalochordates, tunicata or invertebrates; thus endothelin was a vertebrate innovation (Hyndman and Evans 2007; Braasch et al. 2009). Moreover, in the killifish, *Fundulus heteroclitus*, two *EDN1* transcripts are expressed, *EDN1a* and *EDN1b*; (Hyndman and Evans 2007). At the sequence level, *EDN1a* is a longer transcript, likely encoding the full-length prepro-endothelin-1 peptide. It translates into a preproET-1a that is 189 aa long. *EDN1b* is a transcript most likely from a duplicate gene, and not an alternative splice, that has a deletion of AC at base pair position 510 resulting in a stop codon. This translates into a preproET-1 that is 143 aa long. However, the predicted active 21-amino acid ET-1 protein is homologous between *EDN1a* and *EDN1b* (depicted in Fig. 6.1). These two mRNAs were differentially expressed in the killifish, with *EDN1a* expressed highly in the osmoregulatory tissues (gill, operculum, intestine, kidney) while *EDN1b* is most highly expressed in the brain and gill (Hyndman and Evans 2007). *EDN1* duplications are also present in other teleosts (Hyndman and Evans 2007), and evidence suggests that the ancestor to the teleosts underwent a whole genome duplication event, after the split of their common ancestor with the tetrapods (Taylor et al. 2003; Christoffels et al. 2004; Hoegg et al. 2004; Volff 2005; Brunet et al. 2006). Much like duplications of the *EDN1* gene, duplications of *EDN2* and *EDN3* have also been discovered in the teleosts (Braasch et al. 2009), and a teleost specific, *EDN4*, has also been described (Braasch et al. 2009). Currently, the functional and expression patterns of these duplicate and novel endothelins have not been determined.

In contrast to the *EDN*s, the endothelin converting enzymes (*ECE*s) are highly conserved among all organisms. Ancestral *ECEs* are found in Archaea, bacteria, fungi and invertebrate genomes and they appear to be ancestral to *ECE1* and *ECE2* found in the vertebrates (Hyndman and Evans 2007). Evolutionary studies suggest that there was a genome duplication event in the common ancestor between the chordates and vertebrates but before the teleost radiation (Ohno et al. 1968; Canestro et al. 2013). We speculate that these ancestral ECEs play a more generalized proteolytic role that during the evolution of organisms became more selective for cleaving the endothelins in vertebrates (Hyndman and Evans 2007).

6.5 Evolution of the Endothelin Receptors

Similar to the evolutionary history of the *EDN*s, the *EDN* receptors also have an interesting history. In 2009, the phylogenetic relationship among the *EDN* receptors was independently determined by two labs (Braasch et al. 2009; Hyndman et al. 2009). Both studies determined that there are three paralogs of *EDN* receptors: the *EDNRA*, *EDNRB*, and *EDNRC* (called *EDNRB2* by Braasch et al. 2009) (Fig. 6.2). As mentioned above, the *EDNRB2* bird specific receptor, was actually found in all vertebrates basal to the therian mammals (marsupial), and was a distinct gene, thus we recommended it be termed *EDNRC* (Hyndman et al. 2009) although *EDNRB2* is still commonly used (Braasch et al. 2009; Braasch and Schartl 2014). Just like the *EDN* genes, only *EDN* receptor sequences from vertebrates and not cephalochordates, tunicata, or invertebrates were found, suggesting the *EDN* receptors are a vertebrate innovation.

There are duplicate *EDNRA* sequences found in the fish and frog genomes, and are likely the result of genome duplication events that occurred over their evolutionary history (Braasch et al. 2009; Hyndman et al. 2009). As shown in Fig. 6.2, the frog *EDNRA* duplication likely occurred independently in the African clawed frog, *Xenopus laevis*, where the ancestor to the African clawed frog was believed to have undergone a whole genome duplication about 40 million years ago (Hellsten et al. 2007). Thus, the suggested nomenclature is *EDNRA1* and *EDNRA2* for the frog specific duplications, and *EDNRA3* and *EDNRA4* for the fish specific duplications (Hyndman et al. 2009). The *EDNRB* gene also has a teleost specific duplicate, that was suggested to be called *EDNRB2* (Hyndman et al. 2009). One of the most striking findings from the phylogenetic analyses of the ET receptors was that the ET_C (or the gene *EDNRC*) was not found in therian mammals (Braasch et al. 2009; Hyndman et al. 2009; Braasch and Schartl 2014) (Fig. 6.2). Moreover, synteny (the physical co-localization of genes on the chromosome) is conserved between the *EDNRA* and *EDNRB* in all vertebrates. For example, *EDNRA* is located between the *TTC29* and *TMEM184C* genes on the chromosome, and this order of genes is conserved all the way between humans and the elephant shark (about 500 million years of evolutionary history) (Braasch et al. 2009; Hyndman et al. 2009; Braasch and Schartl 2014). In contrast, the *EDNRC* is found between the genes *SPRY3* and *VAMP7* in the fishes, frogs, and platypus, but it is missing from the opossum, rodents, and humans (examples of therian mammals). Interestingly, *SPRY3* and *VAMP7* are located in the therian mammals on the pseudoautosomal region 2 (PAR2) portion of the X chromosome. This area of the X chromosome has undergone numerous chromosomal rearrangements (D'Esposito et al. 1997; Graves et al. 1998; Charchar et al. 2003; Waters et al. 2005). Thus, it seems that in the common ancestor between the monotremes and therian mammals lost the *EDNRC* likely through a chromosomal deletion event and coincides with the origin of the therian X and Y chromosomes (Braasch et al. 2009; Hyndman et al. 2009; Braasch and Schartl 2014). The *EDNs* and *EDNRs* have a very interesting history of gene/genome duplications and deletions, and although therian mammals

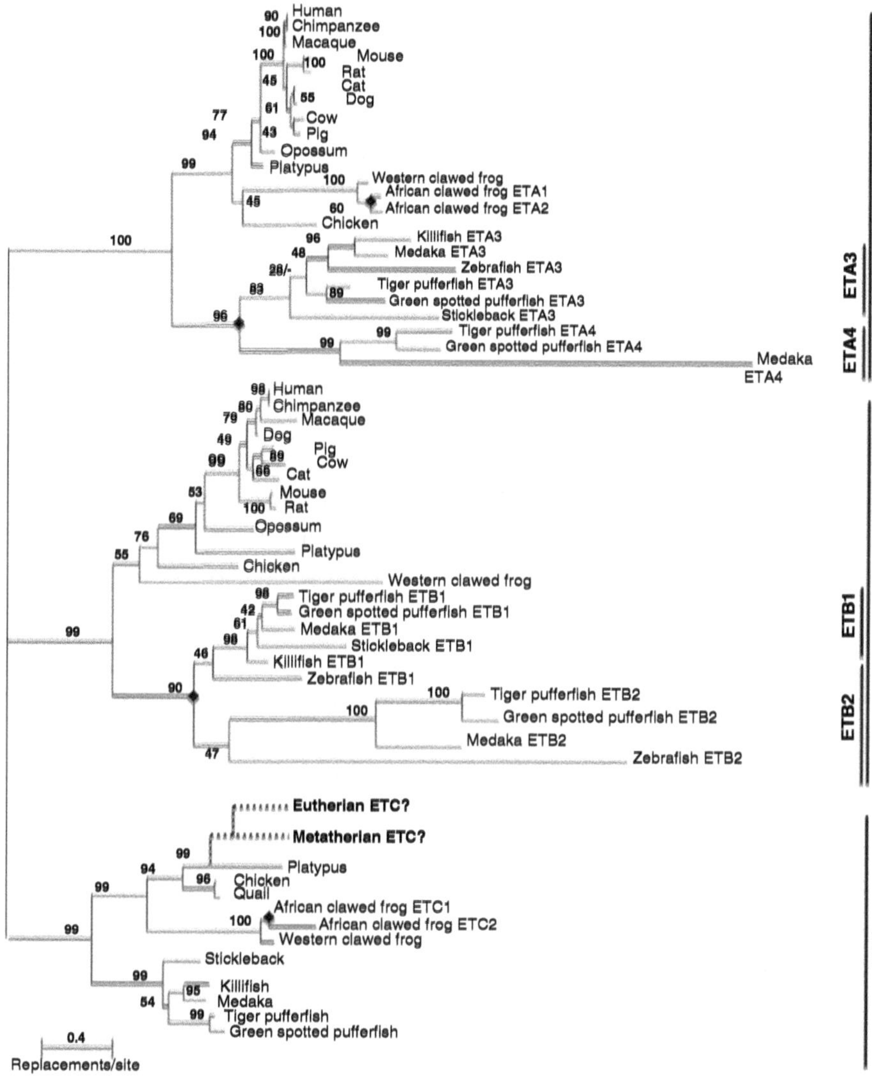

Fig. 6.2 An unrooted, maximum likelihood phylogenetic analysis of the endothelin receptor family based upon their amino acid sequences. The *numbers at the nodes* represent the maximum likelihood bootstrap scores. *Diamonds* refer to ancestral gene duplications, and branch lengths are drawn proportional to their expected numbers of replacements per site. The *dotted branches* and *question marks* highlight the absences of therian ET_C. Modified from Hyndman et al. (2009)

only have the *EDNRA* and *EDNRB*, all other vertebrates have *EDNRC* and multiple copies of these receptors (Fig. 6.2). What the physiological role of these multiple copies is remains for the most part to be determined.

6.6 Sodium Homeostasis in Teleosts

Teleost fishes are the most specious group of vertebrates, and they have a diverse range of habitat use. Some are stenohaline freshwater or stenohaline marine, and these fishes can only maintain ion balance over a small range of environmental salinities. A third broad category is the euryhaline fishes, which can tolerate changes in environmental salinity. The osmoregulation of teleost fishes has been studied for over the past 150 years, and due to space limitations, I will briefly review teleost osmoregulation here [see these reviews: Karnaky (1998), Evans et al. (2005), Marshall and Grossell (2006), Hiroi and McCormick (2012)].

Depending on the salinity of their environment, fishes must overcome osmoregulatory challenges in order to maintain fluid-electrolyte balance. For example, freshwater fishes are hyperosmotic to their environment, and consequently they tend to gain water and lose ions (Evans et al. 2005) (Fig. 6.3a). To compensate for this water load, they produce a dilute urine and absorb ions at the gill. In contrast, seawater fishes are hypo-osmotic to their environment, and they lose water and gain excess salts (Fig. 6.3b). To combat this problem, marine fishes drink the seawater and transport salts across the gut epithelium, creating an osmotic gradient favorable for water to follow (Evans et al. 2005). These excess salts are transported in the blood to the gills where they are actively excreted (Fig. 6.3c). Euryhaline fishes like the killifish, experience large changes in environmental salinity over the course of a day (Marshall 2003), and they must regulate ion uptake or excretion to maintain

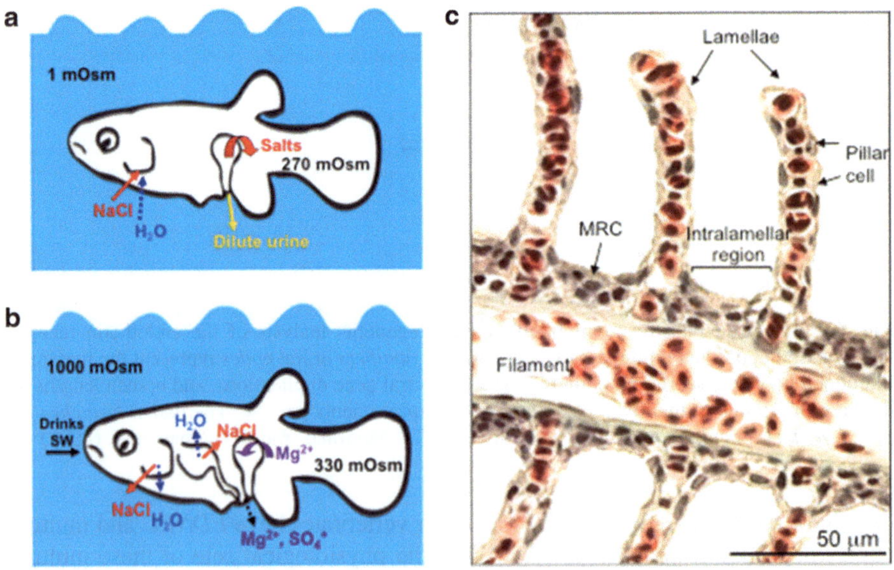

Fig. 6.3 A diagram of teleost ion transport processes in (**a**) fresh water or (**b**) seawater (SW). (**c**) A sagittal section through one filament of the killifish gill. MRC-mitochondion-rich cell

proper ion balance. The killifish does this by rapidly up- or down-regulating gill ion transporter density (Marshall et al. 1999; Scott et al. 2004; Scott and Schulte 2005; Choe et al. 2006).

The fish gill is a multifunctional organ that is the main site of ion and acid/base regulation, gas exchange, and nitrogenous waste excretion (Evans et al. 2005). The fish gill is composed of a series of filaments attached to the gill arch. Perpendicular to the filaments, are the lamella, which are sheets of epithelial cells separated by contractile cells called pillar cells (Fig. 6.3c). The gill epithelium is in direct contact with the environment, and it receives 100 % of the cardiac output (Olson 2002). In the gill, there are specialized cells that are involved in ion transport termed the mitochondrion-rich cell (MRC, also called the chloride cell and ionocyte). These cells are characterized by many mitochondria and ion transporting proteins, and an extensive basolateral membrane. The MRCs have distinct freshwater- and seawater-type morphologies (Perry 1997; Katoh and Kaneko 2003; Hsu et al. 2014). Freshwater MRCs have a flattened apical membrane with many microvilli. In contrast, the seawater MRC has an apical crypt, with no microvilli, and they are generally larger in size (Perry 1997; Katoh and Kaneko 2003). These cells also have different distributions of ion transporters. Freshwater MRCs function in ion uptake and this involves ion transporters such as, the apically expressed V-ATPase, HCO_3^-/Cl^-, Na^+/H^+ exchangers, and a hypothesized Na^+ channel, and the basolaterally expressed Na^+, K^+-ATPase (Evans et al. 2005; Hsu et al. 2014). Recent work has also identified acid-sensing ion channel-4 (ASIC4) in the apical membrane of the FW MRC, where it is hypothesized to function in sodium reabsorption (Dymowska et al. 2014). The seawater MRC functions in ion excretion and apically expresses the cystic fibrosis transmembrane conductance regulator (a chloride channel), and basolaterally expresses the $Na^+/K^+/2Cl^-$ cotransporter, inward rectifier K^+ channel (eKir), and Na^+, K^+-ATPase (Evans et al. 2005; Hsu et al. 2014).

The hormonal regulation of osmoregulation has been extensively studied (see these reviews McCormick 1995; McCormick and Bradshaw 2006). Generally, prolactin is the hormone necessary for freshwater acclimation, and cortisol is the hormone necessary for seawater acclimation (McCormick and Bradshaw 2006). In fishes acclimating to fresh water, prolactin stimulates an increase in the freshwater-type MRC morphology, and stimulates an increase in Na^+, and Cl^- uptake (Evans 2002; Lee et al. 2006; McCormick and Bradshaw 2006). Although cortisol is classified as the "seawater-adapting hormone" (McCormick and Bradshaw 2006), it is also important in freshwater acclimation. Cortisol injections in freshwater-acclimated fishes leads to Na^+ and Cl^- uptake (Perry 1997), freshwater-type MRC morphology and an increase in Na^+, K^+-ATPase activity (Dang et al. 2000). During seawater acclimation, the cortisol and growth hormone/insulin-like growth factor axis is stimulated. In the killifish, 1 h post a fresh water to seawater transfer, there is a sevenfold increase in plasma cortisol concentration (Marshall et al. 1999), and in the tilapia (*Oreochromis mossambicus*) it is released within minutes of a transfer to a higher salinity (Hegab and Hanke 1984) (it may be released within minutes in the killifish but this has not been tested). In addition, growth hormone is released and works synergistically with cortisol to increase gill Na^+, K^+, $2Cl^-$ density and increase gill Na^+, K^+-ATPase activity (Pelis and McCormick 2001). Also, these hormones stimulate development of the seawater-type MRC (Sakamoto and McCormick 2006). A new "osmosensing" transcription factor, OSTF1, was cloned and characterized from the tilapia (Fiol and Kultz 2005) and now many other fishes (Choi and An 2008; Tse et al. 2008, 2011, 2012). Although there is little evidence that the OSTF1 actually senses changes in osmolality, it is rapidly upregulated within minutes of transferring the tilapia to seawater. Fiol et al. (2006) determined that the rapid increase in OSTF1 in response to changes in salinity is dependent on hypertonicity but independent of cortisol. Whether paracrine signals (ET-1, for instance) are involved in the hypertonicity-dependent OSTF1 induction or if *EDN* genes are downstream targets of OSTF1 remains to be determined.

In addition to the few hormones that were described above, there are many other hormones and paracrine/autocrine factors that are believed to be involved in the osmoregulation of fishes. These include endothelin, natriurectic peptides, urotensin II, thyroid hormones, arginine vasotocin, angiotensin II, insulin, vasoactive intestinal peptide (and likely others that are yet to be tested) (Evans 2002; McCormick and Bradshaw 2006; Evans et al. 2011). We will focus on what has been determined with respect to the endothelin system in the context of cardiovascular effects and direct effects on sodium transport.

6.7 Cardiovascular Effects of ET-1

6.7.1 Amphibians and Reptiles

The regulation of vessel tension and blood flow is very important for not only cardiovascular health, but also is critically involved with the regulation of sodium homeostasis. Actions of ET-1 on the cardiovascular system have been determined for non-mammalian vertebrates. In the frog lateral and dorsal aorta, exogenous ET-1 significantly increased vessel tension (Broughton and Donald 2002). It was not determined which receptors mediated these effects. Likewise in the American Alligator, *Alligator mississippiensis*, ET-1 increased vessel tension of intrapulmonary arteries (Skovgaard et al. 2008). This ET-1 mediated constriction was likely via the ET_A receptor, as ET_A blockade with BQ123 attenuated the ET-1-mediated increase in vessel tension, while ET_B blockade with BQ788 had no significant effect (Skovgaard et al. 2008).

The effects of mammalian ET-1 on the systemic and pulmonary circulatory systems in the freshwater turtle, *Trachemys scripta*, have also been determined. Bolus intra-arterial injections of ET1 caused a dose-dependent dilation that was associated with a decrease in systolic pressure (Skovgaard et al. 2005). Unlike mammals where a bolus injection results in an initial vasodilation followed by a prolonged vasoconstriction (Yanagisawa and Masaki 1989), only vasodilation was observed in the anesthetized turtle. Moreover, there was no significant effect of ET-1 on the pulmonary circulation of the turtle. When the turtle was treated with a bolus of the ET_B agonist, BQ-3020, the systemic circulation again dilated to a similar degree as the ET-1 bolus (Skovgaard et al. 2005). Blocking the ET_A receptor with BQ123, did not prevent the ET-1-dependent vasodilation but blocking of the ETB receptor with BQ788 abolished the ET-1-dependent vasodilation (Skovgaard et al. 2005). Thus, the authors conclude that it is the ET_B receptor that dominates in the turtle systemic circulation and mediates the effects of ET-1 (Skovgaard et al. 2005). From these studies, the common theme is that ET-1 can act to regulate vessel tension, and depending on the receptor, it can lead to vasoconstriction or vasodilation similarly to what has been reported for mammals.

6.7.2 *Fishes*

Some of the first studies of ET-1 in non-mammalian vertebrates were performed in fishes. Olson et al. (1991) evaluated the cardiovascular effects of ET-1 in trout using cannulated dorsal aortas, in situ perfused hearts, isolated perfused gills, perfused trunks, and isolated systemic vascular rings. They determined that intra-arterial bolus injections of mammalian ET-1 (667 pmol kg^{-1}) produced a triphasic (increase-decrease-increase) response in dorsal aorta pressure (P_{DA}), and continuous infusions produced an increased mean perfusion pressure in a dose-dependent manner. Hoagland et al. (2000) replicated this study using ET-1 isolated from the trout. They also showed that intra-arterial bolus injections of trout ET-1 (667 pmol kg¯1) produced a triphasic response in P_{DA}, and that continuous infusions produced a dose-dependent increase in P_{DA}. In a similar study, le Mevel et al. (1999) found that intracerebroventricular or intra-arterial injections of mammalian ET-1 (86 pmol kg^{-1}) produced a transient increase in $P_{DA,}$ but a triphasic response was not observed, further suggesting dose-dependent effects of ET-1 on $P_{DA.}$ Additionally, Olson et al. (1991) determined that ET-1 increased mean perfusion pressure in isolated gills in a dose-dependent manner and that the half-maximal effective concentration (EC_{50}) was $<10^{-8}$ M.

Other studies have examined constriction of isolated blood vessels in fishes (Poder et al. 1991; Sverdrup et al. 1994; Evans et al. 1996; Evans and Harrie 2001). In the dogfish shark, mammalian ET-1 caused significant constriction of aortic vascular rings, with an EC_{50} of 10^{-9} M (Evans et al. 1996). Evans and Harrie (2001) showed that ventral aortic rings, from hagfish, lamprey, and eels constricted in response to 0.1 μM mammalian ET1.

Wang et al. (1999) compared the constrictive responses of isolated fish and rat vascular rings to trout ET-1 and mammalian ET-1. They found that trout anterior cardinal veins and branchial arteries were more sensitive to mammalian ET-1 than trout ET-1. The increased sensitivity to mammalian ET-1 was hypothesized to be due to slower degradation of the mammalian ET1 compared to the native trout ET1 (Wang et al. 1999). Degradation of ET-1 has been proposed to be mediated through membrane bound metalloproteinases and internal lysosomes (Jackman et al. 1993). In rats and guinea pigs, the pulmonary circuit rapidly removes 60 % of radioiodinated ET-1 and ET-3 in 1 min (see La and Reid 1995). In humans, 53 % of ET-1 is removed by pulmonary clearance and this is mediated through ET_B (see La and Reid 1995). In fishes, 55 % of an ET-1 bolus was removed during a single pass through the gills (Olson 1998). ET_B receptors are expressed along the vasculature of the filament, which suggests they may function as the clearance receptor in the gill (Hyndman and Evans 2009). Studies to test this hypothesis directly are needed.

Endothelin induced vasoconstriction may have a large effect on gill hemodynamics. The gills are highly vascularized and are perfused by the entire cardiac output (Olson 1998); thus, any change in blood flow or pressure would greatly influence their functions. It has been hypothesized that ET-1 redistributes lamellar blood flow, since a ventral aortic injection of ET-1 caused constriction of pillar cells in the gill lamellae of trout and cod (Nilsson and Sundin 1998; Stenslokken et al. 1999). This contraction resulted in a shift of intralamellar blood flow to the outer marginal channels. Pillar cells contain contractile elements and although they do not appear to be innervated, hormones like ET-1 may signal pillar cells to contract (or dilate depending on the signal, Bettex-Galland and Hughes 1973) (Mistry et al. 2004). There is also no evidence that the lamellar arterioles or filamental arteries (afferent or efferent) were constricted by mammalian ET-1 (Nilsson and Sundin 1998; Stenslokken et al. 1999). Thus, the pillar cells may regulate microcirculation through the gill lamellae. From these studies, it appears that ET-1 may be one of the paracrine/autocrines controlling this system. What these aforementioned studies in fishes, amphibians and reptiles highlight is that ET-1 can alter cardiovascular parameters and this may have important consequences not only for cardiovascular physiology but also for ion transport processes.

6.8 Expression of the Endothelin System in the Gill and Effects of Changing Environmental Salinity

The endothelin system is expressed in the fish gill (Olson et al. 1991; Evans et al. 2004; Hyndman and Evans 2007, 2009; Hyndman et al. 2009). To visualize *EDN1* transcripts in the gill, in situ hybridization experiments were performed with RNA probes specific for *Fundulus EDN1a* and *EDN1b* (Hyndman and Evans 2007). It was determined that *EDN1a* and *EDN1b* in cells of the intralamellar region adjacent to the MRC in the gill, where it is predicted that ET-1 may act in a paracrine fashion to inhibit MRC-dependent salt transport. *EDN1b* was also expressed in the lamellar pillar cells where we speculate that ET-1 may function to regulate pillar cell tone and blood flow through the lamellae (Hyndman and Evans 2007). Big-ET1 was also immunolocalized to cells in the interlamellar region adjacent to the MRC, and to the lamellar pillar cells (Hyndman and Evans

2007). Thus, it appears ET-1 may act as an autocrine to regulate pillar cell tone, and as a paracrine to regulate MRC functions. The ET receptors are also expressed in the killifish gill. Using immunohistochemistry, we localized the ET_A receptor to the MRC, and the ET_B/ET_C (current commercial antibodies cannot differentiate between ET_B and ET_C) was localized to the gill vasculature including the prelamellar arterioles, and the pillar cells (Hyndman and Evans 2007).

Given that killifish live in an estuarine environment where the salinity of the water changes daily with the tides (Marshall et al. 1999; Marshall 2003), it was hypothesized that changes in environmental salinity would regulate gill *EDN*/*EDNR* mRNA and protein expression. Killifish were chronically (30 days) acclimated to fresh water (~0 ppt) or seawater (~32 ppt) and then switched to seawater or freshwater, respectively (Hyndman and Evans 2007). There were only minimal changes to the gill mRNA expression of *EDN1a* and *EDN1b* with either the FW to SW or SW to FW switch; however, there was a significant twofold increase in *ECE1* gill expression when the killifish were switched from FW to SW for 3–24 h (Hyndman and Evans 2007). The conversion of big ET-1 to active 21 aa ET-1 is limited by the enzyme ECE1 (D'Orleans-Juste et al. 2003), therefore it was concluded that acute increases in environmental salinity would increase gill ET-1 (Hyndman and Evans 2007).

The effect of changing environmental salinity on gill *EDN receptor mRNA* expression and receptor localization has also been determined. During the acute (24 h) FW to SW switch, gill *EDNRA*, *EDNRB*, and *EDNRC* mRNAs were significantly higher compared to the control FW to FW switch (Hyndman and Evans 2009). When switched from SW to FW, there were only minimal changes in *EDN receptor* gill expression in the first 24 h; However, after 30 days of acclimation to the FW, there was a significant 50 % reduction in the *EDNRA* gill mRNA, and a 40 % reduction in *EDNRB* mRNA compared to SW acclimated killifish (Hyndman and Evans 2009). Changes in environmental salinity also lead to changes in gill ET receptor expression. In agreement with the gill mRNA study, a switch to FW to SW resulted in a significant two- to threefold increase in ET_A and ET_B/ET_C (Hyndman and Evans 2009). Chronic acclimation to FW resulted in a significant 50 % reduction in ET_A expression, but no statistically significant effect on ETB/ETC. Moreover, chronic FW acclimation resulted in a shift in the ET_A expression in the gill. In SW, the ET_A is expressed in the MRC at high levels and throughout the cell; however, in FW the ET_A expression is reduced and expressed only on the basal side of the MRC (Hyndman and Evans 2009). Changes in environmental salinity did not significantly alter the localization of the ET_B/ET_C receptors (Hyndman and Evans 2009).

6.9 Endothelin Effects on Ion Transport

In addition to cardiovascular responses, ET-1 has been shown to have effects on renal sodium transport in mammals (Garvin and Sanders 1991; Zeidel 1993; Plato et al. 2000; Kohan 2011). In the rat proximal tubule, ET-1 inhibited Na^+, K^+-ATPase activity and bicarbonate transport (Garvin and Sanders 1991). In the rabbit inner medullary collecting duct, ET1 inhibited Na^+-K^+-ATPase activity and lead to sodium natriuresis (Zeidel et al. 1989). In the rat thick ascending limb of the loop of Henle, ET-1 inhibited net chloride flux and this was mediated by an ET_B (Plato et al. 2000). Over the past decade, collecting duct-specific knockout mice for ET-1, ET_A, ET_B, or ET_A/ET_B have been developed and significantly expanded our understanding of the role of the endothelin system in fluid-electrolyte balance and blood pressure control (Ahn et al. 2004; Ge et al. 2005a, b, 2006, 2008). These studies will be discussed in detail in Chap. 7, but to summarize, they determined that during salt loading (through diet), ET-1 and ET_B are necessary for sodium excretion. As a consequence, ET-1 and ET_B collecting duct-knockout mice were severely hypertensive after salt loading (Ahn et al. 2004; Ge et al. 2006). The ET_A collecting duct-knockouts were no different from control animals, suggesting that ET_A in the mammalian collecting duct is not involved in blood pressure regulation (or sodium excretion) (Ge et al. 2005b).

Endothelin inhibition of solute transport also has been demonstrated in fishes. Although in mammals the kidneys play a critical role in the regulation of sodium reabsorption, it is the gills that are the key organ for regulation of sodium balance in fishes. Fishes do have kidneys, and they function in the regulation of water excretion and divalent ions (Mg and SO_4) (Nishimura and Imai 1982; Nishimura and Fan 2003). Fish kidneys also function in the secretion of polypeptides or lipophilic organic anions by the multidrug resistance protein-2 (MRP-2) into the urine. Mrp-2 is expressed in the killifish proximal tubule, and acute treatment with mammalian ET-1 led to inhibition of transport by the Mrp-2 (Notenboom et al. 2002, 2005). Incubation of tubules with the ET_B antagonist, RES-701–1, prevented the ET-1-mediated inhibition of Mrp-2 transport in the proximal tubules, suggesting that ET_B are involved in the signaling cascade. Moreover, ET-1 (via the ET_B) stimulated nitric oxide production in the killifish proximal tubule, and this nitric oxide is critical for ET-1 dependent inhibition of Mrp-2 transport (Notenboom et al. 2002). Other than these few studies, the effects of ET-1 on ion transport in the fish kidney are relatively unexplored.

The killifish opercular epithelium is a model tissue to test mechanistic questions about fish gill sodium and chloride transport processes. It consists of a single epithelial layer that covers the inside of the operculum, and it expresses high numbers of MRCs (Karnaky and Kinter 1977). This epithelium can be mounted in an Ussing chamber and the change in the short circuit current (I_{sc}) in response to agonists or antagonists recorded (Karnaky et al. 1977). The I_{sc} is the result of net transport of chloride across the basolateral (NKCC) to apical membranes (CFTR) of the opercular epithelium; sodium follows passively via paracellular channels in the epithelium (Evans et al. 2004). Mammalian ET-1 inhibits net chloride transport in the killifish opercular epithelium from fish acclimated to seawater (Evans et al. 2004). In addition, the ET_B agonist, sarafotoxin S6c (SRX S6c), also inhibited the net chloride transport in this preparation, suggesting that ET-1 is acting through an ET_B; however, there are no ET_A-specific agonists, thus it has not been determined if ET_A also affects ion transport in the killifish. Furthermore, ET-1 via ET_B stimulates both nitric oxide production and prostaglandin release that is part of the signaling cascade leading to the inhibition of chloride transport (Evans et al. 2004). Thus, much like mammals (discussed in Chap. 7) the ET-1/ET_B/nitric oxide and prostaglandin signaling cascades regulate ion transport processes and these cascades evolved over 400 million years ago.

To summarize, ET/ECE/ET receptors are expressed in the teleost gill, where they are regulated by environmental salinity. Overall studies suggest that during acclimation to SW, where ion accumulation and dehydration are potential problems, many ion transporters and channels are upregulated and function to excrete excess salt across the gill epithelium back to the external environment. The ET-1 system appears to be upregulated when the fish is switched from fresh water to seawater (in essence during salt loading), suggesting it acts as a "brake" to fine-tune ion excretion to maintain proper ion homeostasis. Moreover, ET-1 may regulate local blood flow through the lamellae by constricting pillar cells to optimize gas exchange and other physiological processes. These hypotheses still need to be tested in vivo and may require a gill ET-1 knockout fish. ET-1 mutant (called *suc*) and ET_B mutant (called *rose*) zebrafish are available (Miller et al. 2000; Parichy et al. 2000), but zebrafish are stenohaline fresh water and cannot survive in seawater. Recent advances have been made in targeting gene deletions in medaka (Taniguchi et al. 2006; Ansai et al. 2014; Ansai and Kinoshita 2014). The "brackish" medaka (*Oryzias dancena*) as its common name implies, lives in brackish water and has some salinity tolerance (Yang et al. 2013) and the Japanese medaka (*Oryzias latipes*) can survive in a wide range of environmental salinities (Hsu et al. 2014). Taken together, in the near future we may have gene knockout euryhaline fishes to test our hypotheses about the role of ET-1 in the regulation of sodium homeostasis.

6.10 Perspectives

To conclude, the ET-1 signaling system evolved in the common ancestor to the vertebrates, and through whole genome duplications, gene duplications and gene loss, has resulted in non-mammalian vertebrates having multiple copies of this system. The physiological function(s) of these duplicates remains to be determined, but given that they have remained over the past 400–500 million years of evolution suggests they play an important role. What is conserved from fishes to man, is that ET-1 is critical for proper development, especially for the development of the jaws, and can regulate vascular tone, blood pressure, and ion transport processes.

Moreover, downstream signaling cascades including nitric oxide and prostaglandins are also evident in non-mammalian vertebrates, suggesting a long evolutionary history among these systems. Still, much remains to be elucidated, including what role changes in ET receptors play in the putative regulation of blood pressure and/or ion transport process in non-mammalian vertebrates. With the advent of tissue-specific knockouts, morpholinos and siRNAs, we will hopefully be able to answer some of these remaining questions.

Bibliography

Ahn D, Ge Y, Stricklett PK, Gill P, Taylor D, Hughes AK, Yanagisawa M, Miller L, Nelson RD, Kohan DE (2004) Collecting duct-specific knockout of endothelin-1 causes hypertension and sodium retention. J Clin Invest 114(4):504–511

Ansai S, Kinoshita M (2014) Targeted mutagenesis using CRISPR/Cas system in medaka. Biol Open 3(5):362–371. doi:10.1242/bio.20148177

Ansai S, Inohaya K, Yoshiura Y, Schartl M, Uemura N, Takahashi R, Kinoshita M (2014) Design, evaluation, and screening methods for efficient targeted mutagenesis with transcription activator-like effector nucleases in medaka. Develop Growth Differ 56(1):98–107. doi:10.1111/dgd.12104

Aparicio S, Chapman J, Stupka E, Putnam N, Chia JM, Dehal P, Christoffels A, Rash S, Hoon S, Smit A, Gelpke MD, Roach J, Oh T, Ho IY, Wong M, Detter C, Verhoef F, Predki P, Tay A, Lucas S, Richardson P, Smith SF, Clark MS, Edwards YJ, Doggett N, Zharkikh A, Tavtigian SV, Pruss D, Barnstead M, Evans C, Baden H, Powell J, Glusman G, Rowen L, Hood L, Tan YH, Elgar G, Hawkins T, Venkatesh B, Rokhsar D, Brenner S (2002) Whole-genome shotgun assembly and analysis of the genome of Fugu rubripes. Science 297(5585):1301–1310. doi:10.1126/science.1072104

Arai H, Hori S, Aramori I, Ohkubo H, Nakanishi S (1990) Cloning and expression of a cDNA encoding an endothelin receptor. Nature 348(6303):730–732

Bettex-Galland M, Hughes GM (1973) Contractile filamentous material in the pillar cells of fish gills. J Cell Sci 13(2):359–370

Braasch I, Schartl M (2014) Evolution of endothelin receptors in vertebrates. Gen Comp Endocrinol. doi:10.1016/j.ygcen.2014.06.028

Braasch I, Volff JN, Schartl M (2009) The endothelin system: evolution of vertebrate-specific ligand-receptor interactions by three rounds of genome duplication. Mol Biol Evol 26 (4):783–799. doi:10.1093/molbev/msp015

Broughton BR, Donald JA (2002) Nitric oxide regulation of the central aortae of the toad Bufo marinus occurs independently of the endothelium. J Exp Biol 205(Pt 19):3093–3100

Brunet FG, Roest Crollius H, Paris M, Aury JM, Gibert P, Jaillon O, Laudet V, Robinson-Rechavi M (2006) Gene loss and evolutionary rates following whole-genome duplication in teleost fishes. Mol Biol Evol 23(9):1808–1816. doi:10.1093/molbev/msl049

Canestro C, Albalat R, Irimia M, Garcia-Fernandez J (2013) Impact of gene gains, losses and duplication modes on the origin and diversification of vertebrates. Semin Cell Dev Biol 24 (2):83–94. doi:10.1016/j.semcdb.2012.12.008

Charchar FJ, Svartman M, El-Mogharbel N, Ventura M, Kirby P, Matarazzo MR, Ciccodicola A, Rocchi M, D'Esposito M, Graves JA (2003) Complex events in the evolution of the human pseudoautosomal region 2 (PAR2). Genome Res 13(2):281–286. doi:10.1101/gr.390503

Choe KP, Havird J, Rose R, Hyndman K, Piermarini P, Evans DH (2006) COX2 in a euryhaline teleost, *Fundulus heteroclitus*: primary sequence, distribution, localization, and potential function in gills during salinity acclimation. J Exp Biol 209(Pt 9):1696–1708

Choi CY, An KW (2008) Cloning and expression of Na+/K+ -ATPase and osmotic stress transcription factor 1 mRNA in black porgy, Acanthopagrus schlegeli during osmotic stress. Comp Biochem Physiol B Biochem Mol Biol 149(1):91–100. doi:10.1016/j.cbpb.2007.08.009

Christoffels A, Koh EG, Chia JM, Brenner S, Aparicio S, Venkatesh B (2004) Fugu genome analysis provides evidence for a whole-genome duplication early during the evolution of ray-finned fishes. Mol Biol Evol 21(6):1146–1151. doi:10.1093/molbev/msh114

Clouthier DE, Schilling TF (2004) Understanding endothelin-1 function during craniofacial development in the mouse and zebrafish. Birth Defects Res C Embryo Today 72(2):190–199

D'Esposito M, Matarazzo MR, Ciccodicola A, Strazzullo M, Mazzarella R, Quaderi NA, Fujiwara H, Ko MS, Rowe LB, Ricco A, Archidiacono N, Rocchi M, Schlessinger D, D'Urso M (1997) Differential expression pattern of XqPAR-linked genes SYBL1 and IL9R correlates with the structure and evolution of the region. Hum Mol Genet 6(11):1917–1923

D'Orleans-Juste P, Plante M, Honore JC, Carrier E, Labonte J (2003) Synthesis and degradation of endothelin-1. Can J Physiol Pharmacol 81(6):503–510. doi:10.1139/y03-032

Dang Z, Balm PH, Flik G, Wendelaar Bonga SE, Lock RA (2000) Cortisol increases Na(+)/K(+)-ATPase density in plasma membranes of gill chloride cells in the freshwater tilapia Oreochromis mossambicus. J Exp Biol 203(Pt 15):2349–2355

De Nucci G, Gryglewski RJ, Warner TD, Vane JR (1988a) Receptor-mediated release of endothelium-derived relaxing factor and prostacyclin from bovine aortic endothelial cells is coupled. Proc Natl Acad Sci USA 85(7):2334–2338

De Nucci G, Thomas R, D'orleansjuste P, Antunes E, Walder C, Warner TD, Vane JR (1988b) Pressor effects of circulating endothelin are limited by its removal in the pulmonary circulation and by the release of prostacyclin and endothelium-derived relaxing factor. Proc Natl Acad Sci USA 85(24):9797–9800

Dymowska AK, Schultz AG, Blair SD, Chamot D, Goss GG (2014) Acid-sensing ion channels (ASICs) are involved in epithelial Na^+ uptake in rainbow trout (Oncorhynchus mykiss). Am J Physiol Cell Physiol 307(3):C255–C265. doi:10.1152/ajpcell.00398.2013

Elshourbagy NA, Adamou JE, Gagnon AW, Wu HL, Pullen M, Nambi P (1996) Molecular characterization of a novel human endothelin receptor splice variant. J Biol Chem 271 (41):25300–25307

Evans DH (2002) Cell signaling and ion transport across the fish gill epithelium. J Exp Zool 293 (3):336–347

Evans DH, Gunderson MP (1999) Characterization of an endothelin ET(B) receptor in the gill of the dogfish shark *Squalus acanthias*. J Exp Biol 202(Pt 24):3605–3610

Evans DH, Harrie AC (2001) Vasoactivity of the ventral aorta of the American eel (Anguilla rostrata), Atlantic hagfish (Myxine glutinosa), and sea lamprey (Petromyzon marinus). J Exp Zool 289(5):273–284

Evans DH, Gunderson M, Cegelis C (1996) ETB-type receptors mediate endothelin-stimulated contraction in the aortic vascular smooth muscle of the spiny dogfish shark, *Squalus acanthias*. J Comp Physiol B 165(8):659–664

Evans DH, Rose RE, Roeser JM, Stidham JD (2004) NaCl transport across the opercular epithelium of *Fundulus heteroclitus* is inhibited by an endothelin to NO, superoxide, and prostanoid signaling axis. Am J Physiol Regul Integr Comp Physiol 286(3):R560–R568

Evans DH, Piermarini PM, Choe KP (2005) The multifunctional fish gill: dominant site of gas exchange, osmoregulation, acid-base regulation, and excretion of nitrogenous waste. Physiol Rev 85(1):97–177

Evans DH, Hyndman KA, Cornwell E, Buchanan P (2011) Urotensin II and its receptor in the killifish gill: regulators of NaCl extrusion. J Exp Biol 214(Pt 23):3985–3991. doi:10.1242/jeb.065243, 214/23/3985 [pii]

Fiol DF, Kultz D (2005) Rapid hyperosmotic coinduction of two tilapia (Oreochromis mossambicus) transcription factors in gill cells. Proc Natl Acad Sci USA 102(3):927–932

Fiol DF, Chan SY, Kultz D (2006) Regulation of osmotic stress transcription factor 1 (Ostf1) in tilapia (Oreochromis mossambicus) gill epithelium during salinity stress. J Exp Biol 209 (Pt 16):3257–3265. doi:10.1242/jeb.02352

Froy O, Gurevitz M (1998) Membrane potential modulators: a thread of scarlet from plants to humans. FASEB J 12(15):1793–1796

Garvin J, Sanders K (1991) Endothelin inhibits fluid and bicarbonate transport in part by reducing Na+/K+ ATPase activity in the rat proximal straight tubule. J Am Soc Nephrol 2(5):976–982

Ge Y, Ahn D, Stricklett PK, Hughes AK, Yanagisawa M, Verbalis JG, Kohan DE (2005a) Collecting duct-specific knockout of endothelin-1 alters vasopressin regulation of urine osmolality. Am J Physiol Renal Physiol 288(5):F912–F920

Ge Y, Stricklett PK, Hughes AK, Yanagisawa M, Kohan DE (2005b) Collecting duct-specific knockout of the endothelin A receptor alters renal vasopressin responsiveness, but not sodium excretion or blood pressure. Am J Physiol Renal Physiol 289(4):F692–F698

Ge Y, Bagnall A, Stricklett PK, Strait K, Webb DJ, Kotelevtsev Y, Kohan DE (2006) Collecting duct-specific knockout of the endothelin B receptor causes hypertension and sodium retention. Am J Physiol Renal Physiol 291(6):F1274–F1280

Ge Y, Bagnall A, Stricklett PK, Webb D, Kotelevtsev Y, Kohan DE (2008) Combined knockout of collecting duct endothelin A and B receptors causes hypertension and sodium retention. Am J Physiol Renal Physiol 295(6):F1635–F1640. doi:10.1152/ajprenal.90279.2008, 90279.2008 [pii]

Graves JA, Wakefield MJ, Toder R (1998) The origin and evolution of the pseudoautosomal regions of human sex chromosomes. Hum Mol Genet 7(13):1991–1996

Hegab SA, Hanke W (1984) The significance of cortisol for osmoregulation in carp (Cyprinus carpio) and tilapia (Sarotherodon mossambicus). Gen Comp Endocrinol 54(3):409

Hellsten U, Khokha MK, Grammer TC, Harland RM, Richardson P, Rokhsar DS (2007) Accelerated gene evolution and subfunctionalization in the pseudotetraploid frog Xenopus laevis. BMC Biol 5:31. doi:10.1186/1741-7007-5-31

Hiroi J, McCormick SD (2012) New insights into gill ionocyte and ion transporter function in euryhaline and diadromous fish. Respir Physiol Neurobiol 184(3):257–268. doi:10.1016/j.resp.2012.07.019

Hoagland TM, Weaver L Jr, Conlon JM, Wang Y, Olson KR (2000) Effects of endothelin-1 and homologous trout endothelin on cardiovascular function in rainbow trout. Am J Physiol Regul Integr Comp Physiol 278(2):R460–R468

Hoegg S, Brinkmann H, Taylor JS, Meyer A (2004) Phylogenetic timing of the fish-specific genome duplication correlates with the diversification of teleost fish. J Mol Evol 59 (2):190–203. doi:10.1007/s00239-004-2613-z

Hsu HH, Lin LY, Tseng YC, Horng JL, Hwang PP (2014) A new model for fish ion regulation: identification of ionocytes in freshwater- and seawater-acclimated medaka (Oryzias latipes). Cell Tissue Res 357(1):225–243. doi:10.1007/s00441-014-1883-z

Hyndman KA, Evans DH (2007) Endothelin and endothelin converting enzyme-1 in the fish gill: evolutionary and physiological perspectives. J Exp Biol 210(Pt 24):4286–4297. doi:10.1242/jeb.009969, 210/24/4286 [pii]

Hyndman KA, Evans DH (2009) Effects of environmental salinity on gill endothelin receptor expression in the killifish, *Fundulus heteroclitus*. Comp Biochem Physiol A Mol Integr Physiol 152(1):58–65. doi:10.1016/j.cbpa.2008.08.036, S1095-6433(08)01124-0 [pii]

Hyndman KA, Miyamoto MM, Evans DH (2009) Phylogeny, taxonomy, and evolution of the endothelin receptor gene family. Mol Phylogenet Evol 52(3):677–687. doi:10.1016/j.ympev.2009.04.015, S1055-7903(09)00145-6 [pii]

Inoue A, Yanagisawa M, Kimura S, Kasuya Y, Miyauchi T, Goto K, Masaki T (1989) The human endothelin family: three structurally and pharmacologically distinct isopeptides predicted by three separate genes. Proc Natl Acad Sci USA 86(8):2863–2867

Jackman HL, Morris PW, Rabito SF, Johansson GB, Skidgel RA, Erdos EG (1993) Inactivation of endothelin-1 by an enzyme of the vascular endothelial cells. Hypertension 21(6 Pt 2):925–928

Karnaky KG Jr (1998) Osmotic and Ionic regulation. In: Evans DH (ed) The physiology of fishes, 2nd edn. CRC Press, Boca Raton, pp 159–178

Karnaky KG Jr, Kinter WB (1977) Killifish opercular skin: a flat epithelium with a high density of chloride cells. J Exp Zool 199(3):355–364. doi:10.1002/jez.1401990309

Karnaky KJ Jr, Degnan KJ, Zadunaisky JA (1977) Chloride transport across isolated opercular epithelium of killifish: a membrane rich in chloride cells. Science 195(4274):203–205

Karne S, Jayawickreme CK, Lerner MR (1993) Cloning and characterization of an endothelin-3 specific receptor (ETC receptor) from *Xenopus laevis* dermal melanophores. J Biol Chem 268 (25):19126–19133

Katoh F, Kaneko T (2003) Short-term transformation and long-term replacement of branchial chloride cells in killifish transferred from seawater to freshwater, revealed by morphofunctional observations and a newly established 'time-differential double fluorescent staining' technique. J Exp Biol 206(Pt 22):4113–4123

Kimura S, Kasuya Y, Sawamura T, Shinimi O, Sugita Y, Yanagisawa M, Goto K, Masaki T (1989) Conversion of big endothelin-1 to 21-residue endothelin-1 is essential for expression of full vasoconstrictor activity: structure-activity relationships of big endothelin-1. J Cardiovasc Pharmacol 13(Suppl 5):S5–S7, discussion S18

Kohan DE (2011) Endothelin and collecting duct sodium and water transport. Contrib Nephrol 172:94–106. doi:10.1159/000328687, 000328687 [pii]

La M, Reid JJ (1995) Endothelin-1 and the regulation of vascular tone. Clin Exp Pharmacol Physiol 22(5):315–323

Kohan DE, Rossi NF, Inscho EW, Pollock DM (2011) Regulation of blood pressure and salt homeostasis by endothelin. Physiol Rev 91(1):1–77. doi:10.1152/physrev.00060.2009

le Mevel JC, Delarue C, Mabin D, Vaudry H (1999) Central and peripheral administration of endothelin-1 induces an increase in blood pressure in conscious trout. Am J Physiol 276(4 Pt 2):R1010–R1017

Lecoin L, Sakurai T, Ngo MT, Abe Y, Yanagisawa M, Le Douarin NM (1998) Cloning and characterization of a novel endothelin receptor subtype in the avian class. Proc Natl Acad Sci USA 95(6):3024–3029

Lee CY, Chiappinelli VA (1989) Sequence homology between sarafotoxins S6 and porcine endothelin. Toxicon 27(3):277–279

Lee KM, Kaneko T, Katoh F, Aida K (2006) Prolactin gene expression and gill chloride cell activity in fugu *Takifugu rubripes* exposed to a hypoosmotic environment. Gen Comp Endocrinol 149(3):285–293

Lodhi KM, Sakaguchi H, Hirose S, Hagiwara H (1995) Localization and characterization of a novel receptor for endothelin in the gills of the rainbow trout. J Biochem 118(2):376–379

Marshall WS (2003) Rapid regulation of NaCl secretion by estuarine teleost fish: coping strategies for short-duration freshwater exposures. Biochim Biophys Acta 1618(2):95–105

Marshall WS, Grossell M (2006) Ion transport, osmoregulation, and acid-base balance. In: Evans DH, Claiborne JB (eds) The physiology of fishes, vol 3, 3rd edn. CRC, Boca Raton, FL, pp 177–230

Marshall WS, Emberley TR, Singer TD, Bryson SE, McCormick SD (1999) Time course of salinity adaptation in a strongly euryhaline estuarine teleost, *Fundulus heteroclitus*: a multivariable approach. J Exp Biol 202(Pt 11):1535–1544

McCormick SD (1995) Hormonal control of gill Na+, K+-ATPase and chloride cell function. In: Wood CM, Shuttleworth TJ (eds) Fish physiology ionoregulation: cellular and molecular approaches, vol XIV. Academic, New York, pp 285–315

McCormick SD, Bradshaw D (2006) Hormonal control of salt and water balance in vertebrates. Gen Comp Endocrinol 147(1):3–8

Medeiros DM, Crump JG (2012) New perspectives on pharyngeal dorsoventral patterning in development and evolution of the vertebrate jaw. Dev Biol 371(2):121–135. doi:10.1016/j.ydbio.2012.08.026

Miller CT, Schilling TF, Lee K, Parker J, Kimmel CB (2000) Sucker encodes a zebrafish Endothelin-1 required for ventral pharyngeal arch development. Development 127 (17):3815–3828

Mistry AC, Kato A, Tran YH, Honda S, Tsukada T, Takei Y, Hirose S (2004) FHL5, a novel actin-binding protein, is highly expressed in eel gill pillar cells and responds to wall tension. Am J Physiol Regul Integr Comp Physiol 287(5):R1141–R1154

Nilsson S, Sundin L (1998) Gill blood flow control. Comp Biochem Physiol A Mol Integr Physiol 119(1):137–147

Nishimura H, Fan Z (2003) Regulation of water movement across vertebrate renal tubules. Comp Biochem Physiol A Mol Integr Physiol 136(3):479–498

Nishimura H, Imai M (1982) Control of renal function in freshwater and marine teleosts. Fed Proc 41(8):2355–2360

Notenboom S, Miller DS, Smits P, Russel FG, Masereeuw R (2002) Role of NO in endothelin-regulated drug transport in the renal proximal tubule. Am J Physiol Renal Physiol 282(3): F458–F464

Notenboom S, Miller DS, Kuik LH, Smits P, Russel FG, Masereeuw R (2005) Short-term exposure of renal proximal tubules to gentamicin increases long-term multidrug resistance protein 2 (Abcc2) transport function and reduces nephrotoxicant sensitivity. J Pharmacol Exp Ther 315(2):912–920

Ohno S, Wolf U, Atkin NB (1968) Evolution from fish to mammals by gene duplication. Hereditas 59(1):169–187

Olson KR (1998) Hormone metabolism by the fish gill. Comp Biochem Physiol A Mol Integr Physiol 119(1):55–65

Olson KR (2002) Vascular anatomy of the fish gill. J Exp Zool 293(3):214–231

Olson KR, Duff DW, Farrell AP, Keen J, Kellogg MD, Kullman D, Villa J (1991) Cardiovascular effects of endothelin in trout. Am J Physiol 260(4 Pt 2):H1214–H1223

Parichy DM, Mellgren EM, Rawls JF, Lopes SS, Kelsh RN, Johnson SL (2000) Mutational analysis of endothelin receptor b1 (rose) during neural crest and pigment pattern development in the zebrafish Danio rerio. Dev Biol 227(2):294–306. doi:10.1006/dbio.2000.9899

Pelis RM, McCormick SD (2001) Effects of growth hormone and cortisol on Na(+)-K(+)-2Cl(-) cotransporter localization and abundance in the gills of Atlantic salmon. Gen Comp Endocrinol 124(2):134–143

Perry SF (1997) The chloride cell: structure and function in the gills of freshwater fishes. Annu Rev Physiol 59:325–347

Plato CF, Pollock DM, Garvin JL (2000) Endothelin inhibits thick ascending limb chloride flux via ETB receptor-mediated NO release. Am J Physiol Renal Physiol 279(2):F326–F333

Platzack B, Wang Y, Crossley D, Lance V, Hicks JW, Conlon JM (2002) Characterization and cardiovascular actions of endothelin-1 and endothelin-3 from the American alligator. Am J Physiol Regul Integr Comp Physiol 282(2):R594–R602. doi:10.1152/ajpregu.00733.2000

Poder TC, Silberberg SD, Rampe D (1991) Contraction of reptile, amphibian, and fish blood vessels by endothelin-1. Can J Physiol Pharmacol 69(2):215–217

Pollock DM, Highsmith RF (1998) Endothelin receptors and signaling mechanisms. Springer/ Landes Bioscience, Berlin/Georgetown, TX

Sakamoto T, McCormick SD (2006) Prolactin and growth hormone in fish osmoregulation. Gen Comp Endocrinol 147(1):24–30

Sakurai T, Yanagisawa M, Takuwa Y, Miyazaki H, Kimura S, Goto K, Masaki T (1990) Cloning of a cDNA encoding a non-isopeptide-selective subtype of the endothelin receptor. Nature 348 (6303):732–735

Schilling T (2003) Evolution and development. Making jaws. Heredity 90(1):3–5. doi:10.1038/sj.hdy.6800205

Scott GR, Schulte PM (2005) Intraspecific variation in gene expression after seawater transfer in gills of the euryhaline killifish *Fundulus heteroclitus*. Comp Biochem Physiol A Mol Integr Physiol 141(2):176–182

Scott GR, Richards JG, Forbush B, Isenring P, Schulte PM (2004) Changes in gene expression in gills of the euryhaline killifish *Fundulus heteroclitus* after abrupt salinity transfer. Am J Physiol Cell Physiol 287(2):C300–C309

Shimada K, Takahashi M, Tanzawa K (1994) Cloning and functional expression of endothelin-converting enzyme from rat endothelial cells. J Biol Chem 269(28):18275–18278

Skovgaard N, Warren DE, Jackson DC, Wang T (2005) Endothelin-1 causes systemic vasodilatation in anaesthetised turtles (Trachemys scripta) through activation of ETB-receptors. J Exp Biol 208(Pt 19):3739–3746. doi:10.1242/jeb.01834

Skovgaard N, Zibrandtsen H, Laursen BE, Simonsen U, Wang T (2008) Hypoxia-induced vasoconstriction in alligator (Alligator mississippiensis) intrapulmonary arteries: a role for endothelin-1? J Exp Biol 211(Pt 10):1565–1570. doi:10.1242/jeb.014662

Stenslokken KO, Sundin L, Nilsson GE (1999) Cardiovascular and gill microcirculatory effects of endothelin-1 in Atlantic cod: evidence for pillar cell contraction. J Exp Biol 202 (Pt 9):1151–1157

Stenslokken KO, Sundin L, Nilsson GE (2006) Endothelin receptors in teleost fishes: cardiovascular effects and branchial distribution. Am J Physiol Regul Integr Comp Physiol 290(3): R852–R860

Sultana N, Nag K, Kato A, Hirose S (2007) Pillar cell and erythrocyte localization of fugu ET (A) receptor and its implication. Biochem Biophys Res Commun 355(1):149–155

Sverdrup A, Kruger PG, Helle KB (1994) Role of the endothelium in regulation of vascular functions in two teleosts. Acta Physiol Scand 152(2):219–233

Taniguchi Y, Takeda S, Furutani-Seiki M, Kamei Y, Todo T, Sasado T, Deguchi T, Kondoh H, Mudde J, Yamazoe M, Hidaka M, Mitani H, Toyoda A, Sakaki Y, Plasterk RH, Cuppen E (2006) Generation of medaka gene knockout models by target-selected mutagenesis. Genome Biol 7(12):R116. doi:10.1186/gb-2006-7-12-r116

Taylor JS, Braasch I, Frickey T, Meyer A, Van de Peer Y (2003) Genome duplication, a trait shared by 22000 species of ray-finned fish. Genome Res 13(3):382–390. doi:10.1101/gr.640303

Tse WK, Chow SC, Wong CK (2008) The cloning of eel osmotic stress transcription factor and the regulation of its expression in primary gill cell culture. J Exp Biol 211(Pt 12):1964–1968. doi:10.1242/jeb.017368

Tse WK, Lai KP, Takei Y (2011) Medaka osmotic stress transcription factor 1b (Ostf1b/TSC22D3-2) triggers hyperosmotic responses of different ion transporters in medaka gill and human embryonic kidney cells via the JNK signalling pathway. Int J Biochem Cell Biol 43 (12):1764–1775. doi:10.1016/j.biocel.2011.08.013

Tse WK, Chow SC, Wong CK (2012) Eel osmotic stress transcriptional factor 1 (Ostf1) is highly expressed in gill mitochondria-rich cells, where ERK phosphorylated. Front Zool 9(1):3. doi:10.1186/1742-9994-9-3

Venkatesh B, Lee AP, Ravi V, Maurya AK, Lian MM, Swann JB, Ohta Y, Flajnik MF, Sutoh Y, Kasahara M, Hoon S, Gangu V, Roy SW, Irimia M, Korzh V, Kondrychyn I, Lim ZW, Tay BH, Tohari S, Kong KW, Ho S, Lorente-Galdos B, Quilez J, Marques-Bonet T, Raney BJ, Ingham PW, Tay A, Hillier LW, Minx P, Boehm T, Wilson RK, Brenner S, Warren WC (2014) Elephant shark genome provides unique insights into gnathostome evolution. Nature 505 (7482):174–179. doi:10.1038/nature12826

Volff JN (2005) Genome evolution and biodiversity in teleost fish. Heredity 94(3):280–294. doi:10.1038/sj.hdy.6800635

Wang Y, Olson KR, Smith MP, Russell MJ, Conlon JM (1999) Purification, structural characterization, and myotropic activity of endothelin from trout, Oncorhynchus mykiss. Am J Physiol 277(6 Pt 2):R1605–R1611

Wang Y, Remy-Jouet I, Delarue C, Letourneau M, Fournier A, Vaudry H, Conlon JM (2000) Structural characterization and effects on corticosteroid secretion of endothelin-1 and endothelin-3 from the frog Rana ridibunda. J Mol Endocrinol 24(2):285–293

Waters PD, Delbridge ML, Deakin JE, El-Mogharbel N, Kirby PJ, Carvalho-Silva DR, Graves JA (2005) Autosomal location of genes from the conserved mammalian X in the platypus (Ornithorhynchus anatinus): implications for mammalian sex chromosome evolution. Chromosome Res 13(4):401–410. doi:10.1007/s10577-005-0978-5

Xu D, Emoto N, Giaid A, Slaughter C, Kaw S, deWit D, Yanagisawa M (1994) ECE-1: a membrane-bound metalloprotease that catalyzes the proteolytic activation of big endothelin-1. Cell 78(3):473–485

Yanagisawa M, Masaki T (1989) Endothelin, a novel endothelium-derived peptide. Pharmacological activities, regulation and possible roles in cardiovascular control. Biochem Pharmacol 38(12):1877–1883

Yanagisawa M, Inoue A, Ishikawa T, Kasuya Y, Kimura S, Kumagaye S, Nakajima K, Watanabe TX, Sakakibara S, Goto K et al (1988a) Primary structure, synthesis, and biological activity of rat endothelin, an endothelium-derived vasoconstrictor peptide. Proc Natl Acad Sci USA 85 (18):6964–6967

Yanagisawa M, Kurihara H, Kimura S, Goto K, Masaki T (1988b) A novel peptide vasoconstrictor, endothelin, is produced by vascular endothelium and modulates smooth muscle Ca2+ channels. J Hypertens Suppl 6(4):S188–S191

Yanagisawa M, Kurihara H, Kimura S, Tomobe Y, Kobayashi M, Mitsui Y, Yazaki Y, Goto K, Masaki T (1988c) A novel potent vasoconstrictor peptide produced by vascular endothelial cells. Nature 332(6163):411–415

Yang WK, Kang CK, Chang CH, Hsu AD, Lee TH, Hwang PP (2013) Expression profiles of branchial FXYD proteins in the brackish medaka Oryzias dancena: a potential saltwater fish model for studies of osmoregulation. PLoS One 8(1):e55470. doi:10.1371/journal.pone.0055470

Zeidel ML (1993) Hormonal regulation of inner medullary collecting duct sodium transport. Am J Physiol 265(2 Pt 2):F159–F173

Zeidel ML, Brady HR, Kone BC, Gullans SR, Brenner BM (1989) Endothelin, a peptide inhibitor of Na(+)-K(+)-ATPase in intact renal tubular epithelial cells. Am J Physiol 257(6 Pt 1):C1101–C1107

Chapter 7
Genetic Manipulation of the Endothelin System

Wararat Kittikulsuth and David M. Pollock

Abstract The endothelin (ET) system consists of the three 21-amino acid isopeptides, ET-1, ET-2, and ET-3 synthesized primarily by a family of unique endothelin converting enzymes (ECE). These peptides exert a wide variety of physiological and pathophysiological effects by activating a pair of classical G-protein coupled receptors, ET_A and ET_B. Genetic manipulation of the ET system in rodents has revealed an important role of this system on fetal organ development, blood pressure regulation, and end-organ damage, especially in lung and kidney. Overexpression of ET-1 in mice demonstrates hypertrophic, fibrotic, and inflammatory effects on vasculature, heart, lung and kidney tissues, while overexpression of human ET-2 in rats shows a primary fibrotic effect in glomeruli. Studies from systemic knockout models of the ET system are uniformly lethal, but reveal that the ET-1/ECE-1/ET_A-mediated signaling pathway is necessary for facial and cardiovascular formation, while ET-3/ECE-1/ET_B signaling pathway is important for creation of neural crest-derived enteric neurons and epidermal melanocytes during embryonic development. Furthermore, cell-specific deletion of the ET system in the renal collecting duct leads to impaired water and sodium excretion, increased epithelial sodium channel activity and hypertension. In summary, it is clear that genetic manipulation of the ET system has been, and will continue to be, a powerful tool to aid our understanding of physiological and pathological actions of this complex autocrine and paracrine system.

Keywords Endothelin-1 • Knockout • Transgenic • Collecting duct • Enteric nerves • Cardiac development

---

W. Kittikulsuth
Department of Pharmacology, School of Medicine, Kagawa University, Kagawa, Japan

D.M. Pollock, Ph.D. (✉)
Division of Nephrology, Department of Medicine, University of Alabama at Birmingham, 720 20th Street South, KAUL 802, Birmingham, AL 35233, USA
e-mail: dpollock@uab.edu

K.A. Hyndman, T.L. Pannabecker (eds.), *Sodium and Water Homeostasis*, Physiology in Health and Disease, DOI 10.1007/978-1-4939-3213-9_7

7.1 Introduction

The endothelin system consists of the three 21-amino acid isopeptides, ET-1, -2 and -3. Synthesis of these three isopeptides occurs from precursors called preproendothelins (prepro-ET-1, -2 and -3). Each prepro-ET undergoes proteolytic cleavage from a furin-like protease to form the big-ET (big ET-1 to big ET-3). Then, each big-ET is cleaved by an isoform of the endothelin-converting enzyme (ECE-1 to ECE-3) to form the active ET peptides. Among the three ET isoforms, ET-1 plays a major role in regulating physiological and pathological effects. Two receptor subtypes, ET_A and ET_B are responsible for the actions of the ET peptides. ET_A receptors are mainly expressed in smooth muscle cells and mediate vasoconstriction. The ET_B receptors are widely expressed in a variety of cell types, including endothelial cells and renal epithelial cells and mediate vasodilation and promote sodium excretion. Both ET receptors also bind their ligands irreversibly and thus can act as clearance receptors (Kohan et al. 2011). This latter function has primarily been attributed to ET_B receptors because specific antagonists consistently raise circulating levels of ET-1, although blockade of ET_A receptors can have a similar effect when ET_B expression is compromised (Elmarakby et al. 2004).

There are several techniques used to investigate the role of ET's actions in healthy or disease states. One way is to use pharmacological agents to increase or inhibit ET effects. By using this approach, it is difficult to control the actions of inhibitors to the specific site-of-interest, which leads to off-target responses. Moreover, antagonists may affect several cell types within a localized region that express ET receptors or ECEs, which does not allow study of ET function in a particular cell type. For these reasons, the genetic manipulation of the ET system was developed to examine the actions of ET peptides and their receptors. This approach has been key to providing a greater understanding of this complex autocrine/paracrine system.

Genetic modification in animals refers to animals containing genetic manipulation of the genes of interest by the germ line. This manipulation causes gain or lack of function of the targeted gene early or later in life. Alteration of gene expression can be induced in a systemic or a tissue-specific manner using either a natural- or cell-specific promoter. Therefore, genetic manipulation is a powerful tool for identifying and understanding physiological or pathological function of the targeted gene *in vivo*. This chapter will review how genetic manipulation of the ET system in rodents affects fetal development and regulation of sodium handling and blood pressure.

7.2 Transgenic Models of the Endothelin System

Creation of transgenic animals is a way to study "gain-of-function" of the gene of interest that can be expressed in specific cell types or a broad spectrum of tissues *in vivo*. It involves a process of introducing foreign DNA into the host's genome. Successful integration of DNA leads to an increase in protein expression of the transgene along with basal protein expression of the native gene. For the ET system, this technique has allowed us to understand the potential pathological effect of increased ET-1 levels in many tissues (Table 7.1).

Mice non-specifically overexpressing the human ET-1 gene have increased ET-1 peptide concentrations in plasma and tissues, particularly in lung, liver and kidney (Hocher et al. 1997, 2000). This chronic elevation of ET-1 in the transgenic mice leads to lung inflammation but with normal pulmonary pressure (Hocher et al. 2000). Moreover, the transgenic mice display renal interstitial fibrosis, renal cyst formation, glomerulosclerosis and reduced glomerular filtration rate, which is an age- and androgen-dependent effect (Hocher et al. 1997; Kalk et al. 2009). The ET-1 transgenic mice have a normal systemic arterial pressure; however, as the animals age, salt-sensitive hypertension develops (Shindo et al. 2002).

Several lines of evidence suggest that an increase in nitric oxide may be involved in maintaining the normal blood pressure observed in ET-1 transgenic mice. ET-1 transgenic mice have increased urinary nitrate-nitrite excretion (Hocher et al. 2004). Endothelium-dependent relaxation is enhanced in aorta from ET-1 transgenic mice, which can be inhibited by the nitric oxide (NO) synthase inhibitor L-NAME (Quaschning et al. 2003). Moreover, the administration of L-NAME produces a greater increase in blood pressure in ET-1 transgenic compared to genetic control mice (Hocher et al. 2004). Cross-breeding between ET-1 transgenic mice and mice lacking the gene for endothelial NO synthase (NOS3 KO) have a further increase in blood pressure as compared to either ET-1 transgenic or NOS3 KO mice alone (Quaschning et al. 2007). Similarly, crossing ET-1 transgenic and NOS2 (or iNOS) knockout (KO) mice have significantly elevated blood pressure (Quaschning et al. 2008).

Mice overexpressing human ET-1 in endothelial cells using the tie-2 promoter exhibit a threefold elevation in preproET-1 mRNA in aorta and a sevenfold increase in plasma ET-1. These mice also express vascular endothelial dysfunction and altered vascular structure of resistance vessels, yet have normal blood pressure (Amiri et al. 2004).

Overexpression of human ET-1 specifically in cardiomyocytes using the α-myosin heavy chain (α-MHC) promoter results in a tenfold increase in ET-1 levels in the heart, with no change in ET-1 concentrations in plasma or other tissues. These mice have normal appearance at birth; however, they rapidly develop pulmonary and hepatic congestion, and die at 5 weeks of age. These findings suggest that ET-1 can promote inflammation and cytokine expression in the heart (Yang et al. 2004).

Table 7.1 Summary of phenotype results for genetic modification of endothelin system

Animal model	Phenotype	References
Transgenic mice		
ET-$1^{+/+}$	Lung: chronic inflammation; normal pulmonary pressure Kidney: increased renal cyst formation; renal interstitial fibrosis; glomerulosclerosis; age-dependent salt-sensitive hypertension	Hocher et al. (1997, 2000), Shindo et al. (2002), Kalk et al. (2009)
ET-$1^{+/+}$/NOS$2^{-/-}$	Increased blood pressure	Quaschning et al. (2008)
ET-$1^{+/+}$/NOS$3^{-/-}$	Elevated blood pressure	Quaschning et al. (2007)
Endothelial ET-$1^{+/+}$	Vascular endothelial dysfunction; vascular remodeling	Amiri et al. (2004)
Cardiomyocyte ET$1^{+/+}$	Pulmonary and hepatic congestion	Yang et al. (2004)
Knockout mice		
ET-$1^{-/-}$	Abnormalities in craniofacial and cardiovascular system, thyroid and thymus gland; death after birth	Kurihara et al. (1994, 1995a, b)
ET-$1^{-/+}$	Mild blood pressure elevation; increased resting renal sympathetic nerve activity	Kurihara et al. (1994), Kuwaki et al. (1996, 1999), Ling et al. (1998), Morita et al. (1998)
Cardiomyocyte ET-$1^{-/-}$	Shorter lifespan; age-associated reduction in cardiac function	Zhao et al. (2006)
Collecting duct ET-$1^{-/-}$	Increased blood pressure; salt-sensitive hypertension; impaired ability to excrete water	Ahn et al. (2004), Ge et al. (2005a)
ET-$2^{-/-}$	Growth retardation; internal starvation; severe hypothermia; lung dysfunction; death at an early age	Chang et al. (2013)
ET-$2^{-/-}$ at adulthood	Reduced weight gain; reduced lipid deposition	Chang et al. (2013)
Intestinal epithelium ET-$2^{-/-}$	Normal growth and blood glucose level	Chang et al. (2013)
Neuron ET-$2^{-/-}$	Normal core temperature	Chang et al. (2013)
ET-$3^{-/-}$	Aganglionic megacolon; coat color spotting; death at an early age	Baynash et al. (1994), Kuwaki et al. (2002)
$ET_A{}^{-/-}$	Craniofacial and cardiovascular defects (similar phenotype to ET-$1^{-/-}$)	Clouthier et al. (1998)
Cardiomyocyte $ET_A{}^{-/-}$	Normal development and cardiovascular function; Angiotensin II or isoproterenol-induced myocardial hypertrophy	Kedzierski et al. (2003)
Smooth muscle $ET_A{}^{-/-}$	Defects of arterial network, mandibular and thymus structure	Donato et al. (2014)
Collecting duct $ET_A{}^{-/-}$	Normal blood pressure; impaired ability to excrete water	Ge et al. (2005b)

(continued)

Table 7.1 (continued)

Animal model	Phenotype	References
Whole nephron $ET_A^{-/-}$	Normal blood pressure; fluid retention during high salt intake	Stuart et al. (2012, 2013)
$ET_B^{-/-}$	Aganglionic megacolon; coat color spotting (similar phenotype to ET-3$^{-/-}$)	Hosoda et al. (1994)
Endothelial $ET_B^{-/-}$	Endothelial dysfunction	Bagnall et al. (2006)
Collecting duct $ET_B^{-/-}$	Elevated blood pressure; increased ENaC activity; salt-sensitive hypertension	Ge et al. (2006), Bugaj et al. (2012)
Collecting duct $ET_A^{-/-}$/ $ET_B^{-/-}$	Elevated blood pressure; increased ENaC activity; salt-sensitive hypertension	Ge et al. (2008), Bugaj et al. (2012)
ECE-1$^{-/-}$	Craniofacial and cardiovascular defects (similar phenotype to ET-1$^{-/-}$); aganglionic megacolon and coat color spotting (similar phenotype to ET-3$^{-/-}$)	Yanagisawa et al. (1998)
ECE-2$^{-/-}$	No detectable abnormalities	Yanagisawa et al. (2000)
Transgenic rat		
ET-2$^{+/+}$	Glomerulosclerosis	Liefeldt et al. (1995, 1999)
Dysfunctional ET system in rat		
Homozygous spotting lethal (*sl*/*sl*)	Aganglionic megacolon; white coat color	Gariepy et al. (1996)
Rescued ET_B-deficient	Coat color spotting; salt-sensitive hypertension	Gariepy et al. (1998, 2000)

Overexpression of human ET-2 in the rat causes a marked elevation in ET-2 levels in kidney, intestines, lung, and brain (Liefeldt et al. 1995, 1999). In the kidney, human ET-2 is predominantly expressed in glomeruli so that renal fibrosis in the ET-2 transgenic rats occurs strictly in glomeruli. Moreover, ET-2 transgenic rats have increased urinary protein excretion and reduced glomerular filtration rate with no change in blood pressure. It is not clear whether these findings reflect any known problem related to renal disease, but suggest that further investigation is needed.

In summary, studies from rodents overexpressing human ET-1 and ET-2 suggest deleterious effects of ET-1 and ET-2 in many organs, but are particularly evident in kidney. ET-1 has hypertrophic, fibrotic, and inflammatory effects on vasculature, heart, lung and kidney, while ET-2 has a major fibrotic effect in glomeruli. These deleterious effects of both ET-1 and ET-2 occur independent from blood pressure elevation, which may appear surprising given the potent vasoconstrictor effects of exogenously administered ET-1. However, these findings are consistent with studies where exogenous ET-1 was infused chronically without any change in blood pressure, yet there are clear signs of inflammation and renal dysfunction (Saleh et al. 2010). We speculate that the lack of hypertension is due to efficient clearance of ET peptides by the ET_B receptor and the vasodilator effects that oppose ET_A dependent vasoconstriction.

7.3 Knockout Models of the Endothelin System

Loss-of-function of the protein of interest is another way to study the role of specific gene products. This approach can be achieved through gene targeting by eliminating a specific gene or deleting a portion of the gene that results in the absence of the functional domain of the protein of interest. Homologous recombination is an important step to completely remove gene loci resulting in the production of a mutated or truncated protein, or no protein production at all. Gene deletion can occur in all or specific cell types. The latter is generated by using site-specific recombination technology, such as Cre-lox. The Cre-lox system is composed of (1) Cre recombinase, an enzyme that induces the recombination between two loxP sites on the gene of interest driven by a cell-specific promoter, and (2) loxP, where the recombination occurs within the gene of interest (Kohan 2008). Normally, the process of gene inactivation occurs during the embryogenic state. However, it also can be induced in adulthood or a certain time period using inducers, such as tetracycline or tamoxifen that can activate specific promoters (Kohan 2008). Using this technique, the importance of the ET system in embryonic development and sodium homeostasis has been revealed.

7.3.1 The Endothelin System in Embryonic Development

Homozygous ET-1 KO mice die immediately after birth due to respiratory failure, which is caused by craniofacial developmental abnormalities (Kurihara et al. 1994). ET-1 KO mice also have abnormalities of cardiovascular system (Kurihara et al. 1995b), thyroid and thymus glands (Kurihara et al. 1995a); however, no abnormalities in the lung, kidney, and central nervous system could be found (Kurihara et al. 1994). Moreover, ET_A receptor or ECE-1 KO mice show craniofacial deformities and defects in the cardiovascular systems leading to death soon after birth (Clouthier et al. 1998; Yanagisawa et al. 1998). These phenotypes from ET_A or ECE-1 KO mice are nearly identical to those found in ET-1 KO mice.

Similar to ET-1 or ET_A receptor KO mice, pharmacological inhibition of the ET_A receptor with either a mixed ET_A and ET_B (Spence et al. 1999) or selective ET_A receptor antagonist (Cross et al. 2012) produces teratogenicity, including malformations of the head, mouth, face, and large blood vessels. Furthermore, neutralizing antibodies to ET-1 or an ET_A antagonist leads to cardiovascular defects in the pups from pregnant heterozygous ET-1 KO mice (Kurihara et al. 1995b). Thus, these results from both genetic deletion and pharmacological inhibition demonstrate an important role of ET-1/ECE-1/ET_A-mediated signaling in facial and cardiovascular formation, and control of the respiratory system after birth.

Heterozygous ET-1 KO mice appear to have normal development and are fertile. Unexpectedly, heterozygous ET-1 KO mice, which have reduced plasma and renal ET-1 levels, display a mild elevation in resting blood pressure suggesting a loss of ET_B receptor activation (Kurihara et al. 1994; Morita et al. 1998), but is not exacerbated by salt (see discussion below on ET_B receptor function) (Morita et al. 1998). The mechanism of blood pressure elevation in these mice is not due to impaired nitric oxide activity (Kurihara et al. 1994). One possible explanation is that these mice have increased resting renal sympathetic nerve activity (RSNA) and maximum RSNA during basal conditions (Kuwaki et al. 1996; Ling et al. 1998; Kuwaki et al. 1999). ET-1 and both its receptors are expressed in the central and peripheral nervous system, but little is known about their physiological role.

Cardiomyocyte specific deletion of ET-1 does not affect cardiac structure or function in young mice (Zhao et al. 2006). However, aged cardiomyocyte ET-1 KO mice display a significant reduction in fractional shortening, reduced left ventricle systolic function, and a dilated left heart ventricle (Zhao et al. 2006). For these reasons, these KO mice have a shorter lifespan than controls (median life expectancy: 11 months for KO mice vs. 2 years for wild-type) (Zhao et al. 2006).

Whole body ET-2 KO mice exhibit severe growth retardation and die at the age of 3–4 weeks. These mice display internal starvation, severe hypothermia, and lung dysfunction. Deletion of ET-2 function at adulthood cause diminished weight gain and reduced lipid deposition (Chang et al. 2013). Unlike whole body ET-2 KO mice, intestinal epithelium–specific ET-2 KO mice have normal growth and blood glucose levels, suggesting the internal starvation cannot be explained by intestinal absorption. Similarly, severe hypothermia in ET-2 KO mice cannot be explained by the lack of ET-2 function in the enteric nervous system since neuron-specific ET-2 KO mice had normal core temperature (Chang et al. 2013). These results indicate that ET-2 is essential for growth regulation and survival of postnatal mice and the maintenance of energy homeostasis even in adulthood. However, the precise origin and localized activity remains somewhat unclear. Interesting, ET-2 has been described in some of the early literature as vasoactive intestinal constrictor (VIC) prior to discovery of its amino acid sequence (Bloch et al. 1991).

Disruption of the gene for ET-3 in mice displays aganglionic megacolon, coat color spotting and no melanin pigment in choroidal layer of the retina. Most of the KO mice became sick and die at about 4 weeks after birth (Baynash et al. 1994). These similar phenotypes can be found in ET_B receptor or ECE-1 KO mice (see below) (Hosoda et al. 1994; Yanagisawa et al. 1998). These findings suggest the role of the ET-3/ECE-1/ET_B signaling pathway in neural crest-derived enteric neurons and epidermal melanocytes. Deletion of the ET-3 gene does not affect blood pressure and heart rate in infant mice (Kuwaki et al. 2002).

As described above, ET_A receptor KO mice die immediately after birth with defects of craniofacial and cardiovascular formation (Clouthier et al. 1998). However, ET_A KO mice using the α-MHC-Cre promoter to target cardiomyocytes specifically are viable and have normal development and function of cardiovascular system. These mice develop myocardial hypertrophy after angiotensin II or isoproterenol infusion. Because cardiomyocyte-specific ET_A KO mice have a twofold increase in ET_B receptor binding in cardiac tissue (Kedzierski et al. 2003), it is possible that ET_B receptors may have hypertrophic effects in cardiomyocytes during pathological situations when ET_A receptors are absent.

Recently, smooth muscle (SM)-specific ET_A KO mice have been generated using the SM22-Cre promoter to drive Cre-recombinase. SM ET_A KO mice display developmental abnormalities of the arterial network mandibular and thymus structure, leading to reduced survival. SM ET_A KO mice also exhibit attenuated ET-1-induced vasoconstriction and blood pressure elevation when given exogenously. Moreover, these mice have lower blood pressure during high salt intake as compared to the controls (Donato et al. 2014). These data again confirm a role for ET-1 and ET_A signaling in SM in vascular, mandibular, neural crest and thymus development as well as a less well-understood influence on blood pressure.

Similar to ET-3 KO animals, mice lacking ET_B receptors are born with white spotting of the coat as a result of abnormal melanocyte function. These mice stay healthy in the first week after birth; however, they become severely ill with aganglionic megacolon (Hosoda et al. 1994).

The homozygous spotting lethal (*sl*/*sl*) rat was originally maintained as a model for Hirschsprung's Disease due to a lack of enteric nerve development. Gariepy and colleagues discovered that this rat has a naturally occurring deletion in exon 1 of ET_B receptor (*EDNRB*) gene, which encodes a transmembrane portion of the ET_B receptor. This results in a non-functional ET_B receptor being expressed and so this model is referred to as the ET_B-deficient rat. These rats also display a white coat color, aganglionic megacolon, and die within the first few weeks after birth (Gariepy et al. 1996). Transgene expression of the full, intact ET_B receptor using human dopamine-β-hydroxylase (DβH) promoter increases the expression of *EDNRB* gene in enteric neurons and the putative embryonic neuroblast (Kapur et al. 1991; Mercer et al. 1991); however, non-adrenergic tissues, such as kidneys, do not have functional ET_B receptors. The transgene increases survival and rescues these animals from the development of megacolon. However, *DβH-ENDRB* transgene is not expressed in vascular endothelial cells or renal tubular epithelium and does not prevent coat color spotting phenotype (Gariepy et al. 1998).

As mentioned above, mice that lack ECE-1 expression display abnormalities of craniofacial tissues, great vessel and cardiac outflow structures, which are also observed in ET-1 and ET_A receptor KO mice. Moreover, ECE-1 KO mice show defects of epidermal and choroidal melanocytes, and enteric formation, which is similar to ET-3 and ET_B receptor KO mice. These data suggest the importance of ECE-1 in the conversion of the ET precursor peptides, big ET-1 and big ET-3, to active ET-1 and ET-3. Even though tissue ET-3 levels are dramatically reduced in ECE-1 KO mice, ET-1 and ET-2 levels in the KO mice are only reduced by about 50 %, suggesting a role for other enzymes, such as ECE-2, in conversion of big ETs to ETs (Yanagisawa et al. 1998).

The deletion of ECE-2 in mice results in no detectable defects in embryonic development. Adult ECE-2 KO mice are healthy and fertile. ECE-2 does not appear to play a role in converting big ETs to active ETs under basal conditions since tissue ET-1 and ET-2 levels are comparable between ECE-2 KO and WT mice. Furthermore, tissue ET-1 and ET-2 levels are comparable between double ECE-1/ECE-2 and ECE-1 KO mice (Yanagisawa et al. 2000).

7.4 The Endothelin System in Water and Sodium Homeostasis

7.4.1 Collecting Duct System and Sodium Homeostasis

Among the three ET peptides, ET-1 is a major isoform that is produced and has actions in the kidney (Kohan and Fiedorek 1991; Ujiie et al. 1992). ET-1 tissue content is the highest in the renal medulla, which is mainly driven by inner medullary collecting duct (CD) expression (Kohan and Fiedorek 1991; Ujiie et al. 1992). Several *in vitro* studies demonstrate that ET-1 enhances sodium excretion in the thick ascending limb (TAL) (Plato et al. 2000; Herrera and Garvin 2004; Herrera et al. 2009) and CD (Edwards et al. 1993; Kohan et al. 1993; Gallego and Ling 1996), which is inhibited by an ET_B receptor antagonist. Similarly, an increase in ET_B activation in the renal medulla using a specific ET_B agonist increases sodium and water excretion in rodents, which is independent in changes in medullary blood flow (Nakano et al. 2008; Nakano and Pollock 2009; Kittikulsuth et al. 2011, 2012). These data suggest that ET_B-induced natriuresis is from renal tubular action. However, these studies do not provide direct evidence that ET-1 regulates sodium excretion in TAL or CD in animals. Therefore, mice lacking ET-1 and its receptors were generated using tissue specific KO by Cre-lox methodology. The cre-specific promotor for TAL (Tamms-Horsfall) (Stricklett et al. 2003) and the CD principal cell (AQP2) (Nelson et al. 1998) have each been developed. To date, however, only mice with CD principal cell KO of the ET system have been generated.

As mentioned above, ET_B deficient rats only express ET_B receptors in adrenergic tissues and lack ET_B receptor expression in other tissues such as kidneys, lung, endothelial cells. These rats have elevated plasma ET-1 because ET_B receptors in the vasculature function to clear circulating ET-1 through their irreversible binding properties (Kohan et al. 2011). ET_B deficient rats have slightly increased blood pressure when on a normal salt diet, which increases further during high salt feeding. The mechanism of salt-induced blood pressure elevation in these rats can be partially explained by the over-activation of ET_A receptors since administration of an ET_A receptor antagonist reduces blood pressure and ameliorates renal injury. Interestingly, amiloride, which can inhibit ENaC, blunts blood pressure elevation in these rats during high salt intake (Gariepy et al. 2000). These data are consistent

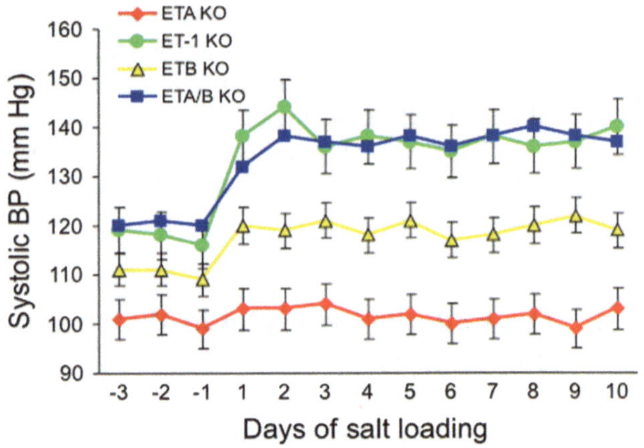

Fig. 7.1 Effect of collecting duct specific deletion of ET-1 system on systolic blood pressure during a normal or high salt diet [with permission from Pollock (2014)]

with observations that ET_B-induced inhibition of ENaC activity (Bugaj et al. 2008) is involved in sodium handling and blood pressure control (Kohan et al. 2011).

Kohan's laboratory has generated a series of mice where components of the ET-1 system have been specifically knocked out of CD principal cells using the Cre-lox system with the AQP2 promoter used to drive Cre-recombinase (Nelson et al. 1998). Mice lacking the ET-1 gene in the CD (CD ET-1 KO) have increased blood pressure during a normal salt diet that increases even more during high salt intake (Fig. 7.1). These mice exhibit an impaired ability to excrete sodium on the first 2 days of a high salt diet. Moreover, administration of diuretics, amiloride or furosemide, ameliorate blood pressure elevation during high salt diet (Ahn et al. 2004). These data suggest the lack of CD ET-1 leads to salt-sensitive hypertension, which is caused by a lack of ability to excrete sodium.

It is quite clear that ET-1 acts in an autocrine or paracrine manner and both ET receptor subtypes, ET_A and ET_B, are expressed in the CD. The Kohan laboratory has further examined which ET receptor subtype is responsible for ET-1-induced sodium excretion. Using similar gene targeting techniques, deletion of ET_A receptors in the CD (CD ET_A KO) does not alter blood pressure (Fig. 7.1) or sodium excretion during normal or high salt feeding (Ge et al. 2005b). Similarly, whole nephron KO mice of ET_A using PAX-8 Cre-promoter have normal blood pressure and sodium excretion during normal and high salt intake, although these mice display mild fluid retention during a high salt diet (Stuart et al. 2012, 2013). Unlike the CD ET_A KO, mice with CD-specific disruption of the ET_B receptor gene (CD ET_B KO) have blood pressure elevation during a normal and high salt diet (Fig. 7.1). CD ET_B KO mice have no change in sodium excretion during chronic sodium load; however, these mice display a reduced ability to excrete an acute sodium load (Ge et al. 2006).

The regulation of sodium excretion is regulated by the balance between tubular transport pathways and renal hemodynamics. ET_B receptors are highly expressed in vascular endothelial cells and cause vasodilation, which helps to excrete sodium (Kohan et al. 2011). Deletion of ET_B receptor in endothelial cells (EC) in mice causes endothelial dysfunction, but how the endothelial ET_B receptor may influence sodium excretion and blood pressure is a bit unclear (Bagnall et al. 2006). When placed on a high salt diet, blood pressure was increased in the EC ET_B KO mouse, but to a similar degree as the control strain, which was salt-sensitive. Thus, further exploration is needed to uncover the full role of EC ET_B receptors.

The degree of blood pressure elevation in CD ET_B KO mice is roughly one-half that observed in CD ET-1 KO mice during a normal and high salt diet (Fig. 7.1). These data suggest that the hypertensive effect in CD ET-1 KO mice is only partially mediated by the lack of CD ET_B receptor action. It is possible that there is a compensatory effect from ET_A receptors in CD ET_B KO mice. For this reason, mice with double deletion of ET_A and ET_B receptor gene in the CD (CD $ET_{A/B}$ KO) were generated. These animals have an identical degree of blood pressure elevation as compared to CD ET-1 KO during a normal and high salt diet (Fig. 7.1) suggesting some sort of receptor crosstalk or cooperation. Since KO of the ET_A receptor from the CD does not affect blood pressure, these data suggest that ET_A receptors also play a role in blood pressure regulation only when ET_B receptors are absent. Unlike CD ET-1 KO mice, however, CD $ET_{A/B}$ KO mice show a slower progression of blood pressure increase (Ge et al. 2008).

Administration of exogenous ET-1 has been shown to inhibit ENaC open probability in the cortical CD, which can be prevented by pharmacological inhibition of the ET_B receptor (Bugaj et al. 2008). Moreover, the patch-clamp technique on isolated split-open cortical CD revealed that ET-1 inhibits ENaC activity in control and CD ET_A KO mice; however, ENaC activity remains after ET-1 stimulation in CD KO mice lacking ET_B or $ET_{A/B}$ (Bugaj et al. 2012). It is well known that ENaC activity is inversely correlated to the amount of salt intake and is a function of circulating aldosterone levels (Stockand et al. 2010; Mironova et al. 2011). ENaC activity in CD ET_A KO mice is low during high salt feeding; however, ENaC activity is inappropriately elevated in CD ET_B or CD $ET_{A/B}$ KO mice (Bugaj et al. 2012).

The full range of studies using CD KO mice of the ET-1 system confirm that ET-1 through ET_B receptors play an important role in control of sodium excretion and blood pressure during high salt intake by inhibiting ENaC activity. Furthermore, ET_A receptors may be involved in sodium handling during high salt feeding if ET_B receptors are dysfunctional, which may account for some of the fluid retention problems observed during administration of ET_A antagonists to subjects with impaired renal function (Mann et al. 2010; Andress et al. 2012).

7.4.2 Collecting Duct System and Water Homeostasis

ET-1 also plays a specific role in regulating water excretion in the CD. CD ET-1 KO mice have reduced plasma vasopressin (AVP) levels with no change in water excretion during normal water intake. Furthermore, CD ET-1 KO mice have an impaired ability to excrete water following an acute water load. Infusion of the AVP receptor 2 agonist, [deamino-Cys1, D-Arg8]-Vasopressin (DDAVP) increases urine osmolality and AQP2 expression in CD ET-1 KO mice as compared to controls. In addition, AVP-stimulated cAMP production in CD ET-1 KO mice is enhanced in the inner medulla compared to controls. These data suggest that the absence of CD ET-1 reduces the ability to excrete water during an acute water load, which may be due to increases in AVP responsiveness (Ge et al. 2005a). It is possible that the diuretic effect of ET-1 on the CD may be a result of ET_B receptor activation since ET_B receptor antagonist inhibits AVP action in rat inner medullary CDs (Edwards et al. 1993).

Knockout of the ET_A receptor from the CD increases plasma AVP with no change in water excretion during a normal water intake. These mice have a modestly enhanced ability to excrete an acute water load. AVP responsiveness is reduced in inner medullary CDs isolated from CD ET_A KO mice (Ge et al. 2005b). These data suggest that while ET-1 induces diuresis via ET_B receptor activation in the CD, data from these KO animals suggest that ET_A receptors may enhance AVP action in the CDs to cause water retention. However, CD ET_A KO mice do not display fluid retention in response to ET_A receptor blockade as do control mice (Stuart et al. 2013) so the precise mechanisms have not been clarified.

7.5 Perspectives

Genetic manipulation of the ET system reveals important developmental, physiological, and pathological actions of this system in many organs. The interaction of ET peptides and its receptors are important in embryonic development. ET-1 via ET_A receptors is critically essential for facial and cardiovascular formation and ET-3 via ET_B signaling pathway is involved in the generation of neural crest-derived enteric neurons and epidermal melanocytes. During the postnatal period, ET-1, possibly through ET_A receptors, has a pro-fibrotic effect on lung, heart and kidney tissue.

In terms of its physiological role, ET-1, primarily via ET_B receptor activation, displays favorable effects on increasing sodium excretion and lowering blood pressure. Cell specific knockouts have been instrumental in elucidating this physiological role. ET_B receptors also appear to provide protection against the profound vasoconstrictor actions that occur with ET_A activation in the vascular system. Disruption of the balance between these two receptor systems can lead to localized tissue inflammation and organ damage that results from unchecked ET_A activation. There is much that has yet to be learned about this complex system and so these and the next generation of genetically manipulated animals will continue to provide insights into this critically important system.

References

Ahn D, Ge Y, Stricklett PK, Gill P, Taylor D, Hughes AK, Yanagisawa M, Miller L, Nelson RD, Kohan DE (2004) Collecting duct-specific knockout of endothelin-1 causes hypertension and sodium retention. J Clin Invest 114(4):504–511. doi:10.1172/JCI21064

Amiri F, Virdis A, Neves MF, Iglarz M, Seidah NG, Touyz RM, Reudelhuber TL, Schiffrin EL (2004) Endothelium-restricted overexpression of human endothelin-1 causes vascular remodeling and endothelial dysfunction. Circulation 110(15):2233–2240. doi:10.1161/01.CIR.0000144462.08345.B9

Andress DL, Coll B, Pritchett Y, Brennan J, Molitch M, Kohan DE (2012) Clinical efficacy of the selective endothelin A receptor antagonist, atrasentan, in patients with diabetes and chronic kidney disease (CKD). Life Sci 91(13–14):739–742. doi:10.1016/j.lfs.2012.01.011

Bagnall AJ, Kelland NF, Gulliver-Sloan F, Davenport AP, Gray GA, Yanagisawa M, Webb DJ, Kotelevtsev YV (2006) Deletion of endothelial cell endothelin B receptors does not affect blood pressure or sensitivity to salt. Hypertension 48(2):286–293. doi:10.1161/01.HYP.0000229907.58470.4c

Baynash AG, Hosoda K, Giaid A, Richardson JA, Emoto N, Hammer RE, Yanagisawa M (1994) Interaction of endothelin-3 with endothelin-B receptor is essential for development of epidermal melanocytes and enteric neurons. Cell 79(7):1277–1285

Bloch KD, Hong CC, Eddy RL, Shows TB, Quertermous T (1991) cDNA cloning and chromosomal assignment of the endothelin 2 gene: vasoactive intestinal contractor peptide is rat endothelin 2. Genomics 10(1):236–242

Bugaj V, Pochynyuk O, Mironova E, Vandewalle A, Medina JL, Stockand JD (2008) Regulation of the epithelial Na+ channel by endothelin-1 in rat collecting duct. Am J Physiol Renal Physiol 295(4):F1063–F1070. doi:10.1152/ajprenal.90321.2008

Bugaj V, Mironova E, Kohan DE, Stockand JD (2012) Collecting duct-specific endothelin B receptor knockout increases ENaC activity. Am J Physiol Cell Physiol 302(1):C188–C194. doi:10.1152/ajpcell.00301.2011

Chang I, Bramall AN, Baynash AG, Rattner A, Rakheja D, Post M, Joza S, McKerlie C, Stewart DJ, McInnes RR, Yanagisawa M (2013) Endothelin-2 deficiency causes growth retardation, hypothermia, and emphysema in mice. J Clin Invest 123(6):2643–2653. doi:10.1172/JCI66735

Clouthier DE, Hosoda K, Richardson JA, Williams SC, Yanagisawa H, Kuwaki T, Kumada M, Hammer RE, Yanagisawa M (1998) Cranial and cardiac neural crest defects in endothelin-A receptor-deficient mice. Development 125(5):813–824

Cross DM, Horsley E, Derzi M, Owen K, Stavros FL (2012) An evaluation of reproductive and developmental toxicity of sitaxentan (thelin) in rats. Birth Defects Res B Dev Reprod Toxicol 95(5):327–336. doi:10.1002/bdrb.21021

Donato AJ, Lesniewski LA, Stuart D, Walker AE, Henson G, Sorensen L, Li D, Kohan DE (2014) Smooth muscle specific disruption of the endothelin-A receptor in mice reduces arterial pressure, and vascular reactivity and affects vascular development. Life Sci 118(2):238–243. doi:10.1016/j.lfs.2013.12.209

Edwards RM, Stack EJ, Pullen M, Nambi P (1993) Endothelin inhibits vasopressin action in rat inner medullary collecting duct via the ETB receptor. J Pharmacol Exp Ther 267 (3):1028–1033

Elmarakby AA, Dabbs Loomis E, Pollock JS, Pollock DM (2004) ETA receptor blockade attenuates hypertension and decreases reactive oxygen species in ETB receptor-deficient rats. J Cardiovasc Pharmacol 44(Suppl 1):S7–S10

Gallego MS, Ling BN (1996) Regulation of amiloride-sensitive Na+ channels by endothelin-1 in distal nephron cells. Am J Physiol 271(2 Pt 2):F451–F460

Gariepy CE, Cass DT, Yanagisawa M (1996) Null mutation of endothelin receptor type B gene in spotting lethal rats causes aganglionic megacolon and white coat color. Proc Natl Acad Sci USA 93(2):867–872

Gariepy CE, Williams SC, Richardson JA, Hammer RE, Yanagisawa M (1998) Transgenic expression of the endothelin-B receptor prevents congenital intestinal aganglionosis in a rat model of Hirschsprung disease. J Clin Invest 102(6):1092–1101. doi:10.1172/JCI3702

Gariepy CE, Ohuchi T, Williams SC, Richardson JA, Yanagisawa M (2000) Salt-sensitive hypertension in endothelin-B receptor-deficient rats. J Clin Invest 105(7):925–933. doi:10.1172/JCI8609

Ge Y, Ahn D, Stricklett PK, Hughes AK, Yanagisawa M, Verbalis JG, Kohan DE (2005a) Collecting duct-specific knockout of endothelin-1 alters vasopressin regulation of urine osmolality. Am J Physiol Renal Physiol 288(5):F912–F920. doi:10.1152/ajprenal.00432.2004

Ge Y, Stricklett PK, Hughes AK, Yanagisawa M, Kohan DE (2005b) Collecting duct-specific knockout of the endothelin A receptor alters renal vasopressin responsiveness, but not sodium excretion or blood pressure. Am J Physiol Renal Physiol 289(4):F692–F698. doi:10.1152/ajprenal.00100.2005

Ge Y, Bagnall A, Stricklett PK, Strait K, Webb DJ, Kotelevtsev Y, Kohan DE (2006) Collecting duct-specific knockout of the endothelin B receptor causes hypertension and sodium retention. Am J Physiol Renal Physiol 291(6):F1274–F1280. doi:10.1152/ajprenal.00190.2006

Ge Y, Bagnall A, Stricklett PK, Webb D, Kotelevtsev Y, Kohan DE (2008) Combined knockout of collecting duct endothelin A and B receptors causes hypertension and sodium retention. Am J Physiol Renal Physiol 295(6):F1635–F1640. doi:10.1152/ajprenal.90279.2008

Herrera M, Garvin JL (2004) Endothelin stimulates endothelial nitric oxide synthase expression in the thick ascending limb. Am J Physiol Renal Physiol 287(2):F231–F235. doi:10.1152/ajprenal.00413.2003

Herrera M, Hong NJ, Ortiz PA, Garvin JL (2009) Endothelin-1 inhibits thick ascending limb transport via Akt-stimulated nitric oxide production. J Biol Chem 284(3):1454–1460. doi:10.1074/jbc.M804322200

Hocher B, Thöne-Reineke C, Rohmeiss P, Schmager F, Slowinski T, Burst V, Siegmund F, Quertermous T, Bauer C, Neumayer HH, Schleuning WD, Theuring F (1997) Endothelin-1 transgenic mice develop glomerulosclerosis, interstitial fibrosis, and renal cysts but not hypertension. J Clin Invest 99(6):1380–1389. doi:10.1172/JCI119297

Hocher B, Schwarz A, Fagan KA, Thöne-Reineke C, El-Hag K, Kusserow H, Elitok S, Bauer C, Neumayer HH, Rodman DM, Theuring F (2000) Pulmonary fibrosis and chronic lung inflammation in ET-1 transgenic mice. Am J Respir Cell Mol Biol 23(1):19–26. doi:10.1165/ajrcmb.23.1.4030

Hocher B, Schwarz A, Slowinski T, Bachmann S, Pfeilschifter J, Neumayer HH, Bauer C (2004) In-vivo interaction of nitric oxide and endothelin. J Hypertens 22(1):111–119

Hosoda K, Hammer RE, Richardson JA, Baynash AG, Cheung JC, Giaid A, Yanagisawa M (1994) Targeted and natural (piebald-lethal) mutations of endothelin-B receptor gene produce megacolon associated with spotted coat color in mice. Cell 79(7):1267–1276

Kalk P, Thöne-Reineke C, Schwarz A, Godes M, Bauer C, Pfab T, Hocher B (2009) Renal phenotype of ET-1 transgenic mice is modulated by androgens. Eur J Med Res 14:55–58

Kapur RP, Hoyle GW, Mercer EH, Brinster RL, Palmiter RD (1991) Some neuronal cell populations express human dopamine beta-hydroxylase-lacZ transgenes transiently during embryonic development. Neuron 7(5):717–727

Kedzierski RM, Grayburn PA, Kisanuki YY, Williams CS, Hammer RE, Richardson JA, Schneider MD, Yanagisawa M (2003) Cardiomyocyte-specific endothelin A receptor knockout mice have normal cardiac function and an unaltered hypertrophic response to angiotensin II and isoproterenol. Mol Cell Biol 23(22):8226–8232

Kittikulsuth W, Pollock JS, Pollock DM (2011) Sex differences in renal medullary endothelin receptor function in angiotensin II hypertensive rats. Hypertension 58(2):212–218. doi:10.1161/HYPERTENSIONAHA.111.172734

Kittikulsuth W, Pollock JS, Pollock DM (2012) Loss of renal medullary endothelin B receptor function during salt deprivation is regulated by angiotensin II. Am J Physiol Renal Physiol 303(5):F659–F666. doi:10.1152/ajprenal.00213.2012

Kohan DE (2008) Progress in gene targeting: using mutant mice to study renal function and disease. Kidney Int 74(4):427–437. doi:10.1038/ki.2008.146
Kohan DE, Fiedorek FT (1991) Endothelin synthesis by rat inner medullary collecting duct cells. J Am Soc Nephrol 2(2):150–155
Kohan DE, Padilla E, Hughes AK (1993) Endothelin B receptor mediates ET-1 effects on cAMP and PGE2 accumulation in rat IMCD. Am J Physiol 265(5 Pt 2):F670–F676
Kohan DE, Inscho EW, Wesson D, Pollock DM (2011) Physiology of endothelin and the kidney. Compr Physiol 1(2):883–919. doi:10.1002/cphy.c100039
Kurihara Y, Kurihara H, Suzuki H, Kodama T, Maemura K, Nagai R, Oda H, Kuwaki T, Cao WH, Kamada N (1994) Elevated blood pressure and craniofacial abnormalities in mice deficient in endothelin-1. Nature 368(6473):703–710. doi:10.1038/368703a0
Kurihara Y, Kurihara H, Maemura K, Kuwaki T, Kumada M, Yazaki Y (1995a) Impaired development of the thyroid and thymus in endothelin-1 knockout mice. J Cardiovasc Pharmacol 26(Suppl 3):S13–S16
Kurihara Y, Kurihara H, Oda H, Maemura K, Nagai R, Ishikawa T, Yazaki Y (1995b) Aortic arch malformations and ventricular septal defect in mice deficient in endothelin-1. J Clin Invest 96 (1):293–300. doi:10.1172/JCI118033
Kuwaki T, Cao WH, Kurihara Y, Kurihara H, Ling GY, Onodera M, Ju KH, Yazaki Y, Kumada M (1996) Impaired ventilatory responses to hypoxia and hypercapnia in mutant mice deficient in endothelin-1. Am J Physiol 270(6 Pt 2):R1279–R1286
Kuwaki T, Ling GY, Onodera M, Ishii T, Nakamura A, Ju KH, Cao WH, Kumada M, Kurihara H, Kurihara Y, Yazaki Y, Ohuchi T, Yanagisawa M, Fukuda Y (1999) Endothelin in the central control of cardiovascular and respiratory functions. Clin Exp Pharmacol Physiol 26 (12):989–994
Kuwaki T, Ishii T, Ju K, Yanagisawa M, Fukuda Y (2002) Blood pressure of endothelin-3 null (-/-) knockout mice and endothelin A receptor null (-/-) knockout mice under anaesthesia. Clin Sci (Lond) 103(Suppl 48):48S–52S. doi:10.1042/CS103S048S
Liefeldt L, Böcker W, Schönfelder G, Zintz M, Paul M (1995) Regulation of the endothelin system in transgenic rats expressing the human endothelin-2 gene. J Cardiovasc Pharmacol 26(Suppl 3):S32–S33
Liefeldt L, Schönfelder G, Böcker W, Hocher B, Talsness CE, Rettig R, Paul M (1999) Transgenic rats expressing the human ET-2 gene: a model for the study of endothelin actions *in vivo*. J Mol Med (Berl) 77(7):565–574
Ling GY, Cao WH, Onodera M, Ju KH, Kurihara H, Kurihara Y, Yazaki Y, Kumada M, Fukuda Y, Kuwaki T (1998) Renal sympathetic nerve activity in mice: comparison between mice and rats and between normal and endothelin-1 deficient mice. Brain Res 808(2):238–249
Mann JF, Green D, Jamerson K, Ruilope LM, Kuranoff SJ, Littke T, Viberti G, ASCEND Study Group (2010) Avosentan for overt diabetic nephropathy. J Am Soc Nephrol 21(3):527–535. doi:10.1681/ASN.2009060593
Mercer EH, Hoyle GW, Kapur RP, Brinster RL, Palmiter RD (1991) The dopamine beta-hydroxylase gene promoter directs expression of E. coli lacZ to sympathetic and other neurons in adult transgenic mice. Neuron 7(5):703–716
Mironova E, Peti-Peterdi J, Bugaj V, Stockand JD (2011) Diminished paracrine regulation of the epithelial Na+ channel by purinergic signaling in mice lacking connexin 30. J Biol Chem 286 (2):1054–1060. doi:10.1074/jbc.M110.176552
Morita H, Kurihara H, Kurihara Y, Shindo T, Kuwaki T, Kumada M, Yazaki Y (1998) Systemic and renal response to salt loading in endothelin-1 knockout mice. J Cardiovasc Pharmacol 31 (Suppl 1):S557–S560
Nakano D, Pollock DM (2009) Contribution of endothelin A receptors in endothelin 1-dependent natriuresis in female rats. Hypertension 53(2):324–330. doi:10.1161/HYPERTENSIONAHA. 108.123687

Nakano D, Pollock JS, Pollock DM (2008) Renal medullary ETB receptors produce diuresis and natriuresis via NOS1. Am J Physiol Renal Physiol 294(5):F1205–F1211. doi:10.1152/ajprenal.00578.2007

Nelson RD, Stricklett P, Gustafson C, Stevens A, Ausiello D, Brown D, Kohan DE (1998) Expression of an AQP2 Cre recombinase transgene in kidney and male reproductive system of transgenic mice. Am J Physiol 275(1 Pt 1):C216–C226

Plato CF, Pollock DM, Garvin JL (2000) Endothelin inhibits thick ascending limb chloride flux via ET(B) receptor-mediated NO release. Am J Physiol Renal Physiol 279(2):F326–F333

Pollock DM (2014) 2013 Dahl Lecture: American Heart Association council for high blood pressure research clarifying the physiology of endothelin. Hypertension 63(5):e110–e117. doi:10.1161/HYPERTENSIONAHA.114.02441

Quaschning T, Koçak S, Bauer C, Neumayer HH, Galle J, Hocher B (2003) Increase in nitric oxide bioavailability improves endothelial function in endothelin-1 transgenic mice. Nephrol Dial Transplant 18(3):479–483

Quaschning T, Voss F, Relle K, Kalk P, Vignon-Zellweger N, Pfab T, Bauer C, Theilig F, Bachmann S, Kraemer-Guth A, Wanner C, Theuring F, Galle J, Hocher B (2007) Lack of endothelial nitric oxide synthase promotes endothelin-induced hypertension: lessons from endothelin-1 transgenic/endothelial nitric oxide synthase knockout mice. J Am Soc Nephrol 18(3):730–740. doi:10.1681/ASN.2006050541

Quaschning T, Voss F, Herzfeld S, Relle K, Kalk P, Godes M, Pfab T, Kraemer-Guth A, Bonz AW, Theuring F, Galle J, Hocher B (2008) Lack of iNOS impairs endothelial function in endothelin-1 transgenic mice. Kidney Blood Press Res 31(2):127–134. doi:10.1159/000124285

Saleh MA, Boesen EI, Pollock JS, Savin VJ, Pollock DM (2010) Endothelin-1 increases glomerular permeability and inflammation independent of blood pressure in the rat. Hypertension 56(5):942–949. doi:10.1161/HYPERTENSIONAHA.110.156570

Shindo T, Kurihara H, Maemura K, Kurihara Y, Ueda O, Suzuki H, Kuwaki T, Ju KH, Wang Y, Ebihara A, Nishimatsu H, Moriyama N, Fukuda M, Akimoto Y, Hirano H, Morita H, Kumada M, Yazaki Y, Nagai R, Kimura K (2002) Renal damage and salt-dependent hypertension in aged transgenic mice overexpressing endothelin-1. J Mol Med (Berl) 80(2):105–116. doi:10.1007/s00109-001-0284-4

Spence S, Anderson C, Cukierski M, Patrick D (1999) Teratogenic effects of the endothelin receptor antagonist L-753,037 in the rat. Reprod Toxicol 13(1):15–29

Stockand JD, Mironova E, Bugaj V, Rieg T, Insel PA, Vallon V, Peti-Peterdi J, Pochynyuk O (2010) Purinergic inhibition of ENaC produces aldosterone escape. J Am Soc Nephrol 21(11):1903–1911. doi:10.1681/ASN.2010040377

Stricklett PK, Taylor D, Nelson RD, Kohan DE (2003) Thick ascending limb-specific expression of Cre recombinase. Am J Physiol Renal Physiol 285(1):F33–F39. doi:10.1152/ajprenal.00366.2002

Stuart D, Rees S, Woodward SK, Koesters R, Strait KA, Kohan DE (2012) Disruption of the endothelin A receptor in the nephron causes mild fluid volume expansion. BMC Nephrol 13:166. doi:10.1186/1471-2369-13-166

Stuart D, Chapman M, Rees S, Woodward S, Kohan DE (2013) Myocardial, smooth muscle, nephron, and collecting duct gene targeting reveals the organ sites of endothelin A receptor antagonist fluid retention. J Pharmacol Exp Ther 346(2):182–189. doi:10.1124/jpet.113.205286

Ujiie K, Terada Y, Nonoguchi H, Shinohara M, Tomita K, Marumo F (1992) Messenger RNA expression and synthesis of endothelin-1 along rat nephron segments. J Clin Invest 90(3):1043–1048. doi:10.1172/JCI115918

Yanagisawa H, Yanagisawa M, Kapur RP, Richardson JA, Williams SC, Clouthier DE, de Wit D, Emoto N, Hammer RE (1998) Dual genetic pathways of endothelin-mediated intercellular signaling revealed by targeted disruption of endothelin converting enzyme-1 gene. Development 125(5):825–836

Yanagisawa H, Hammer RE, Richardson JA, Emoto N, Williams SC, Takeda S, Clouthier DE, Yanagisawa M (2000) Disruption of ECE-1 and ECE-2 reveals a role for endothelin-converting enzyme-2 in murine cardiac development. J Clin Invest 105(10):1373–1382. doi:10.1172/JCI7447

Yang LL, Gros R, Kabir MG, Sadi A, Gotlieb AI, Husain M, Stewart DJ (2004) Conditional cardiac overexpression of endothelin-1 induces inflammation and dilated cardiomyopathy in mice. Circulation 109(2):255–261. doi:10.1161/01.CIR.0000105701.98663.D4

Zhao XS, Pan W, Bekeredjian R, Shohet RV (2006) Endogenous endothelin-1 is required for cardiomyocyte survival *in vivo*. Circulation 114(8):830–837. doi:10.1161/CIRCULATIONAHA.105.577288

Chapter 8
Go with the Flow: Fluid Roles for miRNAs in Vertebrate Osmoregulation

Alex S. Flynt and James G. Patton

Abstract MicroRNAs are a family of small RNAs that regulate gene expression post-transcriptionally. By regulating the expression of multiple genes that mediate salt and water balance, miRNAs enable precise control over osmoregulatory processes in vertebrates. Differential expression of miRNAs and divergent mRNA targeting have allowed for adaptation of osmoregulatory tissues during vertebrate evolution. Interestingly, only a small number of mRNA target relationships have been maintained over the millennia, indicating that gain and loss of miRNA/mRNA networks have enabled species-specific osmoregulation.

Keywords miRNA • Osmoregulation • Evolution • Kidney

8.1 Introduction

Homeostasis is central to survival, and animals have evolved elaborate systems for maintaining water and salt balance. In most vertebrates osmoregulation is, at least in part, carried out by the renal system and is linked to excretion of nitrogenous waste. Configuration of the excretory system varies greatly between different classes of vertebrates reflecting the influence of environmental adaptations. In terrestrial species the renal system is the primary site of osmoregulation while in aquatic vertebrates the renal system functions in tandem with the gills, gut, and skin. In this chapter we will explore how posttranscriptional gene regulation mediated by microRNAs (miRNAs) has contributed to the shaping of water and salt homeostasis systems in vertebrates during their evolution. Understanding the origins of these regulatory relationships will highlight how important mechanisms involved in normal physiology and/or disease processes were established.

A.S. Flynt (✉)
Department of Biological Sciences, University of Southern Mississippi, Johnson Science Tower 512B, 118 College Drive, Box 5018, Hattiesburg, MS 39406, USA
e-mail: Alex.Flynt@USM.edu

J.G. Patton (✉)
Department of Biological Sciences, Vanderbilt University, 2325 Stevenson Center, Box 1820 Station B, Nashville, TN 37235, USA
e-mail: James.G.Patton@Vanderbilt.edu

K.A. Hyndman, T.L. Pannabecker (eds.), *Sodium and Water Homeostasis*, Physiology in Health and Disease, DOI 10.1007/978-1-4939-3213-9_8

8.2 Basic Mechanism of miRNA-Mediate Gene Regulation

A common mode of post-transcriptional gene regulation in multicellular organisms is through the activity of a class of endogenous small regulatory RNA called miRNAs (Bartel 2004). In animal genomes there are hundreds of individual miRNA expressing loci (Griffiths-Jones et al. 2006). Regulation of gene expression by these molecules is widespread, as a majority of protein-coding transcripts are potentially targeted by miRNAs (Friedman et al. 2009). Unsurprisingly, miRNAs contribute significantly to animal development and physiology. Mature miRNAs are ~22 nt long and associate with target mRNAs through partially complementary base-pairing often observed between nucleotides (nt) 2–8 of the miRNA and mRNA (Lagos-Quintana et al. 2001; Lai 2002). This is referred to as the seed region of the miRNA. Typically, miRNA-binding sites are located in the 3′ UTR of mRNAs. Many groups have developed algorithms to locate these sites using genome sequences and gene annotation (Betel et al. 2008; Enright et al. 2003; Friedman et al. 2009; Grun et al. 2005; John et al. 2004; Kiriakidou et al. 2004).

miRNAs originate from longer double-stranded RNA (dsRNA), hairpin precursors (Fig. 8.1) (Kim et al. 2009). Pre-miRNA hairpins are encoded in a variety of genomic contexts, but typically are transcribed by RNA Polymerase II (Lee et al. 2004). For this reason, expression of miRNAs is subject to the transcriptional control schemes typical for mRNAs. Canonical miRNA biogenesis involves sequential cleavages by the dsRNA specific, RNase III enzymes Drosha and Dicer (Murchison and Hannon 2004). Initial nuclear processing of miRNA transcripts involves the "microprocessor complex" which contains Drosha and it's partner Dgcr8 (Gregory et al. 2004; Lee et al. 2003). The microprocessor cleaves the base of the precursor releasing it from the host transcript. Dicer completes maturation of the miRNA in the cytoplasm, producing a small duplex of dsRNA, one strand of which is then loaded into an Argonaute protein (Ago) (Ketting et al. 2001; Zamore et al. 2000).

Binding of a miRNA drives the recruitment of a complex of effector proteins called the RNA induced Silencing Complex (RISC) that inhibits the expression of targeted transcripts (Fig. 8.1) (Martinez and Tuschl 2004; Pham et al. 2004). The core component of RISC are Ago proteins (Sontheimer and Carthew 2004). These proteins directly bind small RNAs when they base-pair with target mRNAs. If an Ago-complexed small RNA pairs perfectly with a transcript, the result is cleavage of the mRNA by the Ago protein piwi domain (Zeng et al. 2003). However, for the majority of miRNAs, pairing is imperfect and does not trigger the immediate destruction of the target. While the precise effect of miRNA targeting is dependent on the life cycle of an mRNA, frequently GW182 proteins are recruited which in turn drive localization of mRNAs to processing bodies (P-bodies) (Liu et al. 2005; Rehwinkel et al. 2005). These cytoplasmic foci are the site of mRNA decay enzymes responsible for decapping and deadenylation. miRNAs also play a role in stress response, associating with stress granules to participate in global changes in translation patterns that occur during stress (Leung et al. 2006).

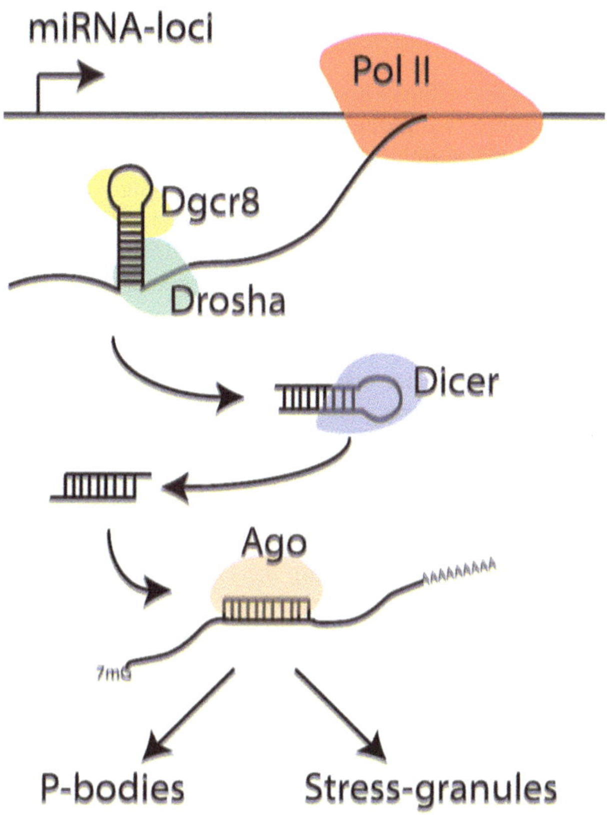

Fig. 8.1 miRNA biogenesis. The miRNA biogenesis pathway and fate of targeted miRNAs (Bartel 2004). miRNA genes are transcribed by DNA Polymerase II and primary transcripts are processed in the nucleus by the Microprocessor Complex containing the RNAse-III like enzyme Drosha and DGCR8 (DiGeorge Syndrome Critical Region 8). Precursor transcripts are then exported to the cytoplasm where processing by the enzyme Dicer releases double stranded RNAs, one strand of which then associates with one or more Argonaute (Ago) proteins. Pairing with target mRNAs leads to degradation, most often associated with Processing bodies (P-bodies)

miRNA biogenesis and function is a deeply conserved process, and appears to be similar in all vertebrates (Carmell et al. 2002; Waterhouse et al. 2011). Sequencing of various vertebrate genomes has revealed that very few changes in the composition of miRNA maturing factors have occurred through evolution. All vertebrates appear to possess a single Dicer, Drosha, and Dgcr8. The number of Ago proteins can be variable between species, but this mostly reflects changes in Ago proteins that bind an unrelated class of small RNAs called piRNAs (Faehnle and Joshua-Tor 2007). The number of miRNA binding Ago proteins is strongly conserved–four homologs have been found in all the sequenced vertebrates. Thus, the rules and mechanisms that govern miRNA function in a vertebrate order will likely be similar in distantly related species.

8.3 miRNA Species Found in Osmoregulatory Cell Types

miRNAs are a fairly heterogeneous group of gene products. Roughly they can be divided into two groups. The first are conserved miRNAs, some of which can be found in all animals and have origins before the divergence of protostomes and deuterostomes (Meunier et al. 2013; Mohammed et al. 2013; Ruby et al. 2007). Conserved miRNAs are exclusively produced by the canonical pathway described above. They are also highly expressed, and/or highly tissue specific. As such, conserved miRNAs are likely to more substantially contribute to regulation of gene expression. The second group consists of newly-evolved miRNAs. These RNAs are generally weakly expressed, and lack refinement in expression patterns or clear target relationships (Farh et al. 2005). In general, they are more diverse in terms of biogenesis and may be produced by alternate maturation pathways (Berezikov et al. 2007).

Both varieties of miRNAs likely contribute to regulation of gene expression in cells carrying out excretory processes. However, conserved miRNAs are probably more critical to maintaining water and salt balance, due to their importance in establishing and maintaining the identities of osmoregulatory cells. Additionally, they may contribute to physiological processes in excretory cells. In addition to conserved functions, these miRNAs may have adopted new regulatory roles due to recently established target gene relationships (Friedman et al. 2009). While newly-evolved miRNAs rarely play significant roles in animal's biology, they might be involved in highly species specific phenomena. Acquisition of a miRNA-mediated regulatory event by a newly evolved miRNA may have been triggered by the need to rapidly adapt to a new environment.

miRNAs are frequently expressed in a cell type specific manner, and there are many cases where restriction of miRNA expression to a particular cell type is highly conserved (Aboobaker et al. 2005; Lagos-Quintana et al. 2002; Sempere et al. 2006). For example, *miR-1* is expressed exclusively in muscles, and *miR-124* in the central nervous system in diverse genera. The same is true of kidney/osmoregulatory tissue. miRNAs have been most extensively profiled in mammals due to multiple groups using high through-put sequencing of miRNA cDNA libraries to identify expressed RNAs (Landgraf et al. 2007). Comparison of expression in humans, mice and rats reveals that roughly 70 miRNAs are highly expressed in the various cell types of the kidney (Fig. 8.2) (Saal and Harvey 2009). While this depth of information is unmatched in such a tissue specific manner in non-mammal vertebrates, efforts have also been made to use *in situ* hybridization to identify miRNA expression patterns. Some of the miRNAs found in the mammalian kidney were profiled in developing chicken embryos (Darnell et al. 2006). A number of these miRNAs were found to also be expressed in the chicken kidney, either specifically or as part of a broader expression pattern: *miR-30*, *miR-34*, *miR-22*, *miR-92*, *miR-107*, *miR-200a/b*, *miR-107*, *miR-130a*, *miR-146*, *miR-16* (Fig. 8.2). Despite the considerable evolutionary distance between mammals and birds (300 million years ago), all these miRNAs are renal tissue expressed. This is

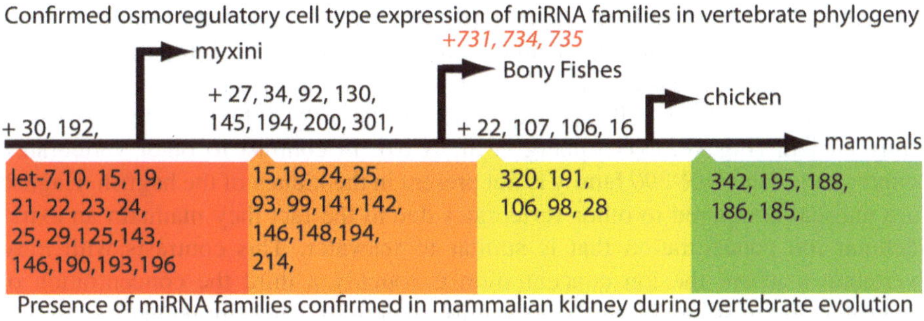

Fig. 8.2 Conservation of kidney-expressed miRNAs. Evolutionary origins of miRNAs expressed in the mammalian kidney in relation to major divergences in the vertebrate lineage (Saal and Harvey 2009). miRNAs listed *above the black line* have been experimentally determined to be expressed in tissues involved in osmoregulation (Darnell et al. 2006; Kloosterman et al. 2006; Wienholds et al. 2005). *Below the black line* are miRNAs expressed in the mammalian kidney that have not been implicated in related cell types in lower vertebrates. Zebrafish specific miRNAs (miR-731, 734, and 735) are indicated above the Bony Fishes in *red*

particularly remarkable considering the different nitrogenous waste excretory systems of uric acid formation in birds and urea in mammals.

In fishes and other lower vertebrates osmoregulation occurs in multiple organs. In adult fishes, maintenance of salt/water balance is maintained through the cooperative action of the gills, gut and renal system (Perry et al. 2003). Environmental pressures are much greater for these animals due to direct exposure to the aquatic environment. Depending on the prevailing concentrations of salts, fishes adjust their physiology. In fresh water environments fishes are constantly losing solutes to the environment. To deal with this, freshwater fishes actively take up ions in their gills in a process that is coupled to exchange of bicarbonate and hydrogen ions. A similar activity occurs in the gut, which is linked to "drinking" behavior to provide an opportunity to acquire additional ions from the environment. The renal system of freshwater fishes produces dilute urine to expel excess water. Saltwater fishes face the opposite problem and excrete excess ions at the gills (Evans et al. 2005). In embryonic fishes osmoregulation is mediated by epidermal cells called ionocytes that are capable of carrying out ion exchange processes of the adult gills (Varsamos et al. 2005).

Interestingly, gill cells and ionocytes resemble the osmoregulatory cells of the kidney of terrestrial vertebrates, suggesting a similar phylogenetic ontogeny (Perry et al. 2003; Varsamos et al. 2005). They are mitochondrial rich, exhibit similar ion transport activities, and have "brush border" structures composed of glycoproteins. Analysis of miRNA expression in the gills, ionocytes, and the renal system of embryonic zebrafish reveals the following RNAs can be found in these tissues: *miR-30a/b*, *miR-34b*, *miR-200 family*, *miR-192*, *miR-194*, *miR-145*, *miR-92b*, *miR-101b*, *miR-130a/b/c*, *miR-301b*, *miR-27c*, *miR-733*, *miR-731*, and *miR-735* (Fig. 8.2) (Kloosterman et al. 2006; Wienholds et al. 2005). While some of these miRNAs are specific to zebrafish such as *miR-731*, *miR-733*, and *miR-735* others are present in higher vertebrates. Indeed, the majority of miRNAs exhibiting osmoregulatory cell expression in zebrafish are shared with higher vertebrates. Furthermore, investigation of miRNA expression in a basal vertebrate, the hagfish, reveals that the kidney/osmoregulatory cell type expression of the *miR-30* family and *miR-192* is truly ancient (Fig. 8.2) (Heimberg et al. 2010). In contrast to other vertebrates, expression of the *miR-200* family is not present in the kidney of the hagfish. Hagfish are unusual compared to other vertebrates due to the fact they maintain an extracellular ion concentration that is similar to seawater. This contrasts with most vertebrates where the ion concentration is roughly a third the concentration of seawater. Acquisition of *miR-200* family expression in osmoregulatory tissues might be a component of the change in gene regulation that led to the establishment of ion concentrations seen in the extracellular fluids of most vertebrates.

8.4 Comparative Genetics of miRNAs in Osmoregulation

Regulation by miRNAs touches many aspects of cell biology. The most relevant aspects of miRNA activity to maintaining water and salt balance are their roles in the specification and maintenance of cell types that carry out osmoregulation and regulation of physiology. miRNAs operate in many ways to affect cell behavior but are frequently found to either target a critical gene to change cell behavior/identity or by affecting a battery of target transcripts to stabilize gene expression profiles (Fig. 8.3) (Flynt and Lai 2008). When interpreting phenotypes caused by manipulation of miRNA function it is important to consider these two modes of action.

The relationship between miRNA and target gene expression is another important aspect of miRNA behavior that informs experiments to discern miRNA function. Gene expression analysis typically shows miRNAs and their target transcripts are rarely co-expressed, as found in different regions of the mammalian kidney (Hornstein and Shomron 2006; Mladinov et al. 2013). Profiling of rat miRNAs and their targets in the proximal convoluted tubule and the medullary thick ascending limb showed a reverse correlation in expression patterns (Mladinov et al. 2013). While these structures are of similar developmental origins, they have distinct roles in maintaining water and salt balance. Here, miRNAs seem to be playing an important role in maintaining cell specific function. Furthermore, the mRNA targets inversely expressed with their respective miRNAs are ion transporters and exchangers, demonstrating the direct integration of miRNAs into the capacity of kidney cell types to direct movement of specific solutes.

The significance of miRNAs to renal development and function came from studying the effects of knocking out the miRNA biogenesis factors: Dicer, Drosha, or Dgcr8 in vertebrates. In mice the use of conditional knockout technology showed that miRNA function was critical in both development and in the maintenance of renal function (Bhatt et al. 2011; Harvey et al. 2008; Shi et al. 2008). Dramatic effects were found in tissue-specific knockouts that targeted developing kidney, podocytes, and juxtaglomerular cells (Chu et al. 2014). When Dicer or Drosha was

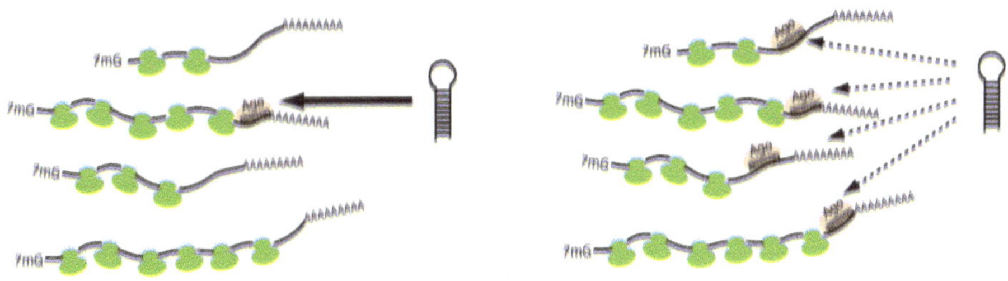

Fig. 8.3 Modes of regulation of gene expression by miRNAs. On the *left* targeting of multiple transcripts is frequently important to the maintenance of cell fate. In dicer mutant kidney glomeruli, many transcripts are derepressed due to loss of *miR-30* function (Shi et al. 2008). The *right panel* shows that miRNAs also act on single targets in a phenotypically significant manner. Regulation of Lhx1 by *miR-30* family members is critical to pronephros development in xenopus (Agrawal et al. 2009)

ablated in the podocytes of adult renal glomeruli, the resulting cells exhibited several morphological defects, suggesting a role for miRNAs in maintenance of tissue specificity and function (Shi et al. 2008). These animals survived postnatally, but succumbed prematurely due to kidney failure. Loss of Dicer function altered the expression of several hundred mRNAs, 25 % of which were predicted targets of *miR-30* family members (Fig. 8.3) (Shi et al. 2008). Most of the defects seen in the mutant podocytes were primarily attributed to the loss of *miR-30* family function. This is one of the most ancient excretory tissue expressed miRNAs, and consistent with observations in mice, appears to be essential for kidney function in other vertebrates. In Xenopus, inhibition of Dicer and Dgcr8 caused pleiotropic defects in pronephros development (Agrawal et al. 2009). Similar developmental abnormalities were phenocopied in treatments that specifically knocked down the *miR-30* family. However, this effect involves the targeting of a single gene, Lhx1. This demonstrates that single miRNA families may act in multiple, context-dependent ways (Fig. 8.3). In some settings, like the adult podocyte, they are responsible for calibrating gene expression, while at other times targeting a single transcript. Interestingly, *miR-30* targets Lhx1 in both mice and frogs, suggesting a conserved role for this family during nephrogenesis. *miR-30* family members are also expressed in the excretory tissues of fishes, including the nephros, gills, and ionocytes. In zebrafish Lhx1 is a predicted target of *miR-30* (Friedman et al. 2009). It remains to be seen in which tissue (s) this interaction affects development.

Another highly conserved kidney expressed miRNA, *miR-192*, is likewise involved in kidney homeostasis (Table 8.1) (Mladinov et al. 2013). Inhibition of *miR-192* in mice prevented adjustment of urine output in mice fed a high salt diet. Furthermore, when mice were fed a low salt diet, expression of *miR-192* decreased by half. *miR-192* targeting of Na+/K+-ATPase beta1 (Atp1b1) appears to be responsible for mediating the effects of this miRNA on water and salt homeostasis. Unlike the situation with the *miR-30* family, potential regulation of Atp1b1 is only conserved to chicken (Friedman et al. 2009). This highlights a common theme with

Table 8.1 miRNA families that show conserved expression in osmoregulatory tissues in humans and zebrafish that have been implicated in water and salt balance by functional studies

miRNA family	Total predicted targets (fishes/ human)	Deeply conserved targets	Role in osmoregulation
miR-30	789/1040	LHX1, PPARGC, GRM3, CUL2, NKX2.2a, ACTC1, MEX3b, EPC2, GALNT1, DAGLA	Essential role in nephrogenesis in Xenopus, responsible for maintenance of mammalian renal glomeruli, participates in osmotic stress in tilapia
miR-192	790/982	ARIADNE	Targets ATP1B1 in mammalian kidney tubules
miR-200	2016/2370	ZFPM2a	Regulates NFAT5, involved in osmotic stress in zebrafish

The total miRNA targets predicted by the microcosm algorithm for zebrafish and humans are shown in column 2. The handful of deeply conserved miRNA-target interactions present in zebrafish and humans listed in column 3

miRNA regulation that target gene regulation seems to be readily rewired by evolution. In this way, despite ancient tissue specificity, miRNAs can take on new functions to suit changing environments and needs. Thus, it would seem that miRNAs are on the forefront of evolution, providing organisms with genetic novelty. Cogent to this point, investigation of *miR-200* function in zebrafish ionocytes showed a role in responding to osmotic stress by regulation of a cyto-skeletal adapter protein (Table 8.1) (Flynt et al. 2009). This particular regulatory interaction however does not extend to other organisms, including other fish species (Friedman et al. 2009). The speed at which these regulatory events arise during evolution will likely mirror the adaptation of individual species to environmental pressures.

The inconsistency in the conservation of miRNA function complicates comparisons of studies from lower animals and mammals. While potential targeting relationships can be predicted by computational methods, functional studies will likely be necessary to draw precise conclusions concerning the exact behavior of miRNAs. Regulation of a transcript in one species might be critical to the role played by a given miRNA, but might be reduced in significance in a different species. To highlight the dramatic changes in miRNA function that have occurred during vertebrate evolution we compared target gene predictions conserved from fishes to man for miRNAs with known osmoregulatory function (Table 8.1) (Betel et al. 2008; Enright et al. 2003; John et al. 2004). Contrary to expectations, among all the targets of osmoregulation-involved miRNAs shared between fishes and man, none have been implicated directly in maintenance of salt and water balance. Lhx1 is the one exception due to its role in guiding kidney development. This likely indicates that the primary conserved function of these miRNAs mostly involves cell specification.

Functionally, miRNAs would be expected to play many roles in the regulation of osmoregulatory processes. Computational prediction of target relationships suggests that direct regulation of water and salt balance by miRNAs is not well-conserved between species. This likely is a result of miRNAs being co-opted for the osmoregulatory needs of species in different clades. If there is a conserved function of miRNAs among all vertebrates it likely is in the establishment and maintenance of gene expression profiles in osmoregulatory cell types. Beyond the context of target gene repression, miRNAs also are deeply involved in the post-transcriptional response to stress.

8.5 miRNAs in Stress

A common impression of regulation of gene expression by miRNAs is that they are fast-acting, compared to production of new proteins by translation. Transcription and maturation of miRNAs may occur more quickly than mRNAs; however, the effects on the steady-state levels of the protein encoded by miRNA-target genes may take several days to manifest (Hausser et al. 2013). There are two major barriers that limit the rate of change in gene activity caused by miRNA targeting (Hausser et al. 2013). The first is the loading of miRNAs into Ago proteins. Several studies have suggested that the abundance of Ago proteins is tightly equilibrated to the presence of miRNAs. Agos that are not associated with a mature small RNA become destabilized (Smibert et al. 2013). The consequence of tying Ago levels to the amount of small RNA present is that "empty" Ago proteins will not be available to accept newly generated small RNAs. In order for newly matured miRNA to associate with an Ago protein, a previously bound small RNA must be turned over. The other major barrier to manifestation of the effects of miRNA gene regulation is protein stability. If a protein's turnover rate is slow, it may not be possible to observe a decrease in expression triggered by miRNA targeting. Thus, transcription and maturation of miRNAs in response to acute stress is probably unlikely.

Despite the fact that miRNAs are not fast acting when produced *de novo*, they have been implicated as important to mediating response to stress. This likely is a consequence of the involvement of miRNAs with stress granules (Fig. 8.4) (Leung et al. 2006). When exposed to osmotic stress, cells alter their ion channel activity and cytoskeletal dynamics (Nishimura and Fan 2003). These changes lead to disruption of intercellular ion concentrations, and membrane ionic strength. The longer term response involves the activity of NFAT5/TonEBP (Miyakawa et al. 1999). Prolonged osmotic stress induces translocation of the transcription factor NFAT5/TonEBP, which triggers production of enzymes that produce osmolytes, such as sorbitol or glycerophosphorylcholine. Sufficient accumulation of osmolytes to balance external solute concentrations may take 12–24 h. During this time the cell shuts down translation, leading to altered trafficking of mRNAs to stress granules (Anderson and Kedersha 2008). Many varieties of RNA binding proteins accumulate in stress granules in response to stress, some of which are

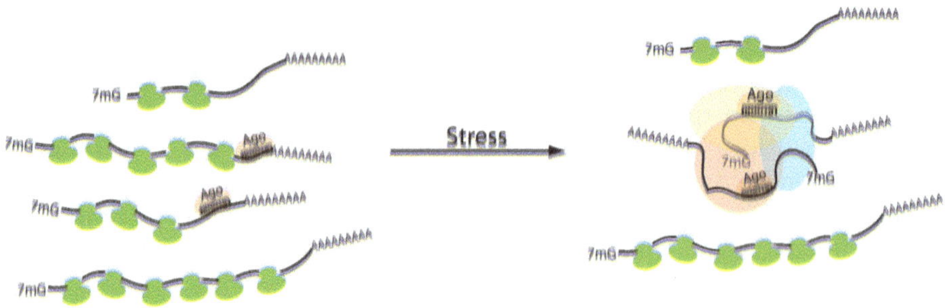

Fig. 8.4 miRNAs and stress. During stress, miRNA targeted transcripts are trafficked to stress granules (Anderson and Kedersha 2008; Leung et al. 2006)

members of the Ago family. This suggests that mRNAs actively targeted by a miRNA will be selectively trafficked to stress granules to eliminate production of new proteins (Fig. 8.4) (Leung et al. 2006). This also predicts that genes in the stress response will likely have fewer miRNA binding sites compared to other transcripts.

The immediate response to stress involves proteins of the Aquaporin (AQP) family (Arima et al. 2003; Nishimura and Fan 2003). These transmembrane channels insert into membranes, and are selective to the passive diffusion of water. Investigation of the abundance of miRNA binding sites in members of this gene family in humans and zebrafish reveals that many members have fewer binding sites compared to average (Fig. 8.5). This does not appear to be simply a function of this gene family having shorter 3′UTRs. The median 3′UTR length of AQPs in both humans and zebrafish is very close to the average length of 3′UTRs (~800 nt) (Mignone and Pesole 2001). These relationships suggest that members of the AQP gene family have selectively avoided accumulation of miRNA binding sites in their 3′UTRs to avoid stress granule trafficking. Unlike the majority of miRNA-target gene relationships this avoidance of target site acquisition is a deeply conserved aspect of the response to osmotic stress. Further investigation of genes involved in solute transport that are depleted of miRNA binding sites may reveal factors involved in the immediate response to osmotic stress. Here, a comparative approach may be particularly fruitful, considering the behavior of the AQP family.

miRNA regulation is also highly influenced by RNA-binding protein dynamics. This can result in selective changes in miRNA targeting. In mammals NFAT5 is typically weakly expressed, but is upregulated in response to stress (Table 8.1) (Huang et al. 2011). This effect is miRNA dependent. Loss of Dicer function or of *miR-200b* and *miR-717* results in elevated levels of NFAT5. Interestingly, the expression of both these miRNAs is modulated by stress, suggesting a role in limiting NFAT5 function in the opposite situation. These target interactions are not conserved indicating that this behavior may be an innovation in higher vertebrates.

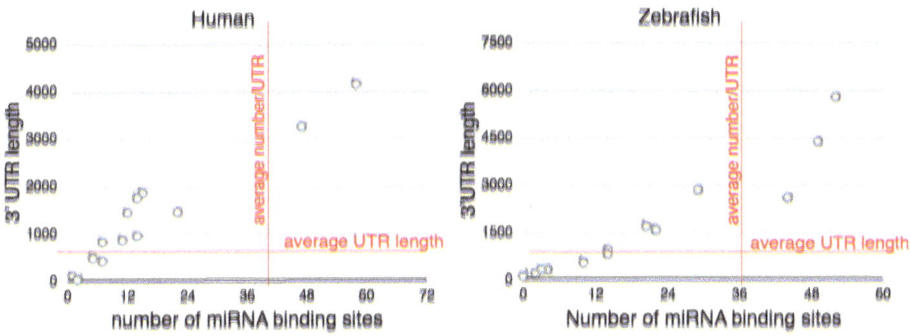

Fig. 8.5 miRNA and 3′ UTRsLength of 3′UTRs and number of miRNA binding sites in members of the human and zebrafish aquaporin (AQP) gene families. The red lines indicate the average length and number of sites in UTRs (Mignone and Pesole 2001; Zare et al. 2014). Note that AQP mRNAs are equally distributed in relation to the mean UTR length. This contrasts with the number of miRNA binding sites, where few AQP transcripts have greater than the average number of miRNA binding sites

8.6 Perspectives

miRNAs are versatile regulators of gene expression, contributing significantly to osmoregulatory processes in vertebrates. In tissues that mediate salt and water balance a host of miRNAs are expressed. Indeed, some of these miRNAs have participated in the shaping of osmoregulatory tissues from early in vertebrate evolution. The precise functions of these genes have become divergent in different classes, where only a handful of target relationships have been maintained over the millennia. The unique environments and stresses faced by lower vertebrates, particularly fishes, likely selected for miRNA functions and targets that appear to not be conserved in humans. Appreciation of where during vertebrate evolution miRNAs gained or lost a regulatory relationship will help identify species-specific osmoregulatory processes.

References

Aboobaker AA, Tomancak P, Patel N, Rubin GM, Lai EC (2005) Drosophila microRNAs exhibit diverse spatial expression patterns during embryonic development. Proc Natl Acad Sci USA 102(50):18017–18022. doi:10.1073/pnas.0508823102

Agrawal R, Tran U, Wessely O (2009) The miR-30 miRNA family regulates Xenopus pronephros development and targets the transcription factor Xlim1/Lhx1. Development 136 (23):3927–3936

Anderson P, Kedersha N (2008) Stress granules: the Tao of RNA triage. Trends Biochem Sci 33 (3):141–150

Arima H, Yamamoto N, Sobue K, Umenishi F, Tada T, Katsuya H, Asai K (2003) Hyperosmolar mannitol stimulates expression of aquaporins 4 and 9 through a p38 mitogen-activated protein kinase-dependent pathway in rat astrocytes. J Biol Chem 278(45):44525–44534

Bartel D (2004) MicroRNAs: genomics, biogenesis, mechanism, and function. Cell 116:281–297
Berezikov E, Chung W-J, Willis J, Cuppen E, Lai EC (2007) Mammalian mirtron genes. Mol Cell 28(2):328–336
Betel D, Wilson M, Gabow A, Marks DS, Sander C (2008) The microRNA.org resource: targets and expression. Nucleic Acids Res 36:D149–D153
Bhatt KM, Mi QS, Dong Z (2011) microRNAs in kidneys: biogenesis, regulation, and pathophysiological roles. Am J Physiol Ren Physiol 300(3):F602–F610
Carmell MA, Xuan Z, Zhang MQ, Hannon GJ (2002) The Argonaute family: tentacles that reach into RNAi, developmental control, stem cell maintenance, and tumorigenesis. Genes Dev 16 (21):2733–2742
Chu JY, Sims-Lucas S, Bushnell DS, Bodnar AJ, Kreidberg JA, Ho J (2014) Dicer function is required in the metanephric mesenchyme for early kidney development. Am J Physiol Ren Physiol 306(7):F764–F772
Darnell DK, Kaur S, Stanislaw S, Konieczka JK, Yatskievych TA, Antin PB (2006) MicroRNA expression during chick embryo development. Dev Dyn 235(11):3156–3165
Enright AJ, John B, Gaul U, Tuschl T, Sander C, Marks DS (2003) MicroRNA targets in drosophila. Genome Biol 5(1):R1
Evans DH, Piermarini PM, Choe KP (2005) The multifunctional fish gill: Dominant site of gas exchange, osmoregulation, acid-base regulation, and excretion of nitrogenous waste. Physiol Rev 85:97–177
Faehnle CR, Joshua-Tor L (2007) Argonautes confront new small RNAs. Curr Opin Chem Biol 11 (5):569–577
Farh KK, Grimson A, Jan C, Lewis BP, Johnston WK, Lim LP, Burge CB, Bartel DP (2005) The widespread impact of mammalian MicroRNAs on mRNA repression and evolution. Science 310(5755):1817–1821
Flynt AS, Lai EC (2008) Biological principles of microRNA-mediated regulation: shared themes amid diversity. Nat Rev Genet 9(11):831–842
Flynt AS, Thatcher EJ, Burkewitz K, Li N, Liu Y, Patton JG (2009) miR-8 microRNAs regulate the response to osmotic stress in zebrafish embryos. J Cell Biol 185(1):115–127
Friedman RC, Farh KK-H, Burge CB, Bartel DP (2009) Most mammalian mRNAs are conserved targets of microRNAs. Genome Res 19(1):92–105
Gregory RI, Yan KP, Amuthan G, Chendrimada T, Doratotaj B, Cooch N, Shiekhattar R (2004) The microprocessor complex mediates the genesis of microRNAs. Nature 432(7014):235–240
Griffiths-Jones S, Grocock RJ, van Dongen S, Bateman A, Enright AJ (2006) miRBase: microRNA sequences, targets and gene nomenclature. Nucleic Acids Res 34(Suppl 1):D140
Grun D, Wang Y-L, Langenberger D, Gunsalus KC, Rajewsky N (2005) microRNA target predictions across seven drosophila species and comparison to mammalian targets. PLoS Comput Biol 1(1), e13
Harvey SJ, Jarad G, Cunningham J, Goldberg S, Schermer B, Harfe BD, McManus MT, Benzing T, Miner JH (2008) Podocyte-specific deletion of dicer alters cytoskeletal dynamics and causes glomerular disease. J Am Soc Nephrol 19(11):2150–2158
Hausser J, Syed AP, Selevsek N, van Nimwegen E, Jaskiewicz L, Aebersold R, Zavolan M (2013) Timescales and bottlenecks in miRNA-dependent gene regulation. Mol Syst Biol 9(1)
Heimberg AM, Cowper-Sal lari R, Sémon M, Donoghue PCJ, Peterson KJ (2010) microRNAs reveal the interrelationships of hagfish, lampreys, and gnathostomes and the nature of the ancestral vertebrate. Proc Natl Acad Sci USA 107:19379–19383
Hornstein E, Shomron N (2006) Canalization of development by microRNAs. Nat Genet 38 (Suppl):S20–S24
Huang W, Liu H, Wang T, Zhang T, Kuang J, Luo Y, Chung SS, Yuan L, Yang JY (2011) Tonicity-responsive microRNAs contribute to the maximal induction of osmoregulatory transcription factor OREBP in response to high-NaCl hypertonicity. Nucleic Acids Res 39 (2):475–485

John B, Enright AJ, Aravin A, Tuschl T, Sander C, Marks DS (2004) Human microRNA targets. PLoS Biol 2(11), e363

Ketting RF, Fischer S, Bernstein E, Sijen T, Hannon GJ, Plasterk R (2001) Dicer functions in RNA interference and in synthesis of small RNA involved in developmental timing in *C. elegans*. Genes Dev 15:2654–2659

Kim VN, Han J, Siomi MC (2009) Biogenesis of small RNAs in animals. Nat Rev Mol Cell Biol 10(2):126–139

Kiriakidou M, Nelson PT, Kouranov A, Fitziev P, Bouyioukos C, Mourelatos Z, Hatzigeorgiou A (2004) A combined computational-experimental approach predicts human microRNA targets. Genes Dev 18(10):1165–1178

Kloosterman WP, Steiner FA, Berezikov E, de Bruijn E, van de Belt J, Verheul M, Cuppen E, Plasterk RHA (2006) Cloning and expression of new microRNAs from zebrafish. Nucleic Acids Res 34(9):2558–2569

Lagos-Quintana M, Rauhut R, Lendeckel W, Tuschl T (2001) Identification of novel genes coding for small expressed RNAs. Science 294(5543):853–858

Lagos-Quintana M, Rauhut R, Yalcin A, Meyer J, Lendeckel W, Tuschl T (2002) Identification of tissue-specific microRNAs from mouse. Curr Biol 12(9):735–739

Lai EC (2002) Micro RNAs are complementary to 3′ UTR sequence motifs that mediate negative post-transcriptional regulation. Nat Genet 30(4):363–364

Landgraf P, Rusu M, Sheridan R, Sewer A, Iovino N, Aravin A, Pfeffer S, Rice A, Kamphorst AO, Landthaler M, Lin C, Socci N, Hermida L, Fulci V, Chiaretti S, Foa R, Schliwka J, Fuchs U, Novosel A, Muller RU, Schermer B, Bissels U, Inman J, Phan Q, Chien M, Weir DB, Choksi R, De Vita G, Frezzetti D, Trompeter H-I, Hornung V, Teng G, Hartmann G, Palkovits M, Di Lauro R, Wernet P, Macino G, Rogler CE, Nagle JW, Ju J, Papavasiliou FN, Benzing T, Lichter P, Tam W, Brownstein MJ, Bosio A, Borkhardt A, Russo JJ, Sander C, Zavolan M, Tuschl T (2007) Mammalian microRNA expression atlas based on small RNA library sequencing. Cell 129(7):1401–1414

Lee Y, Ahn C, Han J, Choi H, Kim J, Yim J, Lee J, Provost P, Radmark O, Kim VN (2003) The nuclear RNase III Drosha initiates microRNA processing. Nature 425(6956):415–419

Lee Y, Kim M, Han J, Yeom KH, Lee S, Baek SH, Kim VN (2004) MicroRNA genes are transcribed by RNA polymerase II. EMBO J 23(20):4051–4060

Leung AKL, Calabrese JM, Sharp PA (2006) Quantitative analysis of Argonaute protein reveals microRNA-dependent localization to stress granules. PNAS 103(48):18125–18130

Liu J, Valencia-Sanchez MA, Hannon GJ, Parker R (2005) MicroRNA-dependent localization of targeted mRNAs to mammalian P-bodies. Nat Cell Biol 7(7):719

Martinez J, Tuschl T (2004) RISC is a 5′ phosphomonoester-producing RNA endonuclease. Genes Dev 18(9):975–980

Meunier J, Lemoine F, Soumillon M, Liechti A, Weier M, Guschanski K, Hu H, Khaitovich P, Kaessmann H (2013) Birth and expression evolution of mammalian microRNA genes. Genome Res 23(1):34–45

Mignone F, Pesole G (2001) mRNA untranslated regions (UTRs). Wiley

Miyakawa H, Woo SK, Dahl SC, Handler JS, Kwon HM (1999) Tonicity-responsive enhancer binding protein, a rel-like protein that stimulates transcription in response to hypertonicity. Proc Natl Acad Sci USA 96(5):2538–2542

Mladinov D, Liu Y, Mattson DL, Liang M (2013) MicroRNAs contribute to the maintenance of cell-type-specific physiological characteristics: miR-192 targets Na+/K+-ATPase β1. Nucleic Acids Res 41(2):1273–1283

Mohammed J, Flynt AS, Siepel A, Lai EC (2013) The impact of age, biogenesis, and genomic clustering on Drosophila microRNA evolution. RNA 19(9):1295–1308

Murchison EP, Hannon GJ (2004) miRNAs on the move: miRNA biogenesis and the RNAi machinery. Curr Opin Cell Biol 16(3):223–229

Nishimura H, Fan Z (2003) Regulation of water movement across vertebrate renal tubules. Comp Biochem Physiol A Mol Integr Physiol 136(3):479–498

Perry SF, Shahsavarani A, Georgalis T, Bayaa M, Furimsky M, Thomas SLY (2003) Channels, pumps, and exchangers in the gill and kidney of freshwater fishes: their role in ionic and acid-base regulation. J Exp Zool A Comp Exp Biol 300A(1):53–62

Pham JW, Pellino JL, Lee YS, Carthew RW, Sontheimer EJ (2004) A Dicer-2-dependent 80s complex cleaves targeted mRNAs during RNAi in Drosophila. Cell 117(1):83–94

Rehwinkel JAN, Behm-Ansmant I, Gatfield D, Izaurralde E (2005) A crucial role for GW182 and the DCP1:DCP2 decapping complex in miRNA-mediated gene silencing. RNA 11 (11):1640–1647

Ruby JG, Stark A, Johnston WK, Kellis M, Bartel DP, Lai EC (2007) Evolution, biogenesis, expression, and target predictions of a substantially expanded set of Drosophila microRNAs. Genome Res 17(12):1850–1864

Saal S, Harvey SJ (2009) MicroRNAs and the kidney: coming of age. Curr Opin Nephrol Hypertens 18(4):317–323

Sempere L, Cole C, McPeek M, Peterson K (2006) The phylogenetic distribution of Meazoan microRNAs: insights into evolutionary complexity and constraint. J Exp Zool 306(B):575–588

Shi S, Yu L, Chiu C, Sun Y, Chen J, Khitrov G, Merkenschlager M, Holzman LB, Zhang W, Mundel P, Bottinger EP (2008) Podocyte-selective deletion of dicer induces proteinuria and glomerulosclerosis. J Am Soc Nephrol 19(11):2159–2169

Smibert P, Yang J-S, Azzam G, Liu J-L, Lai EC (2013) Homeostatic control of Argonaute stability by microRNA availability. Nat Struct Mol Biol 20(7):789–795

Sontheimer EJ, Carthew RW (2004) Molecular biology. Argonaute journeys into the heart of RISC. Science 305(5689):1409–1410

Varsamos S, Nebel C, Charmantier G (2005) Ontogeny of osmoregulation in postembryonic fish: a review. Comp Biochem Physiol A Mol Integr Physiol 141(4):401–429

Waterhouse R, Zdobnov E, Tegenfeldt F, Li J, Kriventseva E (2011) OrthoDB: the hierarchical catalog of eukaryotic orthologs in 2011. Nucleic Acids Res 39:D283–D288

Wienholds E, Kloosterman WP, Miska E, Alvarez-Saavedra E, Berezikov E, de Bruijn E, Horvitz HR, Kauppinen S, Plasterk RHA (2005) MicroRNA expression in zebrafish embryonic development. Science 309(5732):310–311

Zamore PD, Tuschl T, Sharp PA, Bartel DP (2000) RNAi: double-stranded RNA directs the ATP-dependent cleavage of mRNA at 21 to 23 nucleotide intervals. Cell 101(1):25–33

Zare H, Khodursky A, Sartorelli V (2014) An evolutionarily biased distribution of miRNA sites toward regulatory genes with high promoter-driven intrinsic transcriptional noise. BMC Evol Biol 14(1):74

Zeng Y, Yi R, Cullen BR (2003) MicroRNAs and small interfering RNAs can inhibit mRNA expression by similar mechanisms. PNAS 100(17):9779–9784

Chapter 9
MicroRNA and Sodium and Water Balance in Mammals

Maria Angeles Baker, Domagoj Mladinov, and Mingyu Liang

Abstract MicroRNAs are small non-coding RNAs of about 22 nucleotides length that suppress protein expression primarily by complementarity to the 3′ untranslated region of their target messenger RNA. While the 3′ untranslated region is the most known mechanism by which microRNA binds its target messenger RNA, other regions such as the 5′ untranslated region have also been described as microRNA targets. Complementarity to the target is achieved at the seed region of only seven or eight nucleotides, which makes it possible for a single microRNA to target many different messenger RNAs.

MicroRNAs can be expressed in a tissue-specific manner providing a mechanism for regulation of cell type-specific gene expression. In the kidney, for example, microRNAs can regulate kidney epithelial transmembrane transport.

The maintenance of fluid and electrolyte homeostasis is primarily the function of ion channels and transporters that mediate movement of ions across the plasma membrane. MicroRNAs influence fluid and electrolyte balance by regulation of genes involved in transport of electrolytes. In addition to channels and transporters, there are also neurohumoral factors that can be regulated by microRNAs to affect fluid and electrolyte balance.

Keywords MicroRNA • Transport • Epithelium • Fluid and electrolyte homeostasis • Blood pressure

9.1 MicroRNA Biology

9.1.1 Biogenesis of MicroRNAs

MicroRNAs (miRNAs) are a class of endogenous and conserved small RNA molecules that regulate gene expression (Eulalio et al. 2008; Filipowicz

M.A. Baker (✉) • D. Mladinov • M. Liang
Center of Systems Molecular Medicine, Medical College of Wisconsin, Milwaukee, WI, USA

Department of Physiology, Medical College of Wisconsin, 8701 Watertown Plank Road, Milwaukee, WI 53226, USA
e-mail: mliang@mcw.edu

K.A. Hyndman, T.L. Pannabecker (eds.), *Sodium and Water Homeostasis*, Physiology in Health and Disease, DOI 10.1007/978-1-4939-3213-9_9

et al. 2008; Bartel 2009; Krol et al. 2010). Genes encoding miRNAs can reside in intergenic regions (between genes) or within genes. miRNAs within genes can be found in introns of protein-coding RNA or non-coding RNAs, as well as exons of protein-coding RNAs or non-coding RNAs (Krol et al. 2010; Sayed and Abdellatif 2011). That indicates that miRNA genes can be transcribed as independent transcription units or their transcription can be under regulation of transcriptional factors of other genes.

The biogenesis of miRNAs consists of processes that take place both in the nucleus and cytoplasm and is schematically described in Fig. 9.1. MicroRNAs are transcribed from the genomic DNA predominantly by the RNA polymerase II, resulting in a transcript called pri-miRNA. Pri-miRNA has a length of few kilobases and contains one or more stem-loop structures. Subsequently, pri-miRNA is trimmed into pre-miRNA by a nuclear complex involving the enzyme RNase III Drosha. Some pre-miRNAs, called mirtrons, originate from short introns as a result of splicing. The resulting pre-miRNA is 60–70 nucleotides long and has a stem-loop (single hairpin) structure with a 3′ overhang. Pre-miR is then transported into the cytoplasm via the nuclear transport receptor Exportin-5, and further processed by Dicer RNase into a mature miRNA. Dicer is assisted by the transactivation-responsive (TAR) RNA-binding protein (TRBP). The mature miRNA is ~22 nucleotides long. One strand of the miRNA (the guide strand) is loaded into the miRNA-induced silencing complex (miRISC), whereas the other strand (miRNA* or passenger strand) is degraded. The two key factors in the assembly and function of miRISC are Argonaute (AGO) proteins and the glycine-tryptophan protein of 182 kDa (GW182). AGO proteins directly interact with the miRNA and together with the GW182 act as downstream effectors in the repression of the miRNA target. The miRNA sequence enables recognition of the target mRNA through imperfect base pairing. Binding of the miRISC complex promotes degradation and/or translational suppression of the target mRNA and reduces the protein abundance (Eulalio et al. 2008; Filipowicz et al. 2008; Bartel 2009; Krol et al. 2010; Sayed and Abdellatif 2011).

There are several basic principles in miRNA nomenclature (Ambros et al. 2003). miRNA genes are named with sequential numbering. As an example, the names are in the form of hsa-mir-192, where the first three letters identify the organism (in this case, human). The miRNA gene and the predicted stem-loop sequence of the primary transcript are designated mir-192, whereas the mature miRNA is designated miR-192. Multiple precursor sequences that would express identical mature sequences would have additional numbering added to the end of their gene names. Lettered suffixes denote mature sequences that are closely related. The two strands of a mature microRNA are named 3p and 5p to indicate the arm of their origin. In some cases, the strand that is present at lower abundance is called the minor strand and denoted with an asterisk.

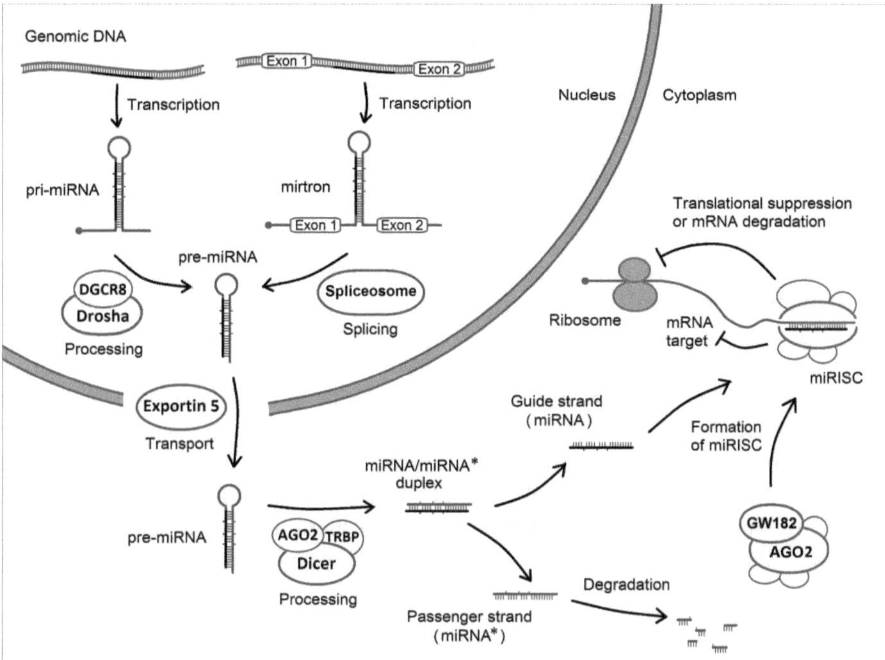

Fig. 9.1 Biogenesis of microRNA. MicroRNAs are transcribed from the genomic DNA resulting in a transcript called pri-miRNA. Pri-miRNA is trimmed into pre-miRNA by a nuclear complex involving the enzyme RNase III Drosha. Some pre-miRNAs called mirtrons originate from short introns as a result of splicing. Pre-miRNA is transported into the cytoplasm via the transport receptor Exportin-5, and further processed by Dicer RNase into a mature miRNA. Dicer is assisted by the protein TRBP. One strand of the mature miRNA (the guide strand) is loaded into the miRISC, whereas the other strand (miRNA* or passenger strand) is degraded. The key factors in the assembly and function of miRISC are AGO and GW182 proteins, which act as downstream effectors in the repression of the miRNA target. The miRISC complex promotes degradation and/or translational suppression of the target mRNA. See text for details. AGO2, Argonaute2; DGCR8, DiGeorge syndrome critical region gene 8 DiGeorge (RNA-binding protein); GW182, glycine-tryptophan protein of 182 kDa; miRISC, miRNA induced silencing complex; TRBP, transactivation-responsive (TAR) RNA-binding protein

9.1.2 Regulation of Expression of miRNAs

Transcription of a miRNA gene is controlled by similar molecular mechanisms to that of protein-coding mRNAs and is likely to be the major way of controlling its abundance (Krol et al. 2010). Wang et al. (2010) manually identified, by performing extensive literature searches, a large number of transcription factor (TF)-miRNA regulatory relationships. They created a database (TransmiR, http://cmbi.bjmu.edu.cn/transmir) containing 82 TFs and 100 miRNAs, with >240 regulatory TF-miRNA pairs (Wang et al. 2010). Multiple transcription factor-binding sites have been identified in the miRNA promoter regions, and several transcriptional factors have been experimentally proven to be key players within regulatory networks that control miRNA abundance (Bandyopadhyay and Bhattacharyya 2009; Re et al. 2009). Transcription factors can regulate miRNA in a positive or negative manner according to its tissue-specificity or developmental specificity (O'Donnell et al. 2005; Chang et al. 2008; Ma et al. 2010). Interestingly, auto-regulatory feedback loops have also been described in which one or more miRNAs suppress their own transcription factor(s) through direct or indirect interactions (Johnston et al. 2005).

Another possible level of regulation is the miRNA processing. As described above, maturation of miRNAs involves multiple steps, each having the potential to be a target for regulation (Krol et al. 2010). Regulation of Drosha, Dicer or other proteins involved in miRNA maturation and/or stabilization would be an effective mechanism of controlling global miRNA abundance in cells (Fig. 9.1). Although transcriptional regulation is considered to be the major mechanism in most instances, in some instances, post-transcriptional regulation of miRNA expression can be of prime importance (Rybak et al. 2008; Viswanathan and Daley 2010). Finally, other regulatory mechanisms have been identified, such as RNA editing and intracellular localization and compartmentalization (Krol et al. 2010).

9.1.3 Mechanisms of Action

As described above, miRNA takes part in the miRISC complex, which binds to the target mRNA. miRNAs pair imperfectly with their target mRNA, usually with a perfect complementarity of the seed sequence (Eulalio et al. 2008; Filipowicz et al. 2008; Bartel 2009; Krol et al. 2010). The seed sequence is a conserved sequence, which is mostly located at positions 2–7 from the miRNA 5′-end, and is essential for the recognition and binding to an mRNA. A miRNA typically binds to the 3′-untranslated region (UTR) of the target mRNA. In addition, several reports showed that miRNAs can mediate their effects through interaction with the 5′-UTR of a target mRNA (Lytle et al. 2007; Tsai et al. 2009; Moretti et al. 2010; Li et al. 2011). Also, several studies demonstrated that miRNAs can interact with the open reading frame (ORF) of a target mRNA, although this effect may be inhibited by displacement of miRISC by the translating ribosomes (Duursma et al. 2008; Forman et al. 2008; Tay et al. 2008; Elcheva et al. 2009).

The exact mechanism by which the miRISC complex inhibits the targeted protein expression has not been fully understood. Generally, it is considered that there are two main ways of action: suppression of translation and decrease of mRNA stability (increased mRNA degradation). In the first scenario, miRNA would decrease protein expression without decreasing mRNA expression, whereas in the second scenario, it would decrease both protein and mRNA expression. A miRNA can predominantly act in the first way, the second way, or both. There are several possible mechanisms of miRISC mediated suppression: (1) repression of translation by blocking elongation or promoting ribosomal dissociation,

(2) co-translational degradation of the nascent polypeptide chain, (3) inhibition of translational initiation by binding to the cap structure, (4) inhibition of translational initiation by prevention of the ribosomal subunit joining, (5) deadenylation of mRNA and prevention of the closed-loop mRNA configuration, and (6) increased mRNA degradation by deadenylation and decapping (Eulalio et al. 2008).

9.2 miRNAs as Regulators of Kidney Epithelial Transmembrane Transport

miRNAs have been shown to be expressed in a tissue-specific fashion, and they might provide an additional mechanism of regulating cell type-specific gene expression (Sempere et al. 2004; Liang et al. 2007; Kriegel et al. 2012; Mladinov et al. 2013). However, there is little direct evidence for a functional role of miRNAs in maintaining cell-type phenotypes, and the potential of miRNAs in determining cell-specific characteristics in nephron segments remains largely unknown. Tian et al. identified 17 miRNAs that were differentially expressed in the renal cortex and medulla of Sprague-Dawley rats and suggested a possible role of miRNA in contributing to the functional specificity of the two kidney regions (Tian et al. 2008). A study by Mladinov et al. analyzed abundance of 118 miRNAs in glomeruli, PCTs and mTALs and determined their differential miRNA expression profiles (Mladinov et al. 2013). It also indicated the potential of interaction between miR-16 and ATP-sensitive inward rectifier K^+ channel (Romk2), $Na^+/K^+/2Cl^-$ cotransporter (Nkcc2 or Slc12a1), and Uromodulin (also called Tamm-Horsfall) glycoprotein (Umod); miR-195 and Romk2, Nkcc2 and Umod; and miR-382 and Romk2.

Several reports corroborate the role of miRNAs as potential contributors to regulation of transport processes in renal epithelial cells. Huang et al. reported significant and rapid changes in miRNA expression as a response to high NaCl exposure in cultured renal medullary epithelial cells (Huang et al. 2011), demonstrating that miRNAs participate in cellular responses to osmotic stress in kidney cells. They studied the Osmotic Response Element Binding Protein (OREBP), a transcription factor critical for cellular osmoresponses, and demonstrated that global suppression of miRNAs by depletion of the *Dicer* protein (a key enzyme in miRNA maturation) caused great up-regulation of OREBP under both isotonic and hypertonic conditions. These data indicate that OREBP expression could be tightly regulated by miRNA. Further, they profiled miRNA expression in cells subjected to isotonic and hypertonic treatments and found 57 miRNAs to be differentially expressed under hypertonicity and isotonicity. Finally, they showed that miR-200b and miR-171 play important roles in osmoregulation/osmoadaptation by regulating the expression of *OREBP* in vivo in mice.

A study by Lin et al. (2011) provides another example of a physiological role of miRNAs in renal epithelial solute balance. The study demonstrates that high potassium intake increases expression of miR-802 in mouse kidney and cortical collecting ducts. The study provides evidence that miR-802 mediates the high potassium diet suppression of caveolin-1. As caveolin-1 decreases the surface expression and activity of ROMK-1 (inward rectifier K^+ channel), conditions of high potassium intake and up-regulation of miR-802 would increase the activity of ROMK-1 (Lin et al. 2011).

One of the most abundant miRNAs in the kidney cortex and the proximal tubule is miR-192 (Tian et al. 2008; Mladinov et al. 2013). A recent study by Mladinov et al. (2013) showed that miR-192 can regulate the abundance of the β1 subunit of Na^+/K^+-ATPase (Atp1β1) and subsequently regulate the enzymatic activity of the membrane bound Na^+/K^+-ATPase in human renal epithelial cells, indicating a possible novel physiological mechanism contributing to renal handling of solute and fluid balance. In further in vivo experiments in mice, they showed that suppression of miR-192 caused time-dependent up-regulation of Atp1β1 that also correlated with a decrease in urine output. When mice were fed a high-salt diet, the experimental suppression of miR-192 blunted the high salt-induced increase of urine output by about 30%. Importantly, feeding rats with sodium-depleted diet resulted in endogenous suppression of miR-192, which was accompanied by a trend of increased Atp1β1 abundance in kidney. This result seems to be in agreement with the study by Elvira-Matelot et al. (2010) that detected decreased kidney miR-192 levels in condition of sodium depletion.

Altogether, described studies in both animals and cultured cells have demonstrated that miRNAs are involved in acclimation to environmental changes in solute concentration.

9.3 Role of miRs in Fluid and Electrolyte Balance

The steady-state maintenance of fluid and electrolytes is primarily the function of ion channels and active transporters that regulate movement of ions across the plasma membrane (Arcangeli 2011). MicroRNAs influence fluid and electrolyte balance by regulation of genes involved in transport of electrolytes. The number of channels or transporters in the cell and the make-up of the subunits of each channel or transporter can be regulated at the transcription or translation level of the corresponding genes by microRNAs (Elvira-Matelot et al. 2011). In addition to channels and transporters, there are also neurohumoral factors that can be regulated by microRNAs to reflect fluid and electrolyte balance throughout the body. In this portion of the chapter, we will focus on microRNAs that affect electrolyte transporters and neurohumoral factors such as the renin-angiotensin-aldosterone system (RAAS), vasopressin, adrenal system, and atrial natriuretic peptide (ANP). Table 9.1 provides a summary of microRNAs affecting fluid and electrolyte balance.

Table 9.1 Summary of microRNAs affecting fluid and electrolyte balance

Ion transporters	miRs	Method	Cell type
Na^+/K^+ ATPase	miR-192	5′ UTR reporter assay	Human kidney
Na^+/K^+ ATPase and Na^+/H^+ Exchanger	miR-142-3p	3′ UTR reporter assay	CAD
CFTR	miR-138	3′ UTR reporter assay	Calu3
CFTR	miR-145, miR-223, miR-494	3′ UTR reporter assay	HEK-293
AE2 (Cl^-/HCO_3^-)	miR-506	3′ UTR reporter assay	H69 Cells
RAAS			
REN/APOE	miR-663	3′ UTR reporter assay	HEK-293
REN/AIFM1	miR-181	3′ UTR reporter assay	HEK-293
Aldosterone	miR-192	3′ UTR reporter assay	MDCK
Smtn	miR-330, miR-125b-5p	Functional studies	SMC
ACE2	miR-421	3′ UTR reporter assay	Huh7
AGT	miR-483-3p	3′ UTR reporter assay	HEK-293
Ang II	miR-212	3′ UTR reporter assay	RVSMC
Ang II (CREB)	miR-132	3′ UTR reporter assay	RVSMC
AT1R	miR-155	3′ UTR reporter assay	HUVEC
hAGT	miR-31, miR-584	3′ UTR reporter assay	Hep3B
TASK2	has-miR-23, has-miR-34	3′ UTR reporter assay	HEK-293
Vasopressin			
OREBP	miR-200b, miR-717, miR-466a	3′ UTR reporter assay	mIMCD3
Adrenal system			
Beta adrenoreceptor	miR-1	3′ UTR reporter assay	HEK-293
	miR-22	3′ UTR reporter assay	PC12
	miR-133	3′ UTR reporter assay	293
Mineralocorticoid receptor	miR-124, miR-135a	3′ UTR reporter assay	HeLa

(continued)

Table 9.1 (continued)

Ion transporters	miRs	Method	Cell type
Atrial natriuretic peptide			
ANP	miR-21, miR-26b, miR-98, miR-1826	cGMP assay	HVSMC
	miR-30a	3′ UTR reporter assay	HEK-293
	miR-1595, miR-2187, miR-2931	cGMP assay	HEK-293
	miR-425	3′ UTR reporter assay	COS-7

9.3.1 Ion Channels and Transporters

Several studies have demonstrated the role of microRNAs in the regulation of ion transporters in physiological and pathological studies.

9.3.1.1 Na^+/K^+ ATPase

Mladinov discovered that miR-192 targets the ATP1β1 by complementarity to its 5′ UTR. The Na^+/K^+ ATPase is the driving force of tubular transport in the kidney and was found to be inversely proportional to the expression of miR-192 in the cortex and outer medulla. In mice studies using anti-miR-192, a blunted adaptational response to increased loads of sodium and water was observed (Mladinov et al. 2013).

The D1 dopamine receptor in the kidney also modulates the Na^+/K^+ ATPase and the Na^+/H^+ exchanger; therefore, regulating diuresis and natriuresis. Using deletion of the 3′ UTR of the D1 receptor gene and site-directed mutagenesis approaches, Tobon et al. showed that miR-142-3p regulates D1 receptor expression (Tobon et al. 2012).

9.3.1.2 CFTR

Cystic fibrosis transmembrane conductance regulator channel (CFTR) encodes a channel for Cl^-, HCO_3^-, and other anions in epithelia and other cell types. Mutation of CFTR causes cystic fibrosis—an autosomal recessive disease that affects fluid and electrolyte balance in multiple organs including: the lungs, gastrointestinal tract, sweat glands, and reproductive tract. The most common mutation is ΔF508. Ramachendran et al. discovered that miR-138 regulates CFTR expression interacting with SIN3A—a transcriptional regulatory protein. Using a miR-138 mimic to treat airway epithelia resulted in an increase of CFTR mRNA and enhanced chloride permeability. An anti-miR-138 had an opposite effect (Ramachandran et al. 2012). Oglesby et al. found an inverse relation between the expression of miR-145, miR-223, and miR-494 and the expression of CFTR in bronchial epithelium of individuals carrying the ΔF508 CFTR mutation (Oglesby et al. 2013).

9.3.1.3 Cl^-/HCO_3^- Anion Exchanger 2

A study of patients with primary biliary cirrhosis conducted by Banales et al. indicated that the Cl^-/HCO_3^- anion exchanger 2 (AE2) participates in pH homeostasis and secretin-stimulated biliary bicarbonate secretion. Banales et al. found mir-506 to be a predictive candidate to target AE2 mRNA by complementarity to its 3′ UTR preventing protein translation (Banales et al. 2012).

Calcium channels have also been linked to microRNA regulation, but such studies tend to focus on cardiac and smooth muscle effects rather than on fluid balance.

An interesting contrast was found on the views of microRNAs and ion transporters in two review papers. Jiang et al. report that alterations in expression and function of ion channels and transporters in response to alterations in the micro environment result in changes in microRNAs (Jiang et al. 2012). On the other hand, Elvira-Matelot et al. report that ion transport is regulated by multiple factors including microRNAs (Elvira-Matelot et al. 2011). Both views are most likely correct with different pathways leading to one or the other outcome.

9.3.2 Neurohumoral Factors

Neurohumoral factors that affect fluid and electrolyte balance are often regulated by microRNAs.

9.3.2.1 RAAS

MicroRNAs and the RAAS have been extensively studied in the last few years with findings indicating that microRNAs are involved in hypertension, cardiovascular disease, and pre-eclampsia.

The RAAS is a major regulator of fluid and electrolyte balance and, therefore, of blood pressure (Elton et al. 2010). Renin is secreted by the juxtaglomerular cells located in the glomerular afferent arterioles (Sequeira-Lopez et al. 2010). Renin carries out the conversion of liver-produced angiotensinogen into angiotensin I that is then converted by angiotensin converting enzyme (ACE) into the potent vasoconstrictor angiotensin II (Ang II), which acts through the angiotensin receptors 1 or 2 (AT_1 or AT_2) (Elton et al. 2010).

Marques et al. used micro-array technology to find differential expression of miRs in 15 untreated hypertensive and 7 normotensive subjects of white European ancestry. Functional studies using a miR-mimic in HEK-293 cells were performed to demonstrate that hsa-miR-663 regulates the renin (REN) and apolipoprotein E (APOE) mRNA and that hsa-miR-181 regulates REN and apoptosis-inducing factor mitochondrion-associated 1 (AIFM1) mRNA (Marques et al. 2011). Follow-up studies of miR-181 in genetically hypertensive mice (BPH/2J mice) by Jackson et al. suggest that increased sympathetic activity and renal sympathetic innervation decrease miR-181a—resulting in elevation of Ren1 expression that contributes to the hypertension of these mice (Jackson et al. 2013).

In another study, Medrano et al. studied vascular smooth muscle cells of the renin lineage and demonstrated that miR-330 and miR-125-b-5p are markers of the renin-producing juxtaglomerular cells. While miR-330 inhibits smoothelin and the smooth muscle phenotype of the renin lineage cells, miR-125-b-5p has the opposite effect (Medrano et al. 2012).

Sequeira-Lopez et al. concluded that microRNA maintains the juxtaglomerular cells and the kidney morphology and function from studies in which they generated mice with a conditional deletion of Dicer. The result was a severely reduced number of juxtaglomerular cells, a decreased expression of the renin genes (Ren1 and Ren2), low plasma renin, and low blood pressure (Sequeira-Lopez et al. 2010).

Angiotensin converting enzyme 2 (ACE2) has also been studied as a target for microRNAs. ACE2 hydrolyses angiotensin II (Ang II) to angiotensin-(1–7), which opposes the effects of Ang II. Lambert et al. identified miR-421, which was implicated in the development of thrombosis, as a down-regulator of ACE2. They confirmed the target by Western blot of cell cultures transfected with a synthetic miR-421 precursor (Lambert et al. 2014).

Using human subjects, rats, and mice, Kemp et al. discovered that Ang II modulates 22 microRNAs in the vascular smooth muscle cells. In rat aortic smooth muscle cells, they also demonstrated that miR-483-3p overexpression inhibits the expression of angiotensinogen (Agt) and, to a lesser extent, angiotensin converting enzyme 1 (ACE-1) (Kemp et al. 2014). It is worthy to note that their data did not show significance for the inhibition of ACE1.

Two groups that have reported that miR-132 and miR-212 modulate the action of Ang II are Jin et al. and Eskildsen et al. Jin et al. first reported their action through the AT1R. They also reported that inhibition of miR-132 attenuates Ang II-induced cyclic AMP-response element binding protein (CREB) activation (Jin et al. 2012). Later, a human study conducted by Eskildsen et al. found that miR-132 and miR-212 were decreased in arteries from bypass-operated patients treated with AT1R blockers, suggesting that these miRs are involved in ANG II-induced hypertension (Eskildsen et al. 2013). However, looking at their data, the decreased level of miR-212 did not reach significance.

Several groups have studied the role of miR-155 in the RAAS. Ceolotto et al. reported that a single-nucleotide polymorphism in the 3′ UTR of the human AT_1 +1166 A/C may have a role in blood pressure regulation. MiR-155 expression was significantly reduced in patients with the CC genotype, while AT_1 protein expression was markedly increased in those patients (Ceolotto et al. 2011). The authors note a positive correlation of AT_1 protein expression with blood pressure.

However, the table showing the clinical characteristics of the subjects shows no statistically significant difference in any characteristics including blood pressure. The data show that the CC genotype has a significant increase of AT_1 compared with AA and AC genotypes, but microRNA expression is shown only in a correlation graph with AT_1. The publication relies heavily on correlations to show significance, but all the r-values show weak correlations. Other studies of the role of miR-155 in the RAAS include findings from Liu et al. that indicate that miR-155 reduces the effects of Ang II-induced ERK 1/2 activation in human-umbilical vein endothelial cells (HUVECs) attenuating damage and apoptosis (Liu et al. 2013). Further elucidation of the role of miR-155 in the RAAS came from Yang et al., who concluded that miR-155 reduces the proliferation of VSMCs induced by Ang II and that it also decreases the AT1R gene and protein expression (Yang et al. 2014). In addition, the increased frequency of the CC genotype of the +1166 A/C polymorphism has also been correlated with pre-eclampsia (Yang et al. 2013).

Another polymorphism associated with hypertension is located on the 3′ UTR of the human angiotensinogen gene at nucleotide 11525 C/A. Mopidevi et al. show that miR-31 and miR-584 target the 11525 C allele more strongly than the 11525 A allele. They demonstrate that these miRs down-regulate human Agt mRNA using transfected human liver cells (Mopidevi et al. 2013). Several other potential SNPs with miR binding sites on the 3′ UTRs of RAAS-related genes have also been reported (Elton et al. 2010; Nossent et al. 2011).

MicroRNAs affecting aldosterone have also been studied. Elvira-Matelot et al. have shown that miR-192 expression is inhibited by aldosterone. When mice were exposed to salt depletion, potassium load, or chronic aldosterone infusion, miR-192 expression was reduced. One of the predicted targets of miR-192, the serine-threonine kinase (WNK1), was also regulated by the same challenges. Elvira-Matelot suggests that miR-192 could participate in regulation of WNK1 expression by aldosterone (Elvira-Matelot et al. 2010).

Studies of human primary aldosteronism (APA) caused by aldosterone-producing adenoma show that Twik-related acid-sensitive K^+ channel 2 (TASK-2) gene is decreased at the transcript and protein levels in the adrenal cortex of APA patients. Using H295R cell transfection, Lenzini et al. found that hsa-miR-23 and hsa-miR-34 target the 3′ UTR of the TASK-2 gene and reduce TASK-2 expression (Lenzini et al. 2014).

9.3.2.2 Arginine Vasopressin

Arginine vasopressin (AVP), also known as antidiuretic hormone, is a peptide hormone secreted by the neurohypophysis that acts in the kidney increasing the water permeability of the collecting ducts (Nielsen et al. 1995). A study by Luo et al. showed that when mIMCD-3 cells were exposed to an arginine-vasopressin analog, the expression of miR-466 (a/b/c/e/p)-3p and miR-200b-3p was reduced. In animal studies using mice, transgenic overexpression of miR-466a-3p was associated with polydipsia, polyuria, disturbed ion homeostasis, and altered kidney morphology (Luo et al. 2014). In another study, the same group used renal medullary epithelial mIMCD3 cells and mice to show that miR-200b and miR-717 regulate the expression of the OREBP, which is a transcription factor critical for cellular osmoresponses. Exposure to sodium chloride significantly decreased the expression of miR-200b and miR-717 (Huang et al. 2011).

9.3.2.3 Adrenal System

The effects of microRNAs in the β-adrenergic pathway and its cardiovascular effects have been another focus of study. Using male Wister rats and a myocardial ischemia model with a posterior treatment with propranolol, Lu et al. observed that miR-1 was overexpressed in ischemic myocardium and that administration of propranolol returned miR-1 to control levels in addition to reducing myocardial injury and restoring the cardiac conduction. Their conclusion was that β blockers are beneficial by down-regulating miR-1 (Lu et al. 2009). The relation of miR-1 to the sympathetic nervous system (SNS) was later corroborated by another group using isoproterenol (a β-adrenoceptor activator) and propranolol (a β inhibitor). Hou et al. used microRNA array analysis on left ventricle samples from rats treated with isoproterenol or propranolol. Their findings showed that several miRs were up-regulated by isoproterenol (miR-1, miR-21, miR-27b, miR-22, miR-24, miR-199a, miR-212, and miR-214) while propranolol up-regulated miR-30c and down-regulated miR-212 (Hou et al. 2012).

Another microRNA that has been studied in relation to the β-adrenergic receptors is miR-133. After being selected from a bioinformatics analysis, gain-of-function and loss-of-function studies revealed that miR-133 controls multiple components of the β-adrenergic receptor transduction cascade and provides cardio-protection during heart failure (Castaldi et al. 2014).

Studies have also been conducted in relation to mineralocorticoid receptors. Sober et al. found that the mineralocorticoid receptor gene NR3C2 was a potential target for several microRNAs while screening loci involved in renal water-salt balance regulation. Further, luciferase assay studies demonstrated that miR-124 and miR-135a suppress NR3C2 activity and could participate in the regulation of the renin-angiotensin-aldosterone system (Sober et al. 2010).

9.3.2.4 Atrial Natriuretic Peptide

Another way fluid balance is affected is through atrial natriuretic peptide (ANP), which is a potent natriuretic, diuretic, and vasodilator. ANP is synthesized in the heart in response to increased intravascular volume and is released into the circulation to act through its guanylyl cyclase receptor (Arora et al. 2013).

From a miR microarray and quantitative real-time PCR analysis on human vascular smooth muscle cells (HVSMC) that had been treated with atrial natriuretic peptide, Kotlo et al. concluded that ANP up-regulated miR-21, miR-26b, and miR-98 while down-regulating miR-1826 (Kotlo et al. 2011). In another study, Somanna et al. used small interfering RNA (siRNA) to elicit a functional knock-down of the guanylyl cyclase/natriuretic peptide receptor A (GC-A/NPRA) in human embryonic kidney 293 cells (HEK-293). These studies concluded that miR-1595 and miR-2931 repress expression of NPRA (Somanna et al. 2013). In addition, studies by Yin et al. demonstrated that down-regulation of miR-30a lead to increased mRNA expression of atrial natriuretic factor and brain natriuretic peptide in H9C2 cells (Yin et al. 2013).

Lastly, utilizing population studies that suggested that the variant, rs5068 (A/g), in the 3′ UTR of the gene encoding for ANP (NPPA) is associated with blood pressure, Arora et al. conducted human studies comparing the genotypes response to salt loads. Arora et al. concluded that miR-425 specifically targets the A allele and silences NPPA; therefore, regulating the production of ANP (Arora et al. 2013).

9.4 Sex Differences in MicroRNA Expression

MicroRNAs regulate a wide range of physiological activities. The differential expression of microRNAs in males and females can have an impact in these regulatory steps and influence physiology, pathophysiology and disease outcome (Mishima et al. 2008; Murphy et al. 2014; Sharma and Eghbali 2014). Mishima et al. used sequencing techniques to find the differential expression of microRNAs in adult mouse testes and ovaries. Their findings showed that 14 miRs were found exclusively in the testes and 48 miRs were found exclusively in the ovaries (Mishima et al. 2008). In addition, sex steroid hormones can also regulate microRNA expression through their interaction with Argonaut, Drosha and Dicer (Sharma and Eghbali 2014). Lastly, the x-chromosome encodes 118 miRs according to the miRBase database (www.mirbase.org) while the y-chromosome only encodes 2 miRs.

Sex differences in microRNA expression have been studied in relation to several conditions including autoimmune diseases, neurodegenerative disorders, metabolic disorders, cardiovascular disease, cancer and liver disease (Murphy et al. 2014; Sharma and Eghbali 2014). No sex difference studies have been performed to specifically examine any role of microRNAs in the regulation of fluid and electrolytes balance. However, sex difference studies of other conditions have shown that some of the microRNAs that we know of are involved in fluid homeostasis exhibit sex differences. Murphy et al. found a differential expression of miR-200b, miR-142, miR-330, miR-125, miR-421, miR-22, miR-21 and miR-155 while looking at sex differences during the development of the rat cerebral cortex (Murphy et al. 2014). In addition, mir-21 was found to be differentially expressed in males and females in a heart hypertrophy study by Queiros et al. in which they stated that miR-21 is regulated by estrogen (Queiros et al. 2013). A pulmonary hypertension study by Wei et al. also found that miR-1, miR-21 and miR-34 were differentially expressed in males and females suffering from the disease

(Wei et al. 2013). Dai and Ahmed found a differential expression of miR-98 between males and females while studying auto-immune diseases (Dai and Ahmed 2014). Another autoimmune study on mice with lupus by Day et al. showed a sex differential expression of miR-31 and miR-155 (Dai et al. 2013). Lastly a study of neurogenerative disorders by Siegel et al. showed that miR-23a's regulation of the sex-linked inhibitor of apoptosis (XIAP) contributes to sex differences in the response to cerebral ischemia (Siegel et al. 2011).

9.5 Perspectives

Significant progress has been made in the understanding of the role of microRNAs in the regulation of fluid and electrolyte balance in mammalian species. Several microRNAs have been found to regulate channels, transporters, or hormonal factors that mediate or regulate cellular transport activities. However, it is conceivable that additional microRNAs are involved in the regulation of fluid and electrolyte balance and remain to be identified. It would be particularly important to ascertain the in vivo role of specific microRNAs in the regulation of fluid and electrolyte balance in whole organisms including model animals and human subjects. It would also be important to understand how microRNAs contribute to the molecular network that coordinates the regulation of physiological processes including fluid and electrolyte balance (Liang 2009).

References

Ambros V, Bartel B, Bartel DP, Burge CB, Carrington JC, Chen X, Dreyfuss G, Eddy SR, Griffiths-Jones S, Marshall M, Mazke M, Ruvkun G, Tuschl T (2003) A uniform system for microRNA annotation. RNA 9(3):277–279

Arcangeli A (2011) Ion channels and transporters in cancer. 3. Ion channels in the tumor cell-microenvironment cross talk. Am J Physiol Cell Physiol 301(4):C762–C771. doi:10.1152/ajpcell.00113.2011

Arora P, Wu C, Khan AM, Bloch DB, Davis-Dusenbery BN, Ghorbani A, Spagnolli E, Martinez A, Ryan A, Tainsh LT, Kim S, Rong J, Huan T, Freedman JE, Levy D, Miller KK, Hata A, Del Monte F, Vandenwijngaert S, Swinnen M, Janssens S, Holmes TM, Buys ES, Bloch KD, Newton-Cheh C, Wang TJ (2013) Atrial natriuretic peptide is negatively regulated by microRNA-425. J Clin Invest 123(8):3378–3382. doi:10.1172/JCI67383

Banales JM, Saez E, Uriz M, Sarvide S, Urribarri AD, Splinter P, Tietz Bogert PS, Bujanda L, Prieto J, Medina JF, LaRusso NF (2012) Up-regulation of microRNA 506 leads to decreased Cl-/HCO3- anion exchanger 2 expression in biliary epithelium of patients with primary biliary cirrhosis. Hepatology 56(2):687–697. doi:10.1002/hep.25691

Bandyopadhyay S, Bhattacharyya M (2009) Analyzing miRNA co-expression networks to explore TF-miRNA regulation. BMC Bioinf 10:163. doi:10.1186/1471-2105-10-163

Bartel DP (2009) MicroRNAs: target recognition and regulatory functions. Cell 136(2):215–233. doi:10.1016/j.cell.2009.01.002

Castaldi A, Zaglia T, Di Mauro V, Carullo P, Viggiani G, Borile G, Di Stefano B, Schiattarella GG, Gualazzi MG, Elia L, Stirparo GG, Colorito ML, Pironti G, Kunderfranco P, Esposito G, Bang ML, Mongillo M, Condorelli G, Catalucci D (2014) MicroRNA-133 modulates the beta1-adrenergic receptor transduction cascade. Circ Res 115(2):273–283. doi:10.1161/CIRCRESAHA.115.303252

Ceolotto G, Papparella I, Bortoluzzi A, Strapazzon G, Ragazzo F, Bratti P, Fabricio AS, Squarcina E, Gion M, Palatini P, Semplicini A (2011) Interplay between miR-155, AT1R A1166C polymorphism, and AT1R expression in young untreated hypertensives. Am J Hypertens 24(2):241–246. doi:10.1038/ajh.2010.211

Chang TC, Yu D, Lee YS, Wentzel EA, Arking DE, West KM, Dang CV, Thomas-Tikhonenko A, Mendell JT (2008) Widespread microRNA repression by Myc contributes to tumorigenesis. Nat Genet 40(1):43–50. doi:10.1038/ng.2007.30

Dai R, Ahmed SA (2014) Sexual dimorphism of miRNA expression: a new perspective in understanding the sex bias of autoimmune diseases. Ther Clin Risk Manag 10:151–163. doi:10.2147/tcrm.s33517

Dai R, McReynolds S, Leroith T, Heid B, Liang Z, Ahmed SA (2013) Sex differences in the expression of lupus-associated miRNAs in splenocytes from lupus-prone NZB/WF1 mice. Biol Sex Differ 4(1):19. doi:10.1186/2042-6410-4-19

Duursma AM, Kedde M, Schrier M, le Sage C, Agami R (2008) miR-148 targets human DNMT3b protein coding region. RNA 14(5):872–877. doi:10.1261/rna.972008

Elcheva I, Goswami S, Noubissi FK, Spiegelman VS (2009) CRD-BP protects the coding region of betaTrCP1 mRNA from miR-183-mediated degradation. Mol Cell 35(2):240–246. doi:10.1016/j.molcel.2009.06.007

Elton TS, Sansom SE, Martin MM (2010) Cardiovascular disease, single nucleotide polymorphisms; and the renin angiotensin system: is there a microRNA connection? Int J Hypertens 2010. pii:281692. doi:10.4061/2010/281692

Elvira-Matelot E, Zhou XO, Farman N, Beaurain G, Henrion-Caude A, Hadchouel J, Jeunemaitre X (2010) Regulation of WNK1 expression by miR-192 and aldosterone. J Am Soc Nephrol 21 (10):1724–1731. doi:10.1681/ASN.2009111186

Elvira-Matelot E, Jeunemaitre X, Hadchouel J (2011) Regulation of ion transport by microRNAs. Curr Opin Nephrol Hypertens 20(5):541–546. doi:10.1097/MNH.0b013e328348b4aa

Eskildsen TV, Jeppesen PL, Schneider M, Nossent AY, Sandberg MB, Hansen PB, Jensen CH, Hansen ML, Marcussen N, Rasmussen LM, Bie P, Andersen DC, Sheikh SP (2013) Angiotensin II regulates microRNA-132/-212 in hypertensive rats and humans. Int J Mol Sci 14 (6):11190–11207. doi:10.3390/ijms140611190

Eulalio A, Huntzinger E, Izaurralde E (2008) Getting to the root of miRNA-mediated gene silencing. Cell 132(1):9–14. doi:10.1016/j.cell.2007.12.024

Filipowicz W, Bhattacharyya SN, Sonenberg N (2008) Mechanisms of post-transcriptional regulation by microRNAs: are the answers in sight? Nat Rev Genet 9(2):102–114. doi:10.1038/nrg2290

Forman JJ, Legesse-Miller A, Coller HA (2008) A search for conserved sequences in coding regions reveals that the let-7 microRNA targets Dicer within its coding sequence. Proc Natl Acad Sci U S A 105(39):14879–14884. doi:10.1073/pnas.0803230105

Hou Y, Sun Y, Shan H, Li X, Zhang M, Zhou X, Xing S, Sun H, Chu W, Qiao G, Lu Y (2012) Beta-adrenoceptor regulates miRNA expression in rat heart. Med Sci Monit 18(8):BR309–BR314

Huang W, Liu H, Wang T, Zhang T, Kuang J, Luo Y, Chung SS, Yuan L, Yang JY (2011) Tonicity-responsive microRNAs contribute to the maximal induction of osmoregulatory transcription factor OREBP in response to high-NaCl hypertonicity. Nucleic Acids Res 39 (2):475–485. doi:10.1093/nar/gkq818

Jackson KL, Marques FZ, Watson AM, Palma-Rigo K, Nguyen-Huu TP, Morris BJ, Charchar FJ, Davern PJ, Head GA (2013) A novel interaction between sympathetic overactivity and aberrant regulation of renin by miR-181a in BPH/2J genetically hypertensive mice. Hypertension 62 (4):775–781. doi:10.1161/HYPERTENSIONAHA.113.01701

Jiang X, Zhang JT, Chan HC (2012) Ion channels/transporters as epigenetic regulators? A microRNA perspective. Sci China Life Sci 55(9):753–760. doi:10.1007/s11427-012-4369-9

Jin W, Reddy MA, Chen Z, Putta S, Lanting L, Kato M, Park JT, Chandra M, Wang C, Tangirala RK, Natarajan R (2012) Small RNA sequencing reveals microRNAs that modulate angiotensin II effects in vascular smooth muscle cells. J Biol Chem 287(19):15672–15683. doi:10.1074/jbc.M111.322669

Johnston RJ Jr, Chang S, Etchberger JF, Ortiz CO, Hobert O (2005) MicroRNAs acting in a double-negative feedback loop to control a neuronal cell fate decision. Proc Natl Acad Sci U S A $32# 102(35):12449–12454. doi:10.1073/pnas.0505530102

Kemp JR, Unal H, Desnoyer R, Yue H, Bhatnagar A, Karnik SS (2014) Angiotensin II-regulated microRNA 483-3p directly targets multiple components of the renin-angiotensin system. J Mol Cell Cardiol 75C:25–39. doi:10.1016/j.yjmcc.2014.06.008

Kotlo KU, Hesabi B, Danziger RS (2011) Implication of microRNAs in atrial natriuretic peptide and nitric oxide signaling in vascular smooth muscle cells. Am J Physiol Cell Physiol 301(4): C929–C937. doi:10.1152/ajpcell.00088.2011

Kriegel AJ, Mladinov D, Liang M (2012) Translational study of microRNAs and its application in kidney disease and hypertension research. Clin Sci 122(10):439–447. doi:10.1042/CS20110159

Krol J, Loedige I, Filipowicz W (2010) The widespread regulation of microRNA biogenesis, function and decay. Nat Rev Genet 11(9):597–610. doi:10.1038/nrg2843

Lambert DW, Lambert LA, Clarke NE, Hooper NM, Porter KE, Turner AJ (2014) Angiotensin-converting enzyme 2 is subject to post-transcriptional regulation by miR-421. Clin Sci 127 (4):243–249. doi:10.1042/CS20130420

Lenzini L, Caroccia B, Campos AG, Fassina A, Belloni AS, Seccia TM, Kuppusamy M, Ferraro S, Skander G, Bader M, Rainey WE, Rossi GP (2014) Lower expression of the TWIK-related acid-sensitive K+ channel 2 (TASK-2) gene is a hallmark of aldosterone-producing adenoma causing human primary aldosteronism. J Clin Endocrinol Metab 99(4):E674–E682. doi:10.1210/jc.2013-2900

Li YP, Gottwein JM, Scheel TK, Jensen TB, Bukh J (2011) MicroRNA-122 antagonism against hepatitis C virus genotypes 1-6 and reduced efficacy by host RNA insertion or mutations in the HCV 5′ UTR. Proc Natl Acad Sci U S A 108(12):4991–4996. doi:10.1073/pnas.1016606108

Liang M (2009) MicroRNA: a new entrance to the broad paradigm of systems molecular medicine. Physiol Genomics 38(2):113–115. doi:10.1152/physiolgenomics.00080.2009

Liang Y, Ridzon D, Wong L, Chen C (2007) Characterization of microRNA expression profiles in normal human tissues. BMC Genomics 8:166. doi:10.1186/1471-2164-8-166

Lin DH, Yue P, Pan C, Sun P, Wang WH (2011) MicroRNA 802 stimulates ROMK channels by suppressing caveolin-1. J Am Soc Nephrol 22(6):1087–1098. doi:10.1681/ASN.2010090927

Liu T, Shen D, Xing S, Chen J, Yu Z, Wang J, Wu B, Chi H, Zhao H, Liang Z, Chen C (2013) Attenuation of exogenous angiotensin II stress-induced damage and apoptosis in human vascular endothelial cells via microRNA-155 expression. Int J Mol Med 31(1):188–196. doi:10.3892/ijmm.2012.1182

Lu Y, Zhang Y, Shan H, Pan Z, Li X, Li B, Xu C, Zhang B, Zhang F, Dong D, Song W, Qiao G, Yang B (2009) MicroRNA-1 downregulation by propranolol in a rat model of myocardial infarction: a new mechanism for ischaemic cardioprotection. Cardiovasc Res 84(3):434–441. doi:10.1093/cvr/cvp232

Luo Y, Liu Y, Liu M, Wei J, Zhang Y, Hou J, Huang W, Wang T, Li X, He Y, Ding F, Yuan L, Cai J, Zheng F, Yang JY (2014) Sfmbt2 10th intron-hosted miR-466(a/e)-3p are important epigenetic regulators of Nfat5 signaling, osmoregulation and urine concentration in mice. Biochim Biophys Acta 1839(2):97–106. doi:10.1016/j.bbagrm.2013.12.005

Lytle JR, Yario TA, Steitz JA (2007) Target mRNAs are repressed as efficiently by microRNA-binding sites in the 5′ UTR as in the 3′ UTR. Proc Natl Acad Sci U S A 104(23):9667–9672. doi:10.1073/pnas.0703820104

Ma L, Young J, Prabhala H, Pan E, Mestdagh P, Muth D, Teruya-Feldstein J, Reinhardt F, Onder TT, Valastyan S, Westermann F, Speleman F, Vandesompele J, Weinberg RA (2010) miR-9, a MYC/MYCN-activated microRNA, regulates E-cadherin and cancer metastasis. Nat Cell Biol 12(3):247–256. doi:10.1038/ncb2024

Marques FZ, Campain AE, Tomaszewski M, Zukowska-Szczechowska E, Yang YH, Charchar FJ, Morris BJ (2011) Gene expression profiling reveals renin mRNA overexpression in human hypertensive kidneys and a role for microRNAs. Hypertension 58(6):1093–1098. doi:10.1161/HYPERTENSIONAHA.111.180729

Medrano S, Monteagudo MC, Sequeira-Lopez ML, Pentz ES, Gomez RA (2012) Two microRNAs, miR-330 and miR-125b-5p, mark the juxtaglomerular cell and balance its smooth muscle phenotype. Am J Physiol Renal Physiol 302(1):F29–F37. doi:10.1152/ajprenal.00460.2011

Mishima T, Takizawa T, Luo SS, Ishibashi O, Kawahigashi Y, Mizuguchi Y, Ishikawa T, Mori M, Kanda T, Goto T, Takizawa T (2008) MicroRNA (miRNA) cloning analysis reveals sex differences in miRNA expression profiles between adult mouse testis and ovary. Reproduction 136(6):811–822. doi:10.1530/rep-08-0349

Mladinov D, Liu Y, Mattson DL, Liang M (2013) MicroRNAs contribute to the maintenance of cell-type-specific physiological characteristics: miR-192 targets Na+/K+-ATPase beta1. Nucleic Acids Res 41(2):1273–1283. doi:10.1093/nar/gks1228

Mopidevi B, Ponnala M, Kumar A (2013) Human angiotensinogen +11525 C/A polymorphism modulates its gene expression through microRNA binding. Physiol Genomics 45(19):901–906. doi:10.1152/physiolgenomics.00056.2013

Moretti F, Thermann R, Hentze MW (2010) Mechanism of translational regulation by miR-2 from sites in the 5′ untranslated region or the open reading frame. RNA 16(12):2493–2502. doi:10.1261/rna.2384610

Murphy SJ, Lusardi TA, Phillips JI, Saugstad JA (2014) Sex differences in microRNA expression during development in rat cortex. Neurochem Int 77:24–32. doi:10.1016/j.neuint.2014.06.007

Nielsen S, Chou CL, Marples D, Christensen EI, Kishore BK, Knepper MA (1995) Vasopressin increases water permeability of kidney collecting duct by inducing translocation of aquaporin-CD water channels to plasma membrane. Proc Natl Acad Sci U S A 92(4):1013–1017

Nossent AY, Hansen JL, Doggen C, Quax PH, Sheikh SP, Rosendaal FR (2011) SNPs in microRNA binding sites in 3′-UTRs of RAAS genes influence arterial blood pressure and risk of myocardial infarction. Am J Hypertens 24(9):999–1006. doi:10.1038/ajh.2011.92

O'Donnell KA, Wentzel EA, Zeller KI, Dang CV, Mendell JT (2005) c-Myc-regulated microRNAs modulate E2F1 expression. Nature 435(7043):839–843. doi:10.1038/nature03677

Oglesby IK, Chotirmall SH, McElvaney NG, Greene CM (2013) Regulation of cystic fibrosis transmembrane conductance regulator by microRNA-145, -223, and -494 is altered in DeltaF508 cystic fibrosis airway epithelium. J Immunol 190(7):3354–3362. doi:10.4049/jimmunol.1202960

Queiros AM, Eschen C, Fliegner D, Kararigas G, Dworatzek E, Westphal C, Sanchez Ruderisch H, Regitz-Zagrosek V (2013) Sex- and estrogen-dependent regulation of a miRNA network in the healthy and hypertrophied heart. Int J Cardiol 169(5):331–338. doi:10.1016/j.ijcard.2013.09.002

Ramachandran S, Karp PH, Jiang P, Ostedgaard LS, Walz AE, Fisher JT, Keshavjee S, Lennox KA, Jacobi AM, Rose SD, Behlke MA, Welsh MJ, Xing Y, McCray PB Jr (2012) A microRNA network regulates expression and biosynthesis of wild-type and DeltaF508 mutant cystic fibrosis transmembrane conductance regulator. Proc Natl Acad Sci U S A 109(33):13362–13367. doi:10.1073/pnas.1210906109

Re A, Cora D, Taverna D, Caselle M (2009) Genome-wide survey of microRNA-transcription factor feed-forward regulatory circuits in human. Mol BioSyst 5(8):854–867. doi:10.1039/b900177h

Rybak A, Fuchs H, Smirnova L, Brandt C, Pohl EE, Nitsch R, Wulczyn FG (2008) A feedback loop comprising lin-28 and let-7 controls pre-let-7 maturation during neural stem-cell commitment. Nat Cell Biol 10(8):987–993. doi:10.1038/ncb1759

Sayed D, Abdellatif M (2011) MicroRNAs in development and disease. Physiol Rev 91 (3):827–887. doi:10.1152/physrev.00006.2010

Sempere LF, Freemantle S, Pitha-Rowe I, Moss E, Dmitrovsky E, Ambros V (2004) Expression profiling of mammalian microRNAs uncovers a subset of brain-expressed microRNAs with possible roles in murine and human neuronal differentiation. Genome Biol 5(3):R13. doi:10.1186/gb-2004-5-3-r13

Sequeira-Lopez ML, Weatherford ET, Borges GR, Monteagudo MC, Pentz ES, Harfe BD, Carretero O, Sigmund CD, Gomez RA (2010) The microRNA-processing enzyme dicer maintains juxtaglomerular cells. J Am Soc Nephrol 21(3):460–467. doi:10.1681/ASN.2009090964

Sharma S, Eghbali M (2014) Influence of sex differences on microRNA gene regulation in disease. Biol Sex Differ 5(1):3. doi:10.1186/2042-6410-5-3

Siegel C, Li J, Liu F, Benashski SE, McCullough LD (2011) miR-23a regulation of X-linked inhibitor of apoptosis (XIAP) contributes to sex differences in the response to cerebral ischemia. Proc Natl Acad Sci U S A 108(28):11662–11667. doi:10.1073/pnas.1102635108

Sober S, Laan M, Annilo T (2010) MicroRNAs miR-124 and miR-135a are potential regulators of the mineralocorticoid receptor gene (NR3C2) expression. Biochem Biophys Res Commun 391 (1):727–732. doi:10.1016/j.bbrc.2009.11.128

Somanna NK, Pandey AC, Arise KK, Nguyen V, Pandey KN (2013) Functional silencing of guanylyl cyclase/natriuretic peptide receptor-A by microRNA interference: analysis of receptor endocytosis. Int J Biochem Mol Biol 4(1):41–53

Tay Y, Zhang J, Thomson AM, Lim B, Rigoutsos I (2008) MicroRNAs to Nanog, Oct4 and Sox2 coding regions modulate embryonic stem cell differentiation. Nature 455(7216):1124–1128. doi:10.1038/nature07299

Tian Z, Greene AS, Pietrusz JL, Matus IR, Liang M (2008) MicroRNA-target pairs in the rat kidney identified by microRNA microarray, proteomic, and bioinformatic analysis. Genome Res 18(3):404–411. doi:10.1101/gr.6587008

Tobon KE, Chang D, Kuzhikandathil EV (2012) MicroRNA 142-3p mediates post-transcriptional regulation of D1 dopamine receptor expression. PLoS One 7(11):e49288. doi:10.1371/journal.pone.0049288

Tsai NP, Lin YL, Wei LN (2009) MicroRNA mir-346 targets the 5′-untranslated region of receptor-interacting protein 140 (RIP140) mRNA and up-regulates its protein expression. Biochem J 424(3):411–418. doi:10.1042/BJ20090915

Viswanathan SR, Daley GQ (2010) Lin28: a microRNA regulator with a macro role. Cell 140 (4):445–449. doi:10.1016/j.cell.2010.02.007

Wang J, Lu M, Qiu C, Cui Q (2010) TransmiR: a transcription factor-microRNA regulation database. Nucleic Acids Res 38(Database issue):D119–D122. doi:10.1093/nar/gkp803

Wei C, Henderson H, Spradley C, Li L, Kim IK, Kumar S, Hong N, Arroliga AC, Gupta S (2013) Circulating miRNAs as potential marker for pulmonary hypertension. PLoS One 8(5):e64396. doi:10.1371/journal.pone.0064396

Yang J, Shang J, Zhang S, Li H, Liu H (2013) The role of the renin-angiotensin-aldosterone system in preeclampsia: genetic polymorphisms and microRNA. J Mol Endocrinol 50(2):R53–R66. doi:10.1530/JME-12-0216

Yang LX, Liu G, Zhu GF, Liu H, Guo RW, Qi F, Zou JH (2014) MicroRNA-155 inhibits angiotensin II-induced vascular smooth muscle cell proliferation. J Renin Angioten Aldosterone Syst 15(2):109–116. doi:10.1177/1470320313503693

Yin X, Peng C, Ning W, Li C, Ren Z, Zhang J, Gao H, Zhao K (2013) miR-30a downregulation aggravates pressure overload-induced cardiomyocyte hypertrophy. Mol Cell Biochem 379 (1–2):1–6. doi:10.1007/s11010-012-1552-z

Chapter 10
Osmoregulation in Desert-Adapted Mammals

John Donald and Thomas L. Pannabecker

Abstract Hot deserts devoid of drinking water pose significant problems to mammals whose high body temperatures challenge the restriction of evaporative water loss (EWL) in order to maintain water balance, in addition to the obligatory loss of water for excretion. Desert mammals maintain water balance by physiological adaptations that minimise water loss, and by gaining water from food and/or from metabolism that produces oxidation water. Larger desert mammals such as ungulates depend on heterothermy and selective brain cooling to minimise EWL and generally do not excrete highly concentrated urine. Small desert mammals, as exemplified by the rodents, also restrict EWL, but are renowned for their ability to excrete small volumes of very concentrated urine, especially when they do not drink. In addition, the gastrointestinal tract of desert mammals generally has a greater absorptive area for water and nutrient uptake than that of non-desert mammals. To offset water loss, desert mammals gain preformed water from food, and small desert mammals in particular, can produce a significant amount of metabolic water from the oxidative metabolism of food substrates. The latter source of water gives rise to the iconic view of small desert rodents that consume seeds and survive by metabolic water production. This chapter will provide an overview of the accepted physiological paradigms that permit desert mammals, particularly small rodents, to exist in arid environments, and will emphasise how desert mammals are valuable models for comparative research in osmoregulation. In addition, the recent research in transcriptomics and its application to understanding osmoregulation in desert mammals will be discussed.

Keywords Desert • Osmoregulation • Water balance • Kidney • Metabolic water • Desert rodent • Desert adaptation

---

J. Donald (✉)
Faculty of Science, Engineering and Built Environmental, School of Life and Environmental Sciences, Deakin University, 75 Pigdons Road, Waurn Ponds, VIC 3216, Australia
e-mail: john.donald@deakin.edu.au

T.L. Pannabecker
Department of Physiology, University of Arizona Health Sciences Center, AHSC 4128, 1501 N Campbell Avenue, Tucson, AZ 85724-5051, USA

K.A. Hyndman, T.L. Pannabecker (eds.), *Sodium and Water Homeostasis*, Physiology in Health and Disease, DOI 10.1007/978-1-4939-3213-9_10

10.1 Introduction

The ability of animals to cope with the extreme heat and aridity of deserts in which there is generally a lack of drinking water has fascinated biologists for many years. In particular, the high ambient temperatures of desert regions pose significant problems to mammals, as their high body temperatures challenge restricting the rate of evaporative water loss (EWL) that is required for cooling, in addition to the obligatory loss of water for excretion of waste products. Pioneering work by Schmidt-Nielsen and others revealed some of the physiological mechanisms that permit mammals of varying body mass to minimise water loss in order to survive in deserts, and many of these paradigms have persisted today (Schmidt-Nielsen and Schmidt-Nielsen 1952; Schmidt-Nielsen 1979; Degen 1997; Walsberg 2000).

The avenues for water loss are primarily EWL by cutaneous and respiratory routes, and excretory water loss from the kidney and the gastrointestinal tract. Small mammals can minimise EWL by behaviours such as burrowing and group nesting to create humid environments (Baudinette 1972; Degen 1997), by seeking micro-habitat refugia during the day, or by being nocturnal (Degen 1997). Larger desert mammals that live in environments with little or no shade cannot seek refuge from the heat of the day and employ heterothermy to reduce EWL (Fuller et al. 2014). Desert mammals can also maximise water recovery from nasal counter-current systems that are also present in many mesic species (Degen 1997; Walsberg 2000), and often have a lower basal metabolic rate to minimise water lost during respiration (Lovegrove 2000, 2003). The loss of water by excretion can be minimised by enhanced water reabsorption in the colon, and the excretion of a small volume of concentrated urine (Degen 1997). The urinary concentrating ability of the kidney scales with body size, and small desert rodents are renowned for their ability to produce a small volume of highly concentrated urine to minimise water loss in comparison to large mammals such as ungulates (Schmidt-Nielsen 1979; Degen 1997). Furthermore, the gastrointestinal tract of desert mammals generally has a greater absorptive area than that of non-desert mammals, which allows for an increased reabsorption of water in food and nutrients (Degen 1997). In the absence of drinking water, desert mammals have strategies to gain water by behavioural and physiological mechanisms that maximise obtaining preformed water from food, and the generation of metabolic water from the oxidative metabolism of food substrates (Degen 1997; Walsberg 2000). Another aspect of desert mammal physiology and water balance is the timing of reproduction and lactation in an arid environment, but this will not be discussed here [see Degen (1997)]. The accepted view of the routes of water loss and gain with a general quantitative comparison between desert and non-desert mammals is shown in Fig. 10.1 (Schmidt-Nielsen 1972; Degen 1997).

This chapter will provide an overview of the accepted physiological paradigms and regulatory systems that permit desert mammals, particularly small rodents, to exist in arid environments, and will emphasise how specific desert mammals are valuable models for comparative research in salt and water homeostasis. In

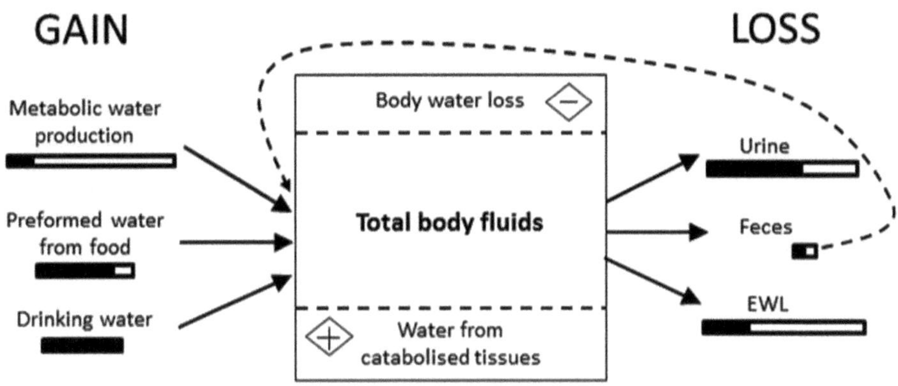

Input (%)	Non-desert (■)	Desert (□)
Drinking	48	0
Preformed water in food	40	10
Metabolic water	12	90
Output (%)		
Urine	60	25
Evaporation	34	70
Feces	6	5

Fig. 10.1 A schematic diagram of the traditional paradigm for water balance in mammals depicting the sources of water input (gain) and water output (loss), and a comparison of the relative proportions of each avenue of water loss and gain between a typical non-desert (*filled square*) and a desert mammal such as kangaroo rat (*open square*). Drawing based on Degen (1997) and water data based on Schmidt-Nielsen (1972, 1975)

addition, the recent research directions in understanding osmoregulation in desert mammals will be reviewed in the final part of the chapter.

10.2 General Physiology of Water Balance in Desert Mammals

The routes of water loss and the physiological mechanisms regulating them have been extensively investigated in desert mammals. Numerous studies have shown that desert mammals have lower rates of water turnover, EWL, and cutaneous water loss (Degen 1997; Bozinovic and Gallardo 2006). Many desert mammals exhibit plasticity in body temperature regulation and utilise heterothermy to reduce water loss by allowing the body temperature to increase during the heat of the day, thus reducing EWL necessitated for evaporative cooling (Fuller et al. 2014). The heat gained during the day is then dissipated at night through thermal conduction that is independent of EWL-driven cooling. Heterothermy has been shown to be an important mechanism in large, free-ranging desert mammals that cannot seek shade from low-lying vegetation (if present), and do not have access to water (Fuller et al. 2014). The degree of heterothermy appears to be regulated in response to the seasonal availability of water and food (Hetem et al. 2010, 2012a; Fuller et al. 2014). For example, in Arabian oryx (*Oryx leucoryx*), the fluctuation in body temperature was highest (7.7 °C) in summer when food and water shortages coincide (Hetem et al. 2010). In the gazelle, *Gazella subgutturosa marica*, heterothermy was used in summer in free-ranging and captive animals that were obtaining water only from food and were not dehydrated as indicated by a stable plasma osmolarity; this indicates the important contribution of heterothermy to maintaining water balance (Ostrowski and Williams 2006). Heterothermic mammals can also suppress EWL by the use of selective brain cooling that protects the brain when the core body temperature is elevated during the day. Specifically, small incremental (tenths of a degree) decreases in carotid blood temperature, compared to the peripheral blood, can reduce the activation of temperature-sensitive neurons in the hypothalamus to a sufficient degree that suppresses water lost by evaporative cooling [see Hetem et al. (2012b)]. EWL can also be minimised by lowering metabolism, and desert mammals generally have lower basal metabolic rate than mesic species (Schwimmer and Haim 2009; Lovegrove 2000, 2003). It is generally accepted that large mammals have lower mass-specific metabolic rate than small mammals, and that small mammals exhibit higher rates of water turnover due to EWL and cutaneous water loss compared to large mammals (Degen 1997).

The loss of free water from the extracellular fluids by evaporation and excretion can lead to an increase in plasma osmolarity and decrease in extracellular fluid volume (ECF), if uncompensated. Generally, non-desert mammals have a plasma osmolarity in the range of 290–310 mOsm kg^{-1}, but many desert mammals have higher baseline plasma osmolarity (300–350 mOsm kg^{-1}) than mesic species (Degen 1997). Furthermore, plasma osmolarity is often unaffected in many small desert mammals in experimental scenarios when they are water-deprived and/or fed a low-protein diet, which indicates that they can compensate to remain in water balance and do not become dehydrated [see Degen (1997), Heimeier et al. (2004), Heimeier and Donald (2006)]. Interestingly, this is not a general observation for all desert mammals, as the plasma osmolarity (indirectly determined from plasma sodium concentration) is increased in water-restricted one-humped dromedary camels (*Camelus dromedaries*; Ali et al. 2012), and was also increased markedly in water-deprived leaf-eared mouse, *Phyllotis darwini*, a South American desert-dwelling rodent (Gallardo et al. 2005). In addition to plasma osmolarity, many desert mammals can defend their ECF volume when water restricted compared to mesic species [see Degen (1997)]. The ability of desert mammals to maintain volume and osmotic homeostasis during water restriction has interesting implications for the endocrine control of osmoregulation and the regulation of renal aquaporins, which is discussed below.

10.3 Renal Physiology in Desert Mammals

One of the most renowned physiological process in desert mammals, especially the small rodents and marsupials, is the production of a small volume of highly concentrated urine, especially when free water is unavailable [see Degen (1997), Bozinovic and Gallardo (2006), Gordge and Roberts (2008)]. For example, the Spinifex hopping mouse of central Australia (*Notomys alexis*; MacMillen and Lee 1967) and the desert mouse possum of South America (*Thylamys pusilla*; Diaz et al. 2001) have been reported to excrete urine with a concentration in excess of 9000 mOsm kg^{-1}. The production of a concentrated urine is dependent on the regulation of two processes; ultrafiltration of plasma in the glomerulus and the reabsorption of water in the renal tubules. Accordingly, many desert mammals have a reduced glomerular filtration rate and enhanced tubular water reabsorption (Degen 1997; Gordge and Roberts 2008).

The ability of mammals to reabsorb water is dependent on the medullary osmotic gradient created by the loop of Henle, which drives the reabsorption of water from the collecting duct. A higher osmotic gradient enables more water to be reabsorbed, thus producing a concentrated urine. The urine concentrating ability is correlated to some degree with the length of the longest loops of Henle, although this is not a perfect relationship. The "relative medullary thickness", or the medullary thickness (and therefore the maximum loop length) relative to kidney volume declines with increasing body mass as shown in studies comparing large numbers of species; however, some of the highest urine concentrations occur in species that do not exhibit the highest relative medullary thickness (Beuchat 1990a). In general though, the increase in concentrating ability with decreasing body size may largely reflect the scaling of the need to conserve water (Beuchat 1990b). Many small desert rodents, faced with large body surface area-to-volume ratio, benefit from exceptionally long loops of Henle that in some way that is not clear leads to the production of highly concentrated urine.

But what is it about the desert rodent kidney that enables it to produce a urine with persistently high solute concentration? Long known as an exemplar (Gottschalk et al. 1963; Schmidt-Nielsen et al. 1948) of the *Krogh Principle* (Krebs 1975), the kangaroo rat, *Dipodomys* spp. is a highly-evolved desert species, which drinks no free water and produces a maximum urine osmolality twice that of the common laboratory rat (6000 vs. 3000 mOsmol kg^{-1}). The day-to-day values are consistently at about 4000 mOsmol kg^{-1} (MacMillen and Lee 1967) compared to about 1500 mOsmol kg^{-1} for the Sprague-Dawley and Munich-Wistar rats. Much of our knowledge on renal function in small desert mammals stems from research on kangaroo rats (*Dipodomys* spp.) of North America as these species has been used for detailed morphological and transporter protein localisation studies in the kidney.

Physiological and molecular characterization of water, sodium, and urea flux pathways in the nephron and blood vessels are of paramount importance to understanding the urine concentrating mechanism, and transepithelial water, sodium, and urea transport in general. It stands to reason that tubular transport mechanisms for sodium, urea and water that are critical for production of a highly concentrated urine will be most evident in kidneys of desert rodents and will provide instructive contrast with rodents such as the Sprague-Dawley rat and Munich-Wistar rat, rodents that produce substantially lower maximum urine concentration than many small desert mammals. We present two examples below that serve to illustrate how transport processes in the kangaroo rat could more effectively concentrate urine, and it is reasonable to assume that these processes would apply to other desert species.

High levels of active sodium transport in the outer medullary thick ascending limb of the kangaroo rat may be one mechanism contributing to the production of highly concentrated urine by this species. Active sodium transport plays a role in generating the corticopapillary solute gradient that increases by nearly twofold between the cortex and the outer medullary-inner medullary boundary, and then increases at a much steeper rate along the axis of the inner medulla (see Chap. 11 for additional information). Na-K-ATPase α-subunit and NKCC2 protein expression and Na-K-ATPase activity in the outer medullary thick ascending limb of the kangaroo rat are substantially higher than they are in the moderately dehydrated Sprague-Dawley rat (Aw et al. 2014). These observations suggest that the thick ascending limb of the kangaroo rat has the capacity for actively reabsorbing sodium at substantially greater rates than in the laboratory rat. A heightened degree of sodium reabsorption would lead to a correspondingly steeper corticopapillary solute gradient, greater fractional water reabsorption from medullary collecting duct segments, and a more concentrated urine. A corresponding observation has been made in the desert jerboa, where Na-K-ATPase activity in the thick ascending limb was shown to decline with hydration, a physiological state that leads to reduction of the corticopapillary solute gradient. These studies further support the hypothesis that a direct correlation exists between active sodium reabsorption and the magnitude of the corticopapillary solute gradient (Doucet et al. 1987). One way to test this hypothesis would be to quantify active sodium transport in the isolated perfused thick ascending limb of desert species, and quantify solute concentrations in the outer medulla and compare these to values determined in the Sprague-Dawley rat.

A greater expression of water channels in nephron segments may also play a role in the ability of desert mammals to produce a more concentrated urine compared to that of non-desert species. The inner medullary thin limb of Henle's loop is considered to function primarily as an equilibrating segment. As fluid descends towards the papilla tip, the osmolality of the luminal fluid gradually increases, equilibrating with the interstitial fluid along the corticopapillary gradient. Then, as fluid rounds the loop bend and ascends towards the outer medulla, the luminal fluid osmolality decreases to approximate that of the interstitial fluid. Expression of aquaporin 1 (AQP1) along the length of the descending thin limb of the kangaroo rat exceeds the length of AQP1 expression along the descending thin limb of the Munich-Wistar rat by 50 % (Urity et al. 2012; Fig. 10.2). A longer water permeable descending thin limb segment in the kangaroo rat possibly leads to a higher degree of water reabsorption than in the laboratory rat and could facilitate osmotic

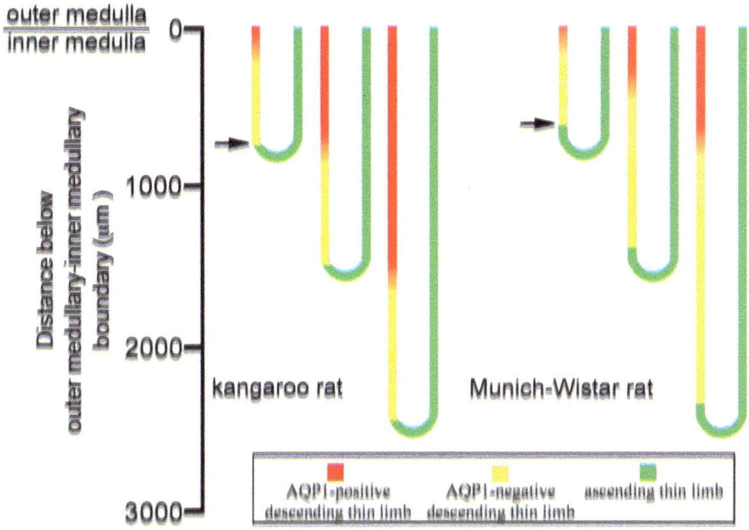

Fig. 10.2 A schematic drawing of the segmentation of inner medullary loops of Henle in kangaroo rat and Munich-Wistar rat. AQP1 is expressed throughout a longer proportion of the inner medullary DTL in the kangaroo rat, compared to the Munich-Wistar rat. The Cl channel ClC-K1 expression begins at a lower level above the bend in the kangaroo rat, compared to the Munich-Wistar rat (*arrows*). From Urity et al. (2012), used with permission

equilibration by water flux between the luminal and interstitial fluid compartments. This may in turn lead to more effective luminal solute concentration and greater driving force for NaCl reabsorption by the kangaroo rat ascending thin limb compared to rat (Urity et al. 2012) and consequently generate higher interstitial solute concentrations.

In the collecting duct principal cell, AQP2 is the primary water pathway for apical water flux, whereas AQP3 and AQP4 are the primary basolateral water channels. In the kangaroo rat inner medullary collecting duct principal cells, the basolateral:apical AQP2 ratio was found to be approximately twofold greater relative to that of Sprague-Dawley rats that were either provided with water or were water-restricted, which may indicate a greater transepithelial water permeability in kangaroo rat than laboratory rat (Espineira and Pannabecker 2014). Furthermore, in the kangaroo rat, AQP4 mRNA is expressed in collecting ducts and S3 segments of proximal tubules, but AQP4 protein is not expressed in the collecting duct, proximal tubules or in any part of the kidney (Huang et al. 2001). Therefore, AQP4 clearly is not required for urinary concentration in kangaroo rat kidney and one or more different aquaporins may provide a pathway for basolateral water flux.

In the desert rodent, *P. darwini*, the renal expression of AQP2, AQP3 and AQP4 in the collecting ducts responded to dehydration, as indicated by an increased plasma osmolarity when water-deprived, but not in a consistent pattern. There was an increase in AQP2 and AQP3 immunolabel and protein abundance

(AQP2 > AQP3) in the collecting ducts, but no change in AQP4 expression (Gallardo et al. 2005). Field studies in degus (*Octodon degus*), a South American desert rodent, found that AQP2 expression responds to seasonal water availability, as greater medullary collecting duct AQP2 expression was observed in animals during summer than in winter, as well as an apparent increase in apical membrane protein expression during the summer. The increased AQP2 expression in summer occurred together with an increase in urine osmolality that was about 1.8-fold higher in summer than in winter (Bozinovic et al. 2003). The data from *P. darwini* and *O. degus* indicate that regulation of AQP2 water channels in desert species may be similar to mesic species in which dehydration induces up-regulation of AQP2 in the collecting duct.

A number of other hormones and transcription factors may play a role in modulating expression and activity of proteins (such as aquaporins) that are integral to producing the highly concentrated urine observed in the kangaroo rat and other desert species such as *N. alexis*. For example, plasma arginine vasopressin (AVP) levels in the kangaroo rat are two- to threefold higher than in the rat (Stallone and Braun 1988), potentially leading to high AQP2 expression and water transport in the kangaroo rat collecting duct. In *N. alexis*, water deprivation increased the expression of TonEBP mRNA and the translocation of TonEBP protein from the cytoplasm into the nucleus of medullary collecting ducts (Bartolo and Donald 2008). In addition, there is an increase in the mRNA expression of aldose reductase and compatible osmolytes, indicative of increased TonEBP transcriptional activity (Bartolo and Donald 2008). On the basis of these dynamic effects and the hypertonic environment of the inner medulla, one might speculate that TonEBP is an important regulator of AQP2 expression and medullary collecting duct water reabsorption in the collecting duct of hopping mouse and other desert species, as shown for laboratory mouse collecting duct cells (Hasler et al. 2006).

The spiny mouse (*Acomys cahirinus*), a rodent native to Egypt and Israel, has also been used for research comparing renal function between a desert species and laboratory mouse (Dickinson et al. 2007). The kidney of the spiny mouse exhibits efficient mechanisms for filtering and excreting high salt concentrations (Dickinson et al. 2007). When the spiny mouse is placed on a high NaCl diet, free water reabsorption exceeds that of the equivalent-sized C57BL/6 mouse by almost 40 %. Water balance, under normal conditions, is similar for both species; however, with a high salt load the spiny mouse maintains its urine concentration but the C57BL/6 mouse does not. The spiny mouse has been hypothesized to express a greater abundance of water channels in the proximal tubule (aquaporins 1–7) and/or collecting duct (aquaporins 2, 3, 4) than C57BL/6 mouse, allowing greater water reabsorption and higher urine concentration (Dickinson et al. 2007). The ability of the spiny mouse to tolerate a larger increase in plasma osmolality compared to the C57BL/6 mouse is one mechanism by which this species adapts to increased NaCl intake. Other adaptations may include significant variations in circulating and renal concentrations of renin, vasopressin and angiotensin II, kidney architecture and intrarenal hemodynamics (Dickinson et al. 2005).

The glycosaminoglycan hyaluronan is the major component of the gel-like renal inner medullary extracellular matrix. In the kidney cortex there is relatively little hyaluronan, although it is present in substantial amounts within the inner medullary interstitium, gradually increasing with depth below the outer medulla with highest concentrations at the papilla tip. Hyaluronan has been hypothesized to function as a mechanical transducer of pelvic wall contractions through repetitive compression and relaxation, and in so doing, influence tubular fluid flows and the urine concentrating mechanism (Knepper et al. 2003). The correlation between increases and decreases of inner medullary hyaluronan with hydration and dehydration, respectively, in the rat and *N. alexis*, further supports a possible role for hyaluronan in water homeostasis (Stridh et al. 2012; Bartolo and Donald 2007).

The kangaroo rat and other desert species offer unique opportunities to investigate the urine concentrating mechanism through multiple approaches including analyses of membrane transporters, cell signalling and regulatory pathways, and integrative aspects arising from kidney architecture and interstitial composition (Issaian et al. 2012; Pannabecker 2013). These combined approaches should provide new insights into mechanisms of Na reabsorption and water transport and their roles in producing the inner medullary solute gradient, and insights into the validity of, and alternatives to, the widely accepted *passive hypothesis* of the urine concentrating mechanism (Sands and Layton 2014).

10.4 Gastrointestinal and Metabolic Physiology in Desert Mammals

The gastrointestinal system plays an important role in osmoregulation in desert mammals due to anatomical specialisations that increase the intestinal surface area to enhance the reabsorption of water and nutrients, the latter providing an important substrate for metabolic water production. Buret et al. (1993) found that the small intestine of the desert-dwelling rodents, *Meriones unguicalatus* and *Psammomys obesus*, had a greater surface area and absorptive capacity compared to rabbits and rats. In addition, the colon of *N. alexis*, is greater in length than other similar-sized mesic rodents, and the mucosa is folded in an oblique manner that increases the surface area for water and nutrient uptake (Murray et al. 1995). Similar adaptations in the anatomy of the large intestine are also observed in desert ungulates, which maximise nutrient and water uptake from the rumen [see Cain et al. (2006)]. The greater ability to reabsorb water in the colon enables desert mammals to reduce the fecal water content to 40–50 % compared with 60–80 % that is commonly observed in mesic species (Degen 1997). In addition, a larger intestinal surface area also serves to increase substrate absorption for metabolic water production, which is discussed below.

At the molecular level, water transport and the distribution of aquaporins have been studied in degus, *O. degus* (Gallardo et al. 2002). Water absorption across the colon of *O. degus* was found to be twice that of water absorption across the rat colon. This is likely to be facilitated by AQP1 and the absorption rate is unaffected by water deprivation (Gallardo et al. 2002). In addition, there were differences in the subcellular distribution of aquaporins when compared to rat. AQP1-immunolabel was found in both basolateral and apical membranes of the epithelial cells but AQP2 was not detected in the colon, which is in contrast to the rat (Gallardo et al. 2001, 2002). In the rat, AQP3 is found in the basolateral membranes of the villus epithelial cells, but in *O. degus*, AQP3 was not observed in the epithelium, but was found in a subepithelial fibroblast layer, pericryptal cells and muscularis mucosae. The distinct distribution of AQP3 may represent an adaptation in desert rodents that plays an important role in colonic fluid reabsorption (Gallardo et al. 2002).

In desert mammals, obtaining preformed water from food and/or metabolic water production are essential in the absence of free water for drinking. Many desert mammals can obtain sufficient water from food by omnivory, and in concert with an increased colonic surface area, reduced EWL, and reduced renal water loss, they can maintain water balance. However, some desert mammals, particularly rodents, can survive on the consumption of food with little or any preformed water (e.g. seeds), and are dependent on the production of metabolic water to maintain water balance. Intriguingly, seed selection trials have shown desert rodents such as kangaroo rats, *D. merriami* (Frank 1988a), and sandy inland mouse, *Pseudomys hermannsburgensis* (Murray and Dickman 1997), can select seeds with higher moisture content and therefore ingest food with more preformed water.

Despite the importance of food as a direct and indirect source of water, there has been little research on how desert mammals that depend on metabolic water can change their food intake strategy or metabolic strategy to enhance the efficiency for metabolic water production. Among substrates for oxidative water production, lipids (fat) produce twice the amount of oxidation water than carbohydrates (starch) per gram, but starch produces 20 % more water than fat per kcal because of a greater demand for oxygen for fat metabolism that then increases EWL. Thus, starch is a preferred substrate for metabolic water production in a dry environment when EWL needs to be minimised. Protein generates the least amount of oxidation water and necessitates obligatory renal water loss to excrete nitrogenous metabolites (Schmidt-Nielsen 1979; Degen 1997; Takei et al. 2012). Interestingly, food selection trials in kangaroo rats showed that they can alter their seed preferences depending on the ambient humidity and therefore the requirement for metabolic water (Frank 1988b). At low humidity, kangaroo rats preferred a diet that was high in lipid and carbohydrate content and intermediate in protein content, but the preference switched to a high lipid diet in high humidity. Thus, kangaroo rats modify their food intake to maximise metabolic water production (Frank 1988b). These types of studies highlight the importance of understanding the foraging ecology of desert mammals with respect to osmoregulation.

Despite the importance of metabolic water production in many desert mammals, research on the mechanisms that regulate it are limited. In particular, metabolic water production is dependent on substrate provision from food and, therefore, it is

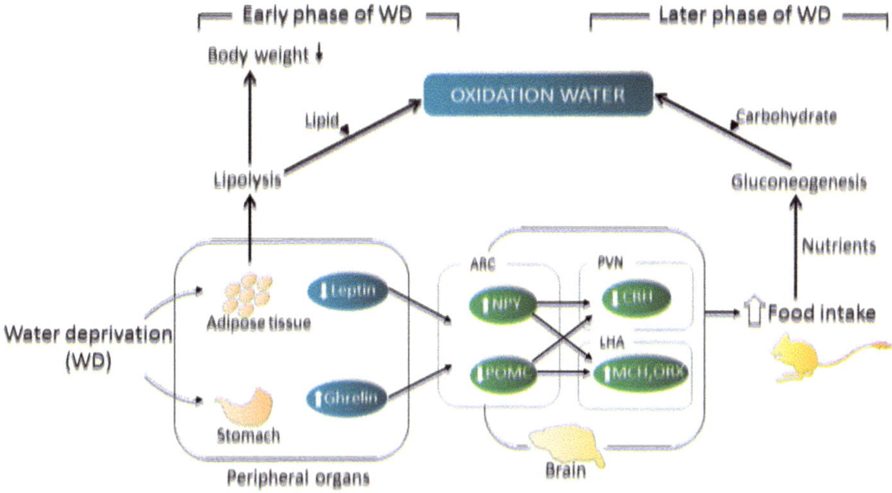

Fig. 10.3 A schema showing the interplay of regulatory systems for appetite and energy metabolism during water deprivation in the Spinifex hopping mouse, *Notomys alexis*. In the initial phase of water deprivation, *N. alexis* catabolises fat stored prior to water deprivation, while sustained food intake occurs in the later phase through integrated actions of the peripheral and central appetite-regulating system, resulting in increased nutrient supply for oxidation water production. The increased nutrient supply also enhanced glycogen deposition in the liver to prepare for prolonged water deficiency. *ARC* arcuate nucleus, *PVN* paraventricular nucleus, *LHA* lateral hypothalamic area, *NPY* neuropeptide Y, *POMC* proopiomelanocortin, *CRH* corticotropin-releasing hormone, *MCH* melanin-concentrating hormone, *ORX* orexin. Redrawn from Takei et al. (2012)

reasonable to predict that there is integration of the appetite control system with osmoregulation. This was examined in *N. alexis*, in which water deprivation for either 12 or 29 days induced a biphasic pattern of food consumption where an initial hypophagia was followed by a sustained increase in appetite for the latter phase of water restriction that was significantly above that of water-replete animals (Takei et al. 2012). The initial changes in food intake were driven by appropriate changes in plasma leptin and ghrelin (Takei et al. 2012), but the sustained food intake observed in the later phase of the 29 water deprivation occurred despite an elevated plasma leptin that should have acted as a satiety signal in the hypothalamus (Donald et al. 2012). Thus, it is possible that the signals controlling energy balance are overridden by the osmoregulatory drive to increase substrate provision for metabolic water production. This may occur centrally in the hypothalamus as the expression of orexigenic and anorectic neuropeptide genes was regulated in a direction that would stimulate appetite (Takei et al. 2012; Donald et al. 2012; summarised in Fig. 10.3). It was also found that as water deprivation was prolonged, body fat disappeared but body weight increased gradually, which was attributed to hepatic glycogen storage (Takei et al. 2012; Fig. 10.3). Switching metabolic strategy from lipids to carbohydrates would enhance metabolic water production per oxygen molecule, thus facilitating lower EWL (Takei et al. 2012). A critical parallel observation in the study with *N. alexis* was that laboratory mice

(*Mus musculus*) subjected to the same water restriction regime but feed a low protein diet maintained a constant food intake and constant lower body mass during the period of the experiment when compared to water-replete mice (Takei et al. 2012). Thus, the food intake pattern in *N. alexis* confronted by an absence of free water could be a key xeric adaptation linked to osmoregulation, and this species provides an excellent animal model to study the integrated control of water and energy balance.

10.5 Endocrinology of Osmoregulation in Desert Mammals

The endocrine regulation of fluid balance in desert mammals has focussed on the roles of AVP and the renin-angiotensin aldosterone system (RAAS), because of their importance in regulating mechanisms that facilitate osmotic and fluid volume homeostasis in desiccating terrestrial environments. This paradigm is primarily based on human and animal studies in which water restriction leads to dehydration and hyperosmolality, which activate AVP signalling and the RAAS, respectively, to increase renal water reabsorption and restore the ECF. However, as indicated above, desert mammals can often tightly regulate plasma osmolarity and ECF in the face of water restriction; therefore classical activation of AVP and RAAS may not be apparent.

A number of studies in desert species have used traditional endocrinological methodologies, such as radioimmunoassay, to examine the hormonal control of osmoregulation. In comparison to non-desert mammals, the plasma levels of AVP in desert rodents such as *A. cahirinus* and *A. russatus* (Castel et al. 1974), *Gerbillus gerbillus* and *Jaculus jaculus* (El Husseini and Haggag 1974), *J. orientalis* (Baddouri et al. 1984), and kangaroo rat (Stallone and Braun 1988) are higher than mesic species. As an example, the plasma AVP levels of *J. orientalis* are 100–200 times greater than those of a laboratory rat and 20 times higher than that of laboratory rats water-deprived for 2 days (Baddouri et al. 1984). Interestingly, an anatomical study of the hypothalamo-neurohypophysial system of two desert rodents, *P. obesus* and *G. tarbuli*, found it to be in a hyperactivated state based on the morphology of the secretory apparatus, suggesting that neurosecretion of AVP was elevated in these species (Ouali-Hassenaoui et al. 2011). Furthermore, *N. alexis* was found to have a large neurohypophysis that contained three times the amount of AVP per unit of body weight compared to a laboratory rat (Bridges and James 1982). These data indicate that desert species may have a greater capacity to synthesise and release AVP than mesic species. However, in *N. alexis* water-deprived for periods up to 28 days, the ECF volume and plasma osmolality were unaffected, and therefore, the hyperosmotic stimulus that triggers AVP release from the posterior pituitary may not be apparent (Heimeier and Donald 2006). Indeed, plasma AVP was found not to change over the course of 28 days of water deprivation in *N. alexis* (Donald et al. unpublished). This observation is supported by a 7 day water deprivation experiment in the same species in which blood pressure was measured and found to be unaffected by DAVP [1-(β-Mercapto-β,β-cyclopenta-methylene-proprionic acid) 8-D-arginine vasopressin], an antagonist of AVP signalling, indicating that tonic AVP-mediated vasoconstriction was not apparent during water deprivation (Weaver et al. 1994). In contrast to hopping mice, plasma AVP increased linearly with an increased plasma osmolarity in kangaroo rats fed a high protein diet (Stallone and Braun 1988). Furthermore, water deprivation for 20 days doubled the plasma AVP concentration in one-humped camels in which plasma Na increased (Ali et al. 2012); a similar observation was made in a short-term water deprivation experiment in the same species (Ben Goumi et al. 1993). An interesting finding in two species of desert-dwelling marsupial wallabies of Western Australia was the contrasting data on the relationship between urine osmolarity and plasma lysine vasopressin (LVP), the antidiuretic hormone of macropodid marsupials (Bradshaw et al. 2001). In the spectacled hare-wallaby, *Lagorchestes conspicillatus*, water deprivation resulted in the production of a concentrated urine and a concomitant increase in plasma LVP. In contrast, Rothschild's rock wallabies, *Petrogale rothschildi*, have a limited renal response to water deprivation that is primarily driven by a reduced glomerular filtration rate that reduces urine volume, and unchanged plasma LVP levels that resulted in negligible up-regulation of tubular water reabsorption (Bradshaw et al. 2001). In summary, AVP is most likely released in response to ECF hyperosmolarity in desert mammals, but many species can maintain plasma osmolarity and ECF when water-restricted and therefore suppress osmotically-driven up-regulation of AVP release.

With respect to the RAAS, the kangaroo rat, *D. spectabilis* and gerbil, *M. unguiculatus*, were found to have higher basal plasma angiotensin II levels than the laboratory rat. However, water deprivation for 4 days caused a 500 % increase in plasma angiotensin II in laboratory rats, but only a 50 % increase in *D. spectabilis and M. unguiculatus* (Wright and Harding 1980). In addition, *N. alexis* was found to have higher basal levels of plasma renin and angiotensinogen than laboratory mice (Weaver et al. 1994). When hopping mice were water-deprived for 7 days, plasma angiotensinogen and renin concentrations increased, but after 28 days of water deprivation, the levels were not different from those measured in control animals with access to water (Weaver et al. 1994). Higher basal plasma angiotensin II levels in desert rodents may serve to reduce water loss by a vasoconstriction-mediated decrease in renal blood flow and a concomitant reduction in the glomerular filtration rate (Weaver et al. 1994). It could also lead to higher plasma aldosterone and AVP levels, thus increasing osmotic and free water reabsorption in the kidney (Weaver et al. 1994). However, in the one-humped camel, two separate water deprivation studies found that plasma aldosterone was not significantly increased despite an increase in plasma osmolality and a decrease in ECF volume (Ben Goumi et al. 1993; Ali et al. 2012); in the earlier study, plasma renin was also elevated indicating that the RAAS had been activated. The absence of an increase in plasma aldosterone was attributed to a feedback response in which the elevated plasma sodium inhibited aldosterone release (Ali et al. 2012). When hypovolemia was induced in camels with furosemide, plasma renin, aldosterone and AVP all increased significantly, indicating a typical volume activation of the fluid-conserving hormone systems (Riad et al. 1994).

In addition to AVP and the RAAS, there are two families of signalling peptides that regulate salt and water handling by the kidney and the intestine by guanosine 3′:5′-cyclic monophosphate (cGMP) signalling (Lucas et al. 2000; Heimeier et al. 2004). The natriuretic peptides (atrial natriuretic peptide [ANP], B-type natriuretic peptide [BNP], and c-type natriuretic peptide [CNP]) and guanylin peptides (guanylin and uroguanylin) are signalling molecules that act by binding to and activating their respective guanylyl cyclase (GC) receptors that increase cGMP production (Takei et al. 2011). ANP, in particular, is released in response to hypervolemia, and primarily mediates diuresis and natriuresis by the kidney to reduce ECF volume. Guanylin has similar effects in the kidney and inhibits the absorption of water and salt by the intestine. Therefore, ANP and guanylin oppose the actions of AVP and RAAS, and it is intriguing to consider how these systems may function in desert mammals that live in environments where they would not normally be confronted with hypervolaemia.

There are several studies that have examined the expressed levels of ANP peptide in the heart and plasma of desert mammals. One study found that the cardiac, kidney and plasma ANP peptide levels in two North African desert rodents, *P. obesus* and *M. libycus*, were lower in comparison to laboratory rats (Lacas et al. 1998; Bachar and Lichtstein 1993). These observations suggested that ANP-mediated diuresis and natriuresis could be lower in desert rodents compared to mesic species. In addition, Osman et al. (2004) cloned both ANP and BNP from the heart of the one-humped camel and provided a detailed description of the distribution of ANP in the heart. The effect of water deprivation on the natriuretic peptide system of desert mammals has been investigated in several species, and there appears to be an inconsistent response (Lacas et al. 2000). Eight days of water deprivation increased cardiac ANP peptide levels, but decreased plasma ANP levels, in the desert rodent *Taterillus gracillus*. However, these parameters were unchanged in *Steatomys caurinus* subjected to the same period of water deprivation (Lacas et al. 2000). In *N. alexis*, the plasma ANP level was significantly reduced after 3 days of water deprivation but was unaffected by 7, 14 and 28 days of water deprivation, respectively, when compared to water-replete animals (Heimeier and Donald 2006). At the genetic level, the same periods of water deprivation had a variable effect on the mRNA expression of ANP, CNP, NPR-A (ANP and BNP receptor), and NPR-B (CNP receptor) and a uniform down-regulation that was predicted was not observed (Heimeier and Donald 2006). Interestingly, after 28 days of water deprivation, the ANP mRNA expression in the heart and kidney was similar to water-replete animals (Heimeier and Donald 2006). Clearer hormonal responses to volume manipulation were found in the one-humped camel, in which water deprivation for 20 days that caused dehydration (as indicated by elevated plasma sodium levels), decreased plasma ANP and BNP levels, respectively, (Adem et al. 2013). However, an earlier study in the same species found that plasma ANP levels were not significantly affected by 14 days of water deprivation (Ben Goumi et al. 1993). In contrast to dehydration, volume loading of camels by saline infusion increased plasma ANP levels (Dahlborn et al. 1992). Taken together, the data in camels indicate that the response of the ANP system is similar in a large desert mammal as it is in mesic species.

In addition to the natriuretic peptide system, the expression of guanylin and uroguanylin mRNAs in the kidney of *N. alexis*, was not affected by water deprivation (Donald and Bartolo 2003). Furthermore, the receptor for guanylin and uroguanylin, GC-C, which is abundantly expressed in the mouse kidney was found not to be expressed in the kidney of *N. alexis* (Donald and Bartolo 2003), possibly indicating that guanylin is not functional in the kidney. In the colon of *N. alexis*, water deprivation increased the mRNA expression of guanylin, uroguanylin and GC-C, which was not predicted as guanylin peptides inhibit fluid reabsorption in the colon (Donald and Bartolo 2003); as discussed above, water reabsorption in the colon is important in desert mammals for water balance.

In summary, the endocrinological contribution to osmoregulation in desert mammals appears to be dependent on the species being studied. In the small desert rodents, the basal plasma levels of AVP and components of the RAAS appear to be higher than those of non-desert mammals, and the expression of ANP is correspondingly lower. However, when small non-desert mammals are water-deprived, they have a much greater capacity to up- or down-regulate these hormone systems to compensate ECF volume and osmolarity than is observed in desert-adapted mammals. The ability of many small desert mammals to tightly regulate plasma osmolarity and ECF volume when water-deprived may mean that the activation of hormonal systems controlling fluid balance is suppressed; hence, clear directional changes are not observed experimentally. In contrast to small mammals, experiments in camels show that plasma osmolarity and ECF volume are clearly affected by water deprivation and that this stimulates the predicted changes in the endocrine control of fluid balance.

10.6 Genomic Approaches to Understanding Osmoregulation in Desert Mammals

Most of the research on the adaptations of desert mammals was performed prior to the advent of comparative genomic and transcriptomic approaches. However, these approaches are now being applied to address questions on the genetic control of osmoregulation, with primary focus on renal function. There are four studies published to date that are technical in nature, therefore only a brief overview will be provided here.

Given its important role as a model desert species, the kangaroo rat was used for transcriptomic analysis to identify osmoregulatory genes that are overexpressed in the kidney compared to a tissue not involved in osmoregulation, such as the spleen (Marra et al. 2012). This study identified 32 genes that were significantly over-expressed in the kidney of which 4 were solute carriers (*Slc12a1*, *Slc13a1*,

Slc22a9, *Slc27a2*) that could be involved in water regulation. In a follow-up study, the kidney transcriptomes of two desert species (*D. spectabilis* and the pocket mouse, *Chaetodipus baileyi*) were compared to a tropical heteromyid species (*Heteromys desmarestianus*). It was found that 1890 genes were differentially expressed between the desert and mesic species and that several osmoregulatory genes were under positive selection (Marra et al. 2014). Within the differentially expressed genes, a range of solute carrier proteins were either up-or down-regulated in desert species relative to mesic species [see Tables S1 and S2, Marra et al. (2014)]. Interestingly, the AVP V_2 receptor had a lower expression in the desert species, which would not be predicted as it may indicate a lower AVP-regulated water reabsorption in the kidney. A similar type of RNA sequencing study that focussed on gene selection in cactus mouse, *Peromyscus eremicus*, found positive selection for the *Slc2a9* gene that encodes a facilitated glucose transporter (MacManes and Eisen 2014). It is intriguing that positive selection for *Slc2a9* gene was also found for the desert species in the study of Marra et al. (2014).

In the Bactrian camel (*C. bactrianus*), transcriptome analysis has been used to determine differential gene expression in water-replete camels compared to those that had been on 29 days of water restriction (Wu et al. 2014); some key findings are summarised here. It was found that the *AQP1*, *AQP2* and *AQP3* genes showed the most significant up-regulation in the kidney of water-restricted camels, indicating the importance of water reabsorption in the kidney. Furthermore, there was a significant down-regulation of TonEBP and its target genes (*TauT*, *BGT1*, *and SMIT*), which raises interesting questions about how camels overcome hypertonic stress in the renal medulla (Wu et al. 2014). It also indicates that transcriptional regulation of the *AQP2* gene is more likely to be by AVP signalling rather than TonEBP (see above). In addition to *AQP*, the genes encoding the epithelial Na channel (ENaC) and Na-K-ATPase were significantly upregulated indicating increased Na reabsorption in the kidney. An intriguing result was the significant up-regulation in the kidney of the *GLUT1* gene and the genes involved in glycolysis (Wu et al. 2014), which together with the data on the glucose transporter in rodents, indicates that renal glucose regulation may be important in osmoregulation in desert mammals.

10.7 Perspective

The accepted paradigms for how mammals survive in deserts that prevail in textbooks have often focussed attention on small rodents such as the kangaroo rat, *Dipodomys* spp., of North America (Walsberg 2000). As discussed by Walsberg (2000), the accepted physiological adaptations of this iconic species to aridity have been based on research that was predicated on the idea that kangaroo rats consume dry carbohydrate-rich seeds and do not drink water, and are therefore dependent on metabolic water production (MacMillen and Hinds 1983). In addition, EWL is minimised by behavioural avoidance of daytime heat, and occupancy of cool (relative to surface temperatures), humid burrows. A similar scenario is also generally proposed for the hopping mouse, *N. alexis*, of central Australia and other Australian small desert mammals. However, in kangaroo rats at least, this paradigm has been challenged by ecophysiological studies that show that they are active during the hottest time of the night (sundown) in surface temperatures that are higher than previously known, and that the burrows had a much lower humidity than what had been reported (Tracy and Walsberg 2002). Therefore, kangaroo rats would be subjected to a much greater water stress than previously thought (Walsberg 2000). However, to offset water loss, ecophysiological studies focussed on foraging found that kangaroo rats consume more preformed water in the diet through selecting insects and succulent vegetation and are actually not strict granivores (Tracy and Walsberg 2002). Interestingly, food preference experiments with Australian hopping mice also found that they are omnivorous and will select food with preformed water (Murray and Dickman 1994). Another important observation in *D. merriami* was that EWL was the primary means of water conservation in conspecifics from locations that had different ambient temperature and aridity, as the urine concentrating ability of kangaroo rats was similar at each location and therefore not correlated with aridity (Tracy and Walsberg 2002). These studies, and those in South American rodents (Bozinovic and Gallardo 2006), highlight the importance of integrating ecology and physiology in understanding how mammals live in deserts.

However, despite potential revisions to the paradigms on how mammals survive in deserts, there remain many fascinating aspects of their ability to survive without water that should be the subject of future research. The ability of the kidney of small desert mammals to produce concentrated urine consistently in the range of 5000–7000 mOsm kg^{-1}, which means that the inner medullary cells are bathed in ECF of a similar concentration, provokes interesting questions in cellular biology about how these cells survive such a harsh hypertonic environment and inhibit apoptosis. In addition, desert rodents are potentially excellent models to study the role of hypertonicity in the regulation of aquaporin expression and trafficking in the renal medulla. Finally, the essential requirement for metabolic water that is dependent on substrate provision from food, provokes interesting questions on the integration of the hypothalamic regulation of energy and water balance.

Acknowledgement The authors acknowledge financial support from Deakin University (JD) and the National Science Foundation and the National Institute of Diabetes and Digestive and Kidney Diseases of the National Institute of Health (TP).

References

Adem A, Al Haj M, Benedict S et al (2013) ANP and BNP Responses to dehydration in the one-humped camel and effects of blocking the renin-angiotensin system. PLoS One 8:1–5

Ali MA, Adem A, Chandranath IS et al (2012) Responses to dehydration in the one-humped camel and effects of blocking the renin-angiotensin system. PLoS One 7:1–4

Aw M, Evans K, Pannabecker TL (2014) Increased expression of sodium transport proteins and Na, K-ATPase activity in the outer medulla of kangaroo rat is related to its greater urine concentrating ability compared to Sprague-Dawley rat. FASEB J 28:1137.7

Bachar H, Lichtstein D (1993) Distribution of atrial natriuretic peptides in the sand rat (*Psammomys obesus*) in comparison to that in the rat. J Basic Clin Physiol Pharmacol 4:47–56

Baddouri K, Butlen D, Imbert-Temboul M et al (1984) Plasma antidiuretic hormone levels and kidney responsiveness to vasopressin in the jerboa, *Jaculus orientalis*. Gen Comp Endocrinol 54:203–215

Bartolo RC, Donald JA (2007) The distribution of renal hyaluronan and the expression of hyaluronan synthases during water deprivation in the Spinifex hopping mouse, *Notomys alexis*. Comp Biochem Physiol A Mol Integr Physiol 148:853–860

Bartolo RC, Donald JA (2008) The effect of water deprivation on the tonicity responsive enhancer binding protein (TonEBP) and TonEBP-regulated genes in the kidney of the Spinifex hopping mouse, *Notomys alexis*. J Exp Biol 211:852–859

Baudinette RV (1972) The impact of social aggregation on the respiratory physiology of Australian hopping mice. Comp Biochem Physiol 41:35–38

Ben Goumi M, Riad F, Giry J et al (1993) Hormonal control of water and sodium in plasma and urine of camels during dehydration and rehydration. Gen Comp Endocrinol 89:378–386

Beuchat CA (1990a) Body size, medullary thickness, and urine concentrating ability in mammals. Am J Physiol 258:R298–R308

Beuchat CA (1990b) Metabolism and the scaling of urine concentrating ability in mammals: resolution of a paradox? J Theor Biol 143:113–122

Bozinovic F, Gallardo P (2006) The water economy of South American desert rodents: from integrative to molecular physiological ecology. Comp Biochem Physiol C Pharmacol Toxicol Endocrinol 142:163–172

Bozinovic F, Gallardo PA, Visser GH et al (2003) Seasonal acclimatization in water flux rate, urine osmolality and kidney water channels in free-living degus: molecular mechanisms, physiological processes and ecological implications. J Exp Biol 206:2959–2966

Bradshaw SD, Morris KD, Bradshaw FJ (2001) Water and electrolyte homeostasis and kidney function of desert-dwelling marsupial wallabies in Western Australia. J Comp Physiol B 171:23–32

Bridges T, James N (1982) The hypothalamo-neurohypophysial system of native Australian desert rodents. The vasopressin and oxytocin contents of hypothalamus and posterior pituitary of *Notomys alexis* and *Pseudomys australis* compared with those of the laboratory rat and mouse in different states of water balance. Aust J Exp Biol Med Sci 60:265–283

Buret A, Hardin J, Olson ME et al (1993) Adaptation of the small intestine in desert-dwelling animals: morphology, ultrastructure and electrolyte transport in the jejunum of rabbits, rats, gerbils and sand rats. Comp Biochem Physiol A Physiol 105:157–163

Cain JW III, Krausman PR, Rosenstock SS et al (2006) Mechanisms of thermoregulation and water balance in desert ungulates. Wildl Soc Bull 34:570–581

Castel M, Borut A, Haines H (1974) Blood titres of vasopressin in various murids (Mammalian: Rodentia). Isr J Zoo 23:208–209

Dahlborn K, Hossainihilali J, Benlamlih S et al (1992) Plasma atrial natriuretic peptide (ANP) concentration and fluid balance during rapid intravenous saline infusions in camels. Acta Physiol Scand 144:485–486

Degen AA (1997) Ecophysiology of small desert mammals. Springer, Berlin

Diaz GB, Ojeda RA, Dacar M (2001) Water conservation in the South American desert mouse opossum, *Thylamys pusilla* (Didelphimorphia, Didelphidae). Comp Biochem Physiol A Mol Integr Physiol 130:323–330

Dickinson H, Walker DW, Cullen-McEwen L et al (2005) The spiny mouse (*Acomys cahirinus*) completes nephrogenesis before birth. Am J Physiol Renal Physiol 289:F273–F279

Dickinson H, Moritz K, Wintour EM et al (2007) A comparative study of renal function in the desert-adapted spiny mouse and the laboratory-adapted C57BL/6 mouse: response to dietary salt load. Am J Physiol Renal Physiol 293:F1093–F1098

Donald JA, Bartolo RC (2003) Cloning and mRNA expression of guanylin, uroguanylin, and guanylyl cyclase C in the Spinifex Hopping mouse, *Notomys alexis*. Gen Comp Endocrinol 132:171–179

Donald JA, Hamid NKA, Horvath P et al. (2012) Appetite regulation in desert-adapted Spinifex hopping mice during water deprivation. FASEB J 26:1069.4

Doucet A, Barlet C, Baddouri K (1987) Effect of water intake on Na-K-ATPase in nephron segments of the desert rodent, *Jaculus orientalis*. Pflugers Arch 408:129–132

El Husseini M, Haggag G (1974) Antidiuretic hormone and water conservation in desert rodents. Comp Biochem Physiol A Mol Integr Physiol 47:347–350

Espineira M, Pannabecker T (2014) Increase in collecting duct basal: apical AQP2 protein expression ratio with increasing depth along the corticopapillary axis positively correlates with maximum urine concentrating capacity in the kangaroo rat and Sprague-Dawley rat. FASEB J 28:1137.2

Frank CL (1988a) The relationship of water content, seed selection, and the water requirements of a heteromyid rodent. Physiol Zool 61:52–534

Frank CL (1988b) Diet selection by a heteromyid rodent—role of net metabolic water production. Ecology 69:1943–1951

Fuller A, Hetem RS, Maloney SK et al (2014) Adaptation to heat and water shortage in large, arid-zone mammals. Physiology (Bethesda) 29:159–167

Gallardo P, Cid LP, Vio CP et al (2001) Aquaporin-2, a regulated water channel, is expressed in apical membranes of rat distal colon epithelium. Am J Physiol Gastrointest Liver Physiol 281: G856–G863

Gallardo P, Olea N, Sepulveda FV (2002) Distribution of aquaporins in the colon of *Octodon degus*, a South American desert rodent. Am J Physiol Regul Integr Comp Physiol 283:R779–R788

Gallardo PA, Cortes A, Bozinovic F (2005) Phenotypic flexibility at the molecular and organismal level allows desert-dwelling rodents to cope with seasonal water availability. Physiol Biochem Zoo 78:145–152

Gordge L, Roberts JR (2008) Kidney function in the Spinifex hopping mouse, *Notomys alexis*. Comp Biochem Physiol A Physiol 150:90–101

Gottschalk CW, Lassiter WE, Mylle M et al (1963) Micropuncture study of composition of loop of Henle fluid in desert rodents. Am J Physiol 204:532–535

Hasler U, Jeon US, Kim JA et al (2006) Tonicity-responsive enhancer binding protein is an essential regulator of aquaporin-2 expression in renal collecting duct principal cells. J Am Soc Nephrol 17:1521–1531

Heimeier RA, Donald JA (2006) The response of the natriuretic peptide system to water deprivation in the desert rodent, *Notomys alexis*. Comp Biochem Physiol A Mol Integr Physiol 143:193–201

Heimeier RA, Bartolo RC, Donald JA (2004) The effect of water deprivation on signalling molecules that utilise cGMP in the Spinifex hopping mouse *Notomys alexis*. Aust Mammal 26:191–198

Hetem RS, Strauss WM, Fick LG et al (2010) Variation in the daily rhythm of body temperature of free-living Arabian oryx (Oryx leucoryx): does water limitation drive heterothermy? J Comp Physiol B 180:1111–1119

Hetem RS, Strauss WM, Fick LG et al (2012a) Does size matter? Comparison of body temperature and activity of free-living Arabian oryx (*Oryx leucoryx*) and the smaller Arabian sand gazelle (*Gazella subgutturosa marica*) in the Saudi desert. J Comp Physiol B 182:437–449

Hetem RS, Strauss WM, Fick LG et al (2012b) Selective brain cooling in Arabian oryx (*Oryx leucoryx*): a physiological mechanism for coping with aridity? J Exp Biol 215:3917–3924

Huang Y, Tracy R, Walsberg GE et al (2001) Absence of aquaporin-4 water channels from kidneys of the desert rodent *Dipodomys merriami merriami*. Am J Physiol Renal Physiol 280:F794–F802

Issaian T, Urity VB, Dantzler WH et al (2012) Architecture of vasa recta in the renal inner medulla of the desert rodent *Dipodomys merriami*: potential impact on the urine concentrating mechanism. Am J Physiol Regul Integr Comp Physiol 303:R748–R756

Knepper MA, Saidel GM, Hascall VC et al (2003) Concentration of solutes in the renal inner medulla: interstitial hyaluronan as a mechano-osmotic transducer. Am J Physiol Renal Physiol 284:F433–F446

Krebs HA (1975) The August Krogh principle: "for many problems there is an animal on which it can be most conveniently studied". J Exp Zool 194:221–226

Lacas S, Bentchikou M, Gabrion J et al (1998) Presence of atrial natriuretic peptide in two desert rodents—comparison with rat. Peptides 19:715–726

Lacas S, Allevard AM, Ag'atteinine S et al (2000) Cardiac natriuretic peptide response to water restriction in the hormonal adaptation of two semidesert rodents from west Africa (*Steatomys caurinus*, *Taterillus gracilis*). Gen Comp Endocrinol 120:176–189

Lovegrove BG (2000) The zoogeography of mammalian basal metabolic rate. Am Nat 156:201–219

Lovegrove BG (2003) The influence of climate on the basal metabolic rate of small mammals: a slow-fast metabolic continuum. J Comp Physiol B 173:87–112

Lucas K, Pitari GM, Kezerounian S, Ruiz-Stewart I et al (2000) Guanylyl cyclases and signalling by cyclic GMP. Pharmacol Rev 52:375–413

MacManes MD, Eisen MB (2014) Characterization of the transcriptome, nucleotide sequence polymorphism, and natural selection in the desert adapted mouse *Peromyscus eremicus*. PeerJ 2:e642

MacMillen RE, Hinds DS (1983) Water regulatory efficiency in heteromyid rodents: a model and its application. Ecology 64:152–164

MacMillen RE, Lee AK (1967) Australian desert mice: independence of exogenous water. Science 158:383–385

Marra NJ, Eo SH, Hale MC et al (2012) *A priori* and *a posteriori* approaches for finding genes of evolutionary interest in non-model species: osmoregulatory genes in the kidney transcriptome of the desert rodent *Dipodomys spectabilis* (banner-tailed kangaroo rat). Comp Biochem Physiol Part D Genomics Proteomics 7:328–339

Marra NJ, Romero A, DeWoody JA (2014) Natural selection and the genetic basis of osmoregulation in heteromyid rodents as revealed by RNA-seq. Mol Ecol 23:2699–2711

Murray BR, Dickman CR (1994) Food preferences and seed selection in two species of Australian desert rodents. Wildl Res 21:647–655

Murray BR, Dickman CR (1997) Factors affecting selection of native seeds in two species of Australian desert rodents. J Arid Environ 35(3):517–525

Murray BR, Hume ID, Dickman CR (1995) Digestive tract characteristics of the Spinifex hopping-mouse, *Notomys alexis* and the sandy inland mouse, *Pseudomys hermannsburgensis* in relation to diet. Aust Mammal 18:93–97

Osman AHK, Yuge S, Hyodo S et al (2004) Molecular identification and immunohistochemical localization of atrial natriuretic peptide in the heart of the dromedary camel (*Camelus dromedarius*). Comp Biochem Physiol A Mol Integr Physiol 139:417–424

Ostrowski S, Williams JB (2006) Heterothermy of free-living Arabian sand gazelles (*Gazella subgutturosa marica*) in a desert environment. J Exp Biol 209:1421–1429

Ouali-Hassenaoui S, Bendjelloul M, Dekar A et al (2011) Distribution of osmoregulatory peptides and neuronal-glial configuration in the hypothalamic magnocellular nuclei of desert rodents. C R Biol 334:855–862

Pannabecker TL (2013) Comparative physiology and architecture associated with the mammalian urine concentrating mechanism: role of inner medullary water and urea transport pathways in the rodent medulla. Am J Physiol Regul Integr Comp Physiol 304:R488–R503

Riad F, Goumi MB, Giry J et al (1994) Renin-aldosterone axis and arginine vasopressin responses to sodium depletion in camels. Gen Comp Endocrinol 95:240–247

Sands JM, Layton HE (2014) Advances in understanding the urine-concentrating mechanism. Annu Rev Physiol 76:387–409

Schmidt-Nielsen K (1972) How animals work. Cambridge University Press, Cambridge

Schmidt-Nielsen K (1975) Animal physiology. Vail-Ballou Press, Binghamton, NY

Schmidt-Nielsen K (1979) Desert animals: physiological problems of heat and water. Dover, New York

Schmidt-Nielsen K, Schmidt-Nielsen B (1952) Water metabolism of desert mammals. Physiol Rev 32:135–166

Schmidt-Nielsen B, Schmidt-Nielsen K, Brokaw A et al (1948) Water conservation in desert rodents. J Cell Comp Physiol 32:331–360

Schwimmer H, Haim A (2009) Physiological adaptations of small mammals to desert ecosystems. Integr Zool 4:357–366

Stallone JN, Braun EJ (1988) Regulation of plasma antidiuretic hormone in the dehydrated kangaroo rat (*Dipodomys spectabilis* M.). Gen Comp Endocrinol 69:119–127

Stridh S, Palm F, Hansell P (2012) Renal interstitial hyaluronan: functional aspects during normal and pathological conditions. Am J Physiol Regul Integr Comp Physiol 302:R1235–R1249

Takei Y, Inoue K, Trajanovska S et al (2011) B-type natriuretic peptide (BNP), not ANP, is the principal cardiac natriuretic peptide in vertebrates as revealed by comparative studies. Gen Comp Endocrinol 171:258–266

Takei Y, Bartolo RC, Fujihara H et al (2012) Water deprivation induces appetite and alters metabolic strategy in *Notomys alexis*: unique mechanisms for water production in the desert. Proc Biol Sci 279:2599–2608

Tracy RL, Walsberg GE (2002) Kangaroo rats revisited: re-evaluating a classic case of desert survival. Oecologia 133:449–457

Urity VB, Issaian T, Braun EJ et al (2012) Architecture of kangaroo rat inner medulla: segmentation of descending thin limb of Henle's loop. Am J Physiol Regul Integr Comp Physiol 302: R720–R726

Walsberg GE (2000) Small mammals in hot deserts: some generalization revisited. Bioscience 50:109–120

Weaver D, Walker L, Alcorn D et al (1994) The contributions of renin and vasopressin to the adaptation of the Australian Spinifex Hopping-mouse (*Notomys alexis*) to free water deprivation. Comp Biochem Physiol A Mol Integr Physiol 108:107–116

Wright JW, Harding JW (1980) Body dehydration in xeric adapted rodents: does the renin-angiotensin system play a role? Comp Biochem Physiol A Mol Integr Physiol 66:181–188

Wu H, Guang X, Al-Fageeh MB et al (2014) Camelid genomes reveal evolution and adaptation to desert environments. Nat Commun 5:1–9

Chapter 11
Renal Medullary Functional Architecture and the Urinary Concentrating Mechanism

Thomas L. Pannabecker

Abstract Most mammals are capable of producing urine having an osmolality that is greater than that of plasma. Production of this concentrated urine is one of the chief mechanisms by which the kidney maintains salt and water balance. Many underlying principles of fluid and solute transport in nephrons and blood vessels are understood to some extent; but how fluid and solute movements throughout the renal medulla are integrated into a unified mechanism for concentrating urine remains poorly understood. During the past two decades we have made important advances in understanding medullary architecture, tubular structure and membrane transporters associated with transepithelial solute and water fluxes within the medulla. These advances have led to identification of significant functional interactions between tubules and interstitium. Defects in expression and function of a number of the medullary membrane transporters are known to underlie several human disorders associated with Na and water imbalance and have provided new insights into the urine concentrating mechanism.

Keywords Aquaporin • NaCl • ClC-K • Renal medulla • Urea • Loops of Henle • Collecting duct • Vasa recta

11.1 Theories That Have Been Proposed to Explain the Urine Concentrating Mechanism

Despite a number of influential though still contested hypotheses (Table 11.1), the mechanism by which mammals concentrate urine remains a mystery. The ability of the kidney to produce a concentrated urine requires a solute gradient of increasing osmolality from the corticomedullary junction to the papilla tip (Fig. 11.1) (Gottschalk and Mylle 1959; Hai and Thomas 1969; Pannabecker 2013). This solute gradient consists chiefly of NaCl and urea. Urine exits the kidney by way of the collecting duct, the final segment of the uriniferous tubule. But in the hydropenic

T.L. Pannabecker (✉)
Department of Physiology, University of Arizona Health Sciences Center, AHSC 4128, 1501 N Campbell Avenue, Tucson, AZ 85724-5051, USA
e-mail: pannabec@email.arizona.edu

K.A. Hyndman, T.L. Pannabecker (eds.), *Sodium and Water Homeostasis*, Physiology in Health and Disease, DOI 10.1007/978-1-4939-3213-9_11

Table 11.1 Commonly accepted hypotheses in support of the urine concentrating mechanism (UCM) and challenges to the hypotheses

UCM Hypothesis	Challenged by:
Countercurrent multiplication[a]	Layton and Layton (2011), Zhai et al. (2007)
Loop of Henle length	Greenwald and Stetson (1988), Beuchat (1990), Schmidt-Nielsen and O'Dell (1961)
Inner medulla sodium chloride/urea mixing[a]	Fenton et al. (2004)
Descending limb of Henle's loop Na+, urea and fluid flux[a]	Nawata et al. (2014), Chou and Knepper (1992, 1993), Urity et al. (2012), Gilbert and Pannabecker (2013)

[a]Tenets of the widely accepted "passive mechanism hypothesis" (Stephenson 1972; Kokko and Rector 1972)

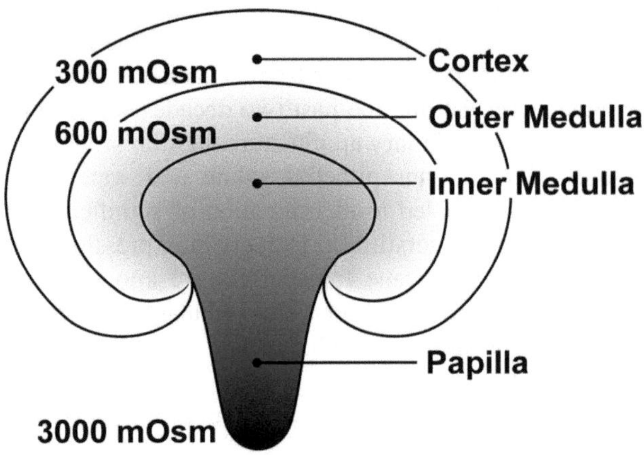

Fig. 11.1 Schematic diagram of a rodent kidney coronal section. The corticopapillary solute gradient that occurs in the kidney of the dehydrated rat is depicted (mosmol/Kg H_2O). The cortex is isosmotic to plasma (300 mosmol), whereas the junction between the cortex and outer medulla is slightly hyperosmotic; the outer and inner medulla become increasingly hyperosmotic with maximal osmolality at the papilla tip (3000 mosmol)

condition, long before urine reaches this final segment to exit the kidney, the antidiuretic hormone vasopressin has increased the osmotic water permeability of the collecting duct epithelium. This increased water permeability leads to increased water movement from the tubule lumen into the more concentrated interstitium (Fig. 11.2). Consequently, as water diffuses across the epithelium and out of the collecting duct the osmolality of the tubular fluid increases, yielding concentrated urine.

One process widely acknowledged to be essential to producing concentrated urine is the energy-requiring step of active sodium reabsorption by the thick ascending limb (Fig. 11.2). Sodium reabsorption from the lumen of the outer medullary thick ascending limb dilutes tubular fluid and is thought to directly raise outer medullary interstitial fluid osmolality. Sodium chloride reabsorption from this segment involves the Na-K-2Cl cotransporter (NKCC2) at the apical membrane and the Na-K-ATPase (the sodium pump) at the basolateral membrane. NKCC2 carries sodium down an electrochemical gradient into the cell and Na-K-ATPase pumps sodium out of the cell against a gradient. The chloride channel ClC-K2 in rodents (and its human ortholog ClCKB) serves as a pathway for efflux of the counterion chloride out of the cell. The potassium channel ROMK serves as an apical channel for recycling potassium from cell to lumen. Active basolateral extrusion of sodium by the sodium pump maintains intracellular sodium concentrations sufficiently low to sustain the inward sodium flux carried by NKCC2 (Mount 2012) (Fig. 11.3). The collecting duct also carries out active sodium reabsorption, particularly in the inner medulla; however, sodium reabsorption in the collecting duct is considered minor compared to that which occurs in the thick ascending limb. Mitochondria are more abundant in thick ascending limbs than in the collecting ducts, but are sparse in thin limbs of Henle's loops and there is no evidence of significant active transport in either descending or ascending thin limb segments; nearly all transepithelial transport in thin limbs of Henle's loops is considered to be passive.

The predominance of passive solute flows and relative absence of active transport in epithelia of the inner medulla, the region of the kidney with the greatest increase in urine osmolality, has led to the hypothesis that the urine concentrating mechanism is primarily a passive process. However, extensive theoretical models have been unsuccessful in explaining how urine is concentrated by the predominantly passive processes existing in the hydropenic kidney (Layton et al. 2009, 2012; Layton 2011; Wexler et al. 1991a, b; Thomas 2000, 2001). On the basis of functional studies it is increasingly clear that the passive mechanism for producing a concentrated urine, at least as originally proposed (Stephenson 1972; Kokko and Rector 1972), is not the correct model (Chou and Knepper 1992, 1993; Fenton et al. 2004; Nawata et al. 2014). These studies and others underlie the motivation to identify alternative processes in the medulla that could participate in producing concentrated urine.

To summarize, although sodium reabsorption by active transport in the thick ascending limb is thought to play a direct role in production of the solute gradient in the outer medulla (Fig. 11.2), it only plays an indirect role in producing the solute gradient in the inner medulla, where the osmotic gradient is steepest (Pannabecker 2013). Because transepithelial solute fluxes across all other medullary tubules are either passive or facilitative, the only active step available for production of the steep osmotic gradient is active transport of sodium by the thick ascending limb; however, the way in which active sodium transport is coupled with fluid and solute flows to produce the inner medullary gradient remains unexplained.

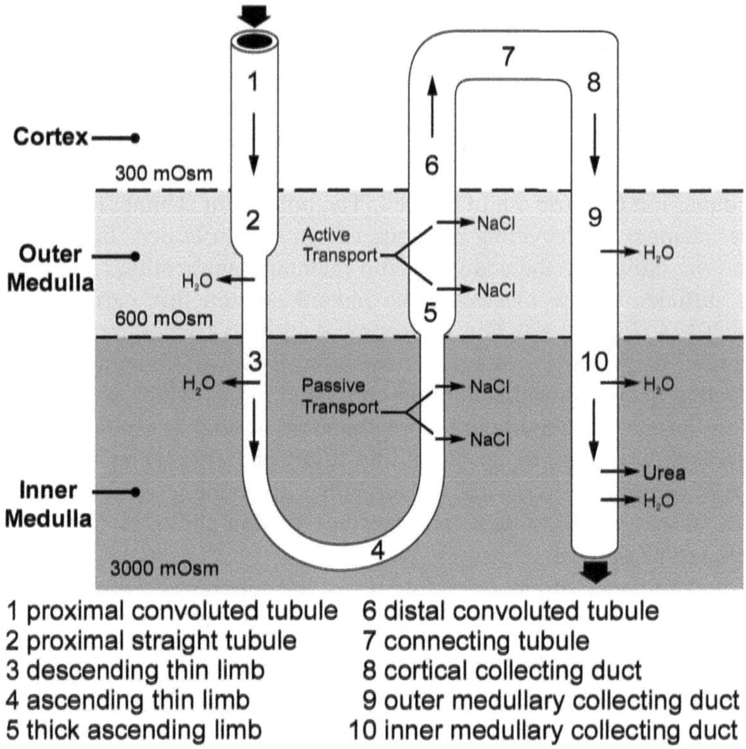

Fig. 11.2 Schematic diagram of a long-looped nephron. *Fluid flows* into the medulla via the proximal tubule and exits the medulla via the *inner medullary* collecting duct (*large arrows*). *Vertical arrows* indicate flow direction through the nephron. *Horizontal arrows* identify tubular segments where medullary NaCl, urea and water reabsorption occur. NaCl reabsorption occurs by active transport in the thick ascending limb and by passive transport in the ascending thin limb. The corticopapillary solute gradient that occurs in the kidney of the dehydrated rat is depicted (mosmol/Kg H_2O)

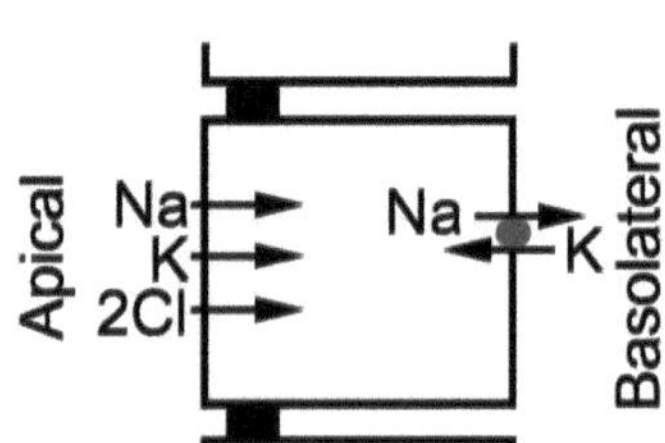

Fig. 11.3 Schematic diagram of Na transport through the thick ascending limb cell. Na, K-ATPase in the basolateral membrane and NKCC2 in the apical membrane mediate active sodium reabsorption in the thick ascending limb

11.2 Medullary Architecture

The medulla is organized into distinct corticopapillary zones and lateral regions that are defined by the configuration of loops of Henle, collecting ducts and blood vessels. The outer medulla of most rodents is defined by two corticopapillary axial zones (the outer stripe and the inner stripe) and two lateral regions (the vascular bundle and the interbundle regions) (Fig. 11.2). The inner medullas of rat and kangaroo rat consist of two axial zones (an outer zone and inner zone) and two lateral regions (the intracluster and intercluster regions) (Pannabecker et al. 2008b; Issaian et al. 2012; Urity et al. 2012). Similar inner medullary subdivisions likely occur in other rodents (Bankir and De Rouffignac 1985; Bankir et al. 1979; Kriz 1981; Jamison and Kriz 1982; Bachmann and Kriz 1982; Lemley and Kriz 1987, 1991; Chou et al. 1993; Pannabecker 2012; Pannabecker et al. 2008a).

The proximity of one structure to another and the expression of segment-specific cell membrane transport pathways influence both axial flows as well as highly organized lateral flows of fluid and solutes that occur between tubules and vessels. These flows impact solute exchange, cycling and sequestration patterns related to the urine concentrating mechanism (Pannabecker et al. 2008a; Layton et al. 2010; Layton 2011; Lemley and Kriz 1987; Thomas and Wexler 1995; Wang et al. 1998; Knepper and Roch-Ramel 1987), oxygen delivery (Chen et al. 2009a, b, 2010; Fry et al. 2014; Pannabecker and Layton 2014) and medullary and renal function more generally. This section briefly summarizes the axial zonation and lateral regionalization of the medulla and tubule morphology found in the rat, mouse, hamster, chinchilla, kangaroo rat and other species.

11.2.1 Architecture of the Outer Medulla

Vascular bundles form the dominant organizing motif in the outer medulla. Outer medullary circulation consists primarily of two spatially distinct vascular arrangements: (1) descending and ascending vasa recta that lie chiefly within vascular bundle regions and (2) venous capillary networks that lie between bundles (the interbundle regions) (Moffat and Fourman 1963; Rollhauser et al. 1964). Descending and ascending vasa recta carry plasma in a papillary or cortical direction, respectively, and interbundle capillaries carry plasma chiefly in a cortical direction (Fig. 11.4). Descending vasa recta connect to venous capillary networks at all levels along the corticopapillary axis (Pallone et al. 2003). Capillary networks in turn, connect to ascending vasa recta at all axial levels. Outer medullary capillary networks are assembled together with thick ascending limbs, descending limbs of long-looped nephrons and collecting ducts in the interbundle region.

Critical matching of blood flow and oxygen supply and demand is essential for optimizing the urine concentrating mechanism. Excessive blood flow tends to

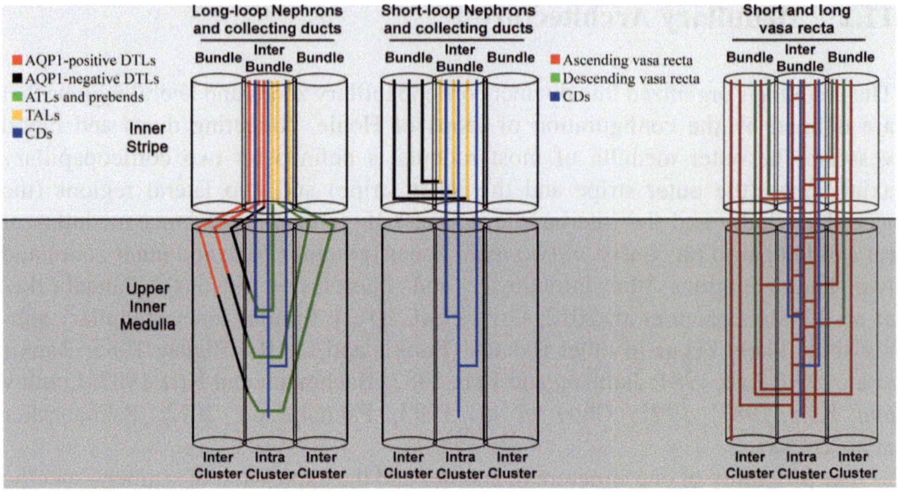

Fig. 11.4 Schematic diagram depicting regional distribution of nephrons and blood vessels of the rat inner stripe of the outer medulla and inner medulla. The inner stripe consists of two regions, the bundle and interbundle regions. The upper inner medulla also consists of two regions, the intercluster and intracluster regions. Collecting ducts, thick ascending limbs, descending thin limbs of long-loop nephrons and fenestrated interconnecting capillaries occupy the outer medullary interbundle region (Bankir and De Rouffignac 1985; Kriz 1981). Collecting ducts, fenestrated interconnecting capillaries, ascending thin limbs, and to a lesser extent aquaporin 1-negative descending thin limbs occupy the inner medullary intracluster region (Pannabecker and Dantzler 2006; Yuan and Pannabecker 2010; Gilbert and Pannabecker 2013)

collapse the osmolality gradient due to solute washout, whereas insufficient blood flow tends to collapse the gradient due to reduced oxygen supply and therefore diminished sodium reabsorption by the thick ascending limb, a critical process for producing concentrated urine. On the basis of functional and theoretical evidence, axial and lateral oxygen tension gradients are believed to exist in the outer medulla (Chen et al. 2009a, b, 2010; Fry et al. 2014). Gradients arise largely as a result of the physical distance between medullary oxygen source (descending vasa recta) and sink. High metabolic requirements exist for carrying out active reabsorption of sodium by the thick ascending limb. A gradient develops because those thick ascending limbs positioned in the interbundle region lie a significant distance from the oxygen-carrying descending vasa recta, which lie in the bundle regions. Medullary hypoxia induces necrosis of cells lining thick ascending limbs and long-loop descending limbs in the outer medulla, with the degree of necrosis being directly proportional to the distance from vascular bundles (Brezis et al. 1984; Kopolovic et al. 1989; Schurek and Kriz 1985). Medullary oxygenation in humans is affected in poorly understood ways by multiple pharmacological and pathophysiological events in the clinical setting (Brezis and Rosen 1995; Sendeski 2011). Increased understanding of blood flow and oxygenation of the renal medulla is critical for more completely understanding renal tubular solute transport and sodium and water homeostasis in general (Pannabecker and Layton 2014).

11.2.2 Architecture of the Inner Medulla

Upon entering the inner medulla the collecting ducts begin to coalesce as they descend towards the papilla, forming distinct groups, or clusters of collecting ducts. Collecting duct clusters form the dominant organizing motif in the inner medulla (Fig. 11.4). Inner medullary circulation consists of two spatially distinct vascular arrangements as in the outer medulla, with the descending and ascending vasa recta lying between collecting duct clusters and venous capillary networks intermixed with the collecting duct clusters. At the outer medullary-inner medullary border most descending thin limbs are positioned between collecting duct clusters in the intercluster region. As each thin limb descends, it gradually approaches the collecting duct cluster and continues its descent within the intracluster region before forming a 180° bend (Westrick et al. 2013) (Fig. 11.4). Distal to the bend, the ascending segment then rises towards the outer medullary-inner medullary border and gradually moves back into the intercluster region. In the lower third of the inner medulla, where collecting ducts are markedly reduced in number due to their coalescences, the collecting duct cluster regions disappear.

11.2.3 Tubule Morphology Defines Multiple Medullary Nephron and Blood Vessel Segments

All medullary nephrons descend from the cortex at least partway through the outer or inner medulla, then form a 180° bend and ascend back to the cortex. Bends are formed at all levels of the inner stripe of the outer medulla and inner medulla; as a consequence, nephron number at each transverse level gradually declines with depth below the cortex. There are two categories of medullary nephrons based on morphological criteria. Short-looped nephrons consist of a thin descending limb and thick ascending limb and form their bends in the outer medulla. Long-looped nephrons consist of thin descending and thin ascending limbs and a thick ascending limb and form their bends in the inner medulla. All thick ascending limbs reside in the outer medulla. Short and long-looped nephrons exhibit functional differences and similarities and a species-dependent ratio of short-to-long looped nephrons is considered to play a role in maximum concentrating capacity (Kriz 1981).

On the basis of cell ultrastructural characteristics observed in numerous rodent species, the short-looped descending thin limbs are known to consist of the type 1 epithelium, the long-looped descending thin limbs consist sequentially of types 2 and 3 epithelia and a short prebend segment and the ascending thin limbs consist of type 4 epithelia (Jamison and Kriz 1982; Kriz 1981; Schwartz and Venkatachalam 1974; Schwartz et al. 1979; Kaissling and Kriz 1992; Bachmann and Kriz 1982; Zhai et al. 2006; Chou et al. 1993). The ascending thin limb makes an abrupt transition to the thick ascending limb of the inner stripe. Thick ascending limbs of the inner stripe are morphologically distinct from those of the outer stripe and cortex, exhibiting greater cell height, more abundant mitochondria, and deeper and more extensive and complex invaginations of the basal plasmalemma (Kone et al. 1984).

As the unbranched collecting duct segments of the outer medulla begin to coalesce in the inner medulla they become larger in diameter. In the rat about 7000 collecting ducts enter the inner medulla (Oliver 1968; Saxen 1987) and coalesce into about 10–15 collecting ducts at the papilla tip (Pannabecker and Dantzler 2007). Collecting ducts consist of two cell types in the outer medulla, the intercalated and principal cells, at about a 2:3 ratio. Only about 10 % of collecting duct cells in the initial third of the inner medulla are intercalated whereas collecting ducts of the terminal two-thirds of the inner medulla consist of a single structurally and functionally distinct cell type, the "IMCD" cell (Christensen et al. 2012; Madsen et al. 1988) (Clapp et al. 1989).

Descending vasa recta arise from efferent arterioles of glomeruli that are positioned deep in the cortex, near the corticomedullary border. There is a gradual transition along the vessel to a point where the smooth muscle cells of the efferent arteriole wall are completely replaced by pericytes. Pericytes disappear from descending vasa recta at levels below the initial third of the inner medulla. Descending vasa recta join to the venous capillaries, which generally exhibit repeated branching and form sparse networks. Capillary networks of the inner stripe of the outer medulla and inner medulla join to unbranched ascending vasa recta that return to the cortex linking to arcuate or interlobular veins. Capillary networks of the outer stripe return directly to arcuate or interlobular veins (Pallone et al. 2003; Moffat and Fourman 1963; Rollhauser et al. 1964).

Descending vasa recta consist of a continuous endothelium and ascending vasa recta and venous capillaries consist of a discontinuous endothelium. The discontinuous endothelia consist of fenestrae arranged in plaques and are considered to have very high permeability to water and small solutes as shown for fenestrae of vessels in other tissues (Pallone et al. 1994, 2003; MacPhee and Michel 1995; Michel and Curry 1999).

11.3 Roles for Water and Sodium Chloride Tubular Transport in Producing and Sustaining the Medullary Solute Gradient

At each level of the medulla the osmolality of all tubules is nearly equivalent to that of the interstitium, but is hypertonic to systemic plasma. This is a fundamental feature of the urine concentrating mechanism. It was shown long ago that luminal fluid near the bend of thin limbs of Henle's loop of hamster, kangaroo rat and the sand rat *Psammomys obesus* is nearly isosmotic with fluid of the inner medullary interstitium, collecting ducts and blood vessels at the same level, implying that solutes are secreted into the tubule lumen, water is reabsorbed from it or both secretion and reabsorption occur (Gottschalk and Mylle 1959). The water channel aquaporin 1 is abundantly expressed in long-looped descending thin limb segments (type 2 segments) lying in the outer medulla and upper 60 % of the inner medulla and is the predominant pathway by which water is reabsorbed from these segments; however, aquaporin 1 expression is relatively low in descending thin limb segments (type 3 segments) in the lower 40 % of the rat, kangaroo rat, and chinchilla inner medulla (Chou et al. 1993; Pannabecker and Dantzler 2007; Urity et al. 2012). The collective histotopography of inner medullary aquaporin 1-positive and aquaporin 1-negative descending thin limbs and their different water permeabilities is important for reducing the fluid load delivered to the deep inner medulla and is related to distinct contributions of these segments to production of the corticopapillary osmotic gradient (Layton et al. 2010).

Type 2 segments (aquaporin 1-positive) in both rat and chinchilla have been shown to exhibit substantial transepithelial water permeability, whereas type 3 segments (aquaporin 1-negative) exhibit little or no transepithelial water permeability (Chou and Knepper 1992; Nawata et al. 2014). Thus, transepithelial water flux could play a major role in osmotic equilibration between the lumen of type 2 segments and the interstitial fluid compartment, but such an equilibration by water flux is unlikely to occur with type 3 segments. Transepithelial permeability for small solutes such as sodium and urea is very high in types 2 and 3 segments (Chou and Knepper 1993; Nawata et al. 2014). Solute secretion into the type 3 segments, and not water reabsorption, likely plays the dominant role in osmotic equilibration of these segments (Marsh 1970; Pennell et al. 1974). By the time luminal fluid reaches the bend of the loop, its sodium chloride concentration is sufficiently high so as to enable passive sodium chloride reabsorption by the ascending thin limb (Layton et al. 2004, 2012). It is important to note at this point that, because loops form bends at all levels of the inner medulla, types 2 and 3 segments co-exist along most levels of the inner medulla (Fig. 11.4).

Sodium chloride reabsorption by the ascending thin limb adds solutes to the inner medullary interstitium and as long as sodium chloride reabsorption by the ascending thin limb exceeds urea secretion into this segment the interstitium will become concentrated and luminal fluid will become dilute (Pannabecker 2013). Sodium is believed to move across the epithelium largely by way of the paracellular pathway (Koyama et al. 1991; Chou and Knepper 1993), although a substantial transcellular flux may also be involved. In contrast, the pathway of the counterion chloride is transcellular, occurring by way of the chloride channel ClC-K1 (Uchida et al. 1993; Vandewalle et al. 1997). The descending thin limb from the outer inner medulla of rat and hamster (likely type 2) and the type 2 descending thin limb of chinchilla are sodium permselective, whereas the chinchilla type 3 descending thin limb and the rat, hamster and chinchilla ascending thin limb (type 4 segment) are significantly more permeable to chloride than to Na (Imai 1977; Chou and Knepper 1993). Possibly these segmental permselective differences are due to segmental differences in solute selectivity of the paracellular pathways.

The thick ascending limb chloride channel ClC-K2 (rodents) and ClC-KB (human) and the chloride channel ClC-K1 expressed in the thin ascending limb of rodents (ClC-KA in humans) play critical roles in NaCl reabsorption by these two segments and in producing a concentrated urine. Expression of ClC-K1 mRNA is upregulated by dehydration (Vandewalle et al. 1997; Uchida et al. 1993) and expression of ClC-K2 mRNA is downregulated by high salt (Vitzthum et al. 2002). There is 81 % homology between ClC-K1 and ClC-K2 amino acid sequences in rat and 91 % homology between ClC-KA and ClC-KB amino acid sequences in human; there are no antibodies that specifically recognize each protein. The ClC-KA and ClC-KB genes lie adjacent to each other on human chromosome 1p36, separated by 11 kb of genomic DNA and likely arose as a result of recent gene duplication. Both Cl^- channels require the subunit Barttin for appropriate membrane trafficking and for full channel activity. Genetic studies have shown that deletion or loss of function of NKCC2 and ROMK1 results in Bartter syndrome types I and II, deletion or loss of function of ClC-KB results in Bartter syndrome type III and deletion or loss of function of the Barttin gene results in Bartter syndrome type IV, which is accompanied with sensorineural deafness.

Genetic variants of the gene *CLCNK* are associated with salt-sensitive hypertension in human. Four single nucleotide polymorphisms within CLCNKA loci were found to be associated with increased blood pressure after Na-loading (Barlassina et al. 2007). Plasma renin activity and heart rate (measured before Na-load) were significantly lower in patients carrying the alleles associated with the larger mean blood pressure increase after Na-load, indicating the alleles are linked to chronic volume expansion. A single nucleotide polymorphism within CLCNKB was found to activate ClC-Kb chloride channel function in vitro and may predispose to the development of essential hypertension in vivo (Jeck et al. 2004).

Cells of thin limbs of the loop of Henle exhibit little or no Na-K-ATPase activity and only weak expression of α, β or γ subunit protein (Garg et al. 1981; Pihakaski-Maunsbach et al. 2006); however, cells of thin limbs from the innermost third of the inner medulla show stronger expression of γ subunit and splice variant γ_a than in the first and second third (Pihakaski-Maunsbach et al. 2006). A role for Na-K-ATPase in transport in thin limbs of the loop of Henle is not clear, Na-K-ATPase may be involved with intracellular sodium, potassium and volume maintenance (Lopes et al. 1988).

As stated above, vasopressin increases the osmotic water permeability of the collecting duct epithelium, leading to increased water movement from the collecting duct into the more concentrated interstitium and this concentrates solutes in the final urine. Aquaporin 2 is the principle water channel in the collecting duct and is expressed in the apical membrane and in subapical endosomes of medullary collecting ducts in kidneys of most, if not all rodents examined, including rats and mice (Fenton and Knepper 2007a; Nielsen et al. 2002), chinchilla (Pannabecker unpublished), kangaroo rat (Urity et al. 2012; Issaian et al. 2012), musk shrew (Maeda et al. 2008), and degus (Bozinovic et al. 2003). Aquaporin 2 is also expressed to a lesser, but significant degree in the inner medullary collecting duct basolateral membrane (see below). Vasopressin regulates aquaporin 2 protein sorting from endosomes to apical membrane and also regulates abundance of total aquaporin 2 protein within the cell (Brown et al. 2012) (see Chap. 3).

Despite its importance in overall collecting duct water reabsorption, there is relatively little known about aquaporin 2 channelopathies in human kidney. A well-known disorder related to aquaporin 2 expression arises from lithium toxicity, a condition that results from lithium treatment for bipolar disorder. Lithium leads to *nephrogenic diabetes insipidus*, which is characterized by reduced aquaporin 2 apical expression, reduced water reabsorption and excessive water loss (for references and further details, see Chap. 3).

Hypertonicity has been shown to serve as a regulator of medullary collecting duct aquaporin 2 protein expression, trafficking and water permeability (Lankford et al. 1991; van Balkom et al. 2003; Yui et al. 2012). The transcription factor tonicity-responsive enhancer binding protein (TonEBP) has been shown to play a role in regulating hypertonicity-dependent aquaporin 2 expression in rat primary cultured collecting duct cells independently of the cyclic AMP response element Cre activation (Storm et al. 2003). In cortical collecting duct cells derived from mouse ($mpkCCDc_{14}$) hypertonicity regulates aquaporin 2 protein abundance independently of vasopressin (Hasler et al. 2005). The transcription factor TonEBP has been shown to play a central role in regulating aquaporin 2 expression in $mpkCCDc_{14}$ cells by enhancing aquaporin 2 gene transcription, and this appears to occur independently of vasopressin (Hasler et al. 2006). Hypertonicity can influence preferential apical or basolateral membrane targeting of aquaporin 2 in the rat collecting duct. When medullary slices are treated with hypertonic media, vasopressin increases the basolateral:apical aquaporin 2 protein expression ratio in the collecting ducts of the outer zone of the inner medulla (van Balkom et al. 2003). It has been suggested that following constitutive sorting to the basolateral membrane in the rat, vasopressin treatment induces transcytotic aquaporin 2 apical membrane expression with the consequent increase in transepithelial water permeability (Yui et al. 2012). Basolateral aquaporin 2 expression also occurs in the kangaroo rat inner medullary collecting duct; however, the basolateral sorting index is nearly twofold higher than in the rat collecting duct (Espineira and Pannabecker 2014). The relative significance of inner medullary collecting duct basolateral and apical aquaporin 2 expression (and membrane water permeabilities) in the process of transepithelial water flows needs further clarification. The mechanism by which hypertonicity influences aquaporin 2 targeting in medullary collecting ducts is not known. Investigations into variable aquaporin 2 regulatory pathways amongst species with high inner medullary interstitial osmolality and high urine concentrating capacity may provide clues to the overall physiological roles of aquaporin 2 membrane sorting.

11.4 Peristaltic Contractions of the Papillary Pelvic Wall and the Urine Concentrating Mechanism

In unipapillate mammals such as rodents, the renal pelvis is encapsulated by two distinct smooth muscle layers that extend from the outer medulla to the tip of the papilla (Lacy and Schmidt-Nielsen 1979; Gosling and Dixon 1971). This muscular wall is a proximal extension of the ureter. In kidneys of multipapillate mammals such as the human, each papilla is surrounded by a funnel-shaped calyx, which is the equivalent of the rodent pelvis and it is also encapsulated by distinct smooth muscle layers (Dwyer and Schmidt-Nielsen 2003). Peristaltic contractions of the papillary pelvic wall markedly influence fluid flow through medullary thin limbs of Henle's loops, collecting ducts and vasa recta and these influences have been proposed to play an important role in producing concentrated urine (Dwyer and Schmidt-Nielsen 2003; Pruitt et al. 2006; Oliver et al. 1982). Both the pelvic wall (and calyces in humans) and the ureter continually contract and relax in a rhythmic fashion. Peristaltic muscle contractions produce a transient narrowing of the papilla that proceeds as a wave along the length of the papilla. Each wave is followed by a brief period of relaxation during which the papilla returns to its expanded state. Papillary contractions are initiated every few seconds. Pressure changes induced by these contraction/relaxation episodes impact fluid flow rates through the papillary tubular segments and ureter. Pelvic wall contractions propel fluid toward the papilla tip in both descending and ascending segments, whereas the subsequent relaxation permits reverse transient reflux. This reflux is followed by a period of orthograde flow that is dominated by filtration pressure; orthograde flow is the flow that we usually associate with steady-state nephron fluid flow (Schmidt-Nielsen et al. 1980; Schmidt-Nielsen and Reinking 1981; Reinking and Schmidt-Nielsen 1981; Schmidt-Nielsen and Graves 1982). Contraction/relaxation cycles thereby lead to bolus flows through tubules and blood vessels as opposed to steady-state flows, low tubule patency relative to high blood vessel patency and variable temporal patterns of fluid contact with epithelia. All of these factors may significantly impact fluid and solute reabsorption rates and the concentrating mechanism. The glycosaminoglycan hyaluronan exists in substantial amounts within the inner medullary interstitium, gradually increasing with depth below the outer medulla with highest concentrations at the papilla tip. Hyaluronan is the chief component of the gel-like interstitial matrix of the inner medulla. Hyaluronan has been hypothesized to function as a mechanical transducer of pelvic wall contractions, by acting as the interface between the pelvic wall and tubular epithelia. Hyaluronan thereby serves to transduce the repetitive compression and relaxation phases of the pelvic wall, and in so doing, influence tubular fluid flows and the urine concentrating mechanism (Knepper et al. 2003) (see Chap. 10 for additional details on medullary hyaluronan).

Elimination of the contractions of the pelvic wall by heat application or xylocaine treatment was found to significantly decrease inner medullary osmolality in one study with hamsters (Pruitt et al. 2006) but had no effect on medullary osmolality in a second study with Munich-Wistar rat (Oliver et al. 1982). The role that pelvic wall contractions may play in producing the corticopapillary osmotic gradient and a concentrated urine remains relatively unstudied and controversial; however, dynamic mathematical models involving peristaltic contractions of the pelvic wall will likely lead to more complete understanding of fluid and solute flows through the medulla and thereby provide new insights into the urine concentrating mechanism.

In-depth studies of the pelvic wall have focused on the cellular mechanisms involved with regulating smooth muscle contractions, the role of the pelvic wall in maintaining peristaltic ureteral urine flow and the impact of impaired pelvic wall peristalsis on hydronephrosis and renal sepsis. Congenital defects in peristalsis are commonly associated with obstructive or nonobstructive ureteral and renal damage in humans, and are particularly important in the developing kidney (Chevalier 2006; Thom and Rosenblum 2013). The morphogenetic signaling pathway *hedgehog* is the only signaling pathway known to regulate development of ureteral peristalsis and function (Herzlinger 2011). The pelvic wall is the site of pacemaker cells that control ureteral peristalsis. Genetic studies have recently identified several proteins associated with the pacemaker mechanism that appear to exert control over ureter peristalsis. These proteins include the hyperpolarizing cation channel 3 (HCN3), C-Kit and GLI3 repressor (Hurtado et al. 2010; Cain et al. 2011; David et al. 2005). Clinical studies have shown lower density of C-Kit positive cells in the muscle bundles from patients with obstructed ureteropelvic junction and from patients with vesicoureteral reflux compared to normal tissue (Solari et al. 2003; Schwentner et al. 2005). The origins of gut and heart pacemaker cells, which utilize similar signaling pathways, may provide insights into regulatory and developmental mechanisms associated with pelvic muscle wall contractions (Feeney and Rosenblum 2014).

11.5 Perspective

Most mammals produce a concentrated urine, a process that lies at the core of the kidney's role in fluid and solute homeostasis. The urine concentrating mechanism has been conceptualized as the countercurrent multiplication of an osmotic pressure difference sustained by active NaCl reabsorption from water-impermeable ascending limbs of Henle's loops (Gottschalk and Mylle 1959; Hai and Thomas 1969; Kuhn and Ryffel 1942). This conceptual model has been widely accepted for the outer medulla where active NaCl reabsorption has been demonstrated (Burg and Green 1973; Rocha and Kokko 1973). However, countercurrent flow coupled with active solute reabsorption does not explain the concentrating process in the inner medulla, where the steepest osmotic gradient is generated.

Genetic models have been profoundly influential in understanding physiological processes associated with the urine concentrating mechanism (Fenton and Knepper 2007a). Notable among these many models include mouse knockouts of AQP1,

ClCK-1, UTA1/3, and UTA2. Akizuki et al (2001) demonstrated that inner medullary urea accumulation and a concentrating effect are impaired by defective ascending thin limb Cl^- transport in a ClC-K1 knockout; a direct implication that inner medullary NaCl and urea hypertonicity arise from interacting transport systems, i.e. countercurrent systems, or perhaps from interacting transport pathways within a nephron segment. A study of UTA1/3 gene deletion (Fenton et al. 2004) clearly demonstrated a role for UTA1 and/or UTA3 in the urine concentrating mechanism and supports a conclusion that urea absorption from collecting ducts has little or no effect on the accumulation of NaCl in the inner medulla. This finding contradicts the passive mechanism as originally described by Kokko and Rector (1972) and Stephenson (1972). Separately, a study of the UTA2 knockout indicates only a minor role for urea flux via this transporter, and therefore for urea secretion, in the descending thin limb (Fenton and Knepper 2007b; Uchida et al. 2005); however, the very high phloretin-insensitive transepithelial urea fluxes of the descending thin limb provide strong evidence of other urea pathways (Chou and Knepper 1993; Nawata et al. 2014).

Genetic models have provided insights into signaling pathways that regulate embryonic medullary development and formation in rodents (Yu et al. 2009; Yu 2011; Pietila et al. 2011). However, although these studies have characterized patterns of medullary development at the gross anatomical level, genetic studies that examine in detail the fine points of medullary architecture in the adult kidney and their relationships to the urine concentrating process are needed. Future animal studies of solute transport and medullary development will drive the next generation of genetic animal models that will elucidate all membrane transporters that underlie fluid and solute flows amongst the many medullary compartments, those compartments that are known and those yet to be identified, and that are essential to forming a concentrated urine.

Acknowledgements The author acknowledges financial support from the National Science Foundation and the National Institute of Diabetes and Digestive and Kidney Diseases of the National Institutes of Health.

References

Akizuki N, Uchida S, Sasaki S, Marumo F (2001) Impaired solute accumulation in inner medulla of Clcnk1-/- mice kidney. Am J Physiol Renal Physiol 280:F79–F87

Bachmann S, Kriz W (1982) Histotopography and ultrastructure of the thin limbs of the loop of Henle in the hamster. Cell Tissue Res 225(1):111–127

Bankir L, De Rouffignac C (1985) Urinary concentrating ability: insights from comparative anatomy. Am J Physiol 249:R643–R666

Bankir L, Kaissling B, Rouffignac CD, Kriz W (1979) Vascular organization of the kidney of psammomys-obesus. Anat Embryol (Berl) 155(2):149–160

Barlassina C, Dal Fiume C, Lanzani C, Manunta P, Guffanti G, Ruello A, Bianchi G, Del Vecchio L, Macciardi F, Cusi D (2007) Common genetic variants and haplotypes in renal

CLCNKA gene are associated to salt-sensitive hypertension. Hum Mol Genet 16 (13):1630–1638
Beuchat CA (1990) Metabolism and the scaling of urine concentrating ability in mammals: resolution of a paradox? J Theor Biol 143:113–122
Bozinovic F, Gallardo PA, Visser GH, Cortes A (2003) Seasonal acclimatization in water flux rate, urine osmolality and kidney water channels in free-living degus: molecular mechanisms, physiological processes and ecological implications. J Exp Biol 206(Pt 17):2959–2966
Brezis M, Rosen S (1995) Hypoxia of the renal medulla—its implications for disease. N Engl J Med 332(10):647–655. doi:10.1056/NEJM199503093321006
Brezis M, Rosen S, Spokes K, Silva P, Epstein FH (1984) Transport-dependent anoxic cell injury in the isolated perfused rat kidney. Am J Pathol 116(2):327–341
Brown D, Bouley R, Paunescu TG, Breton S, Lu HA (2012) New insights into the dynamic regulation of water and acid-base balance by renal epithelial cells. Am J Physiol Cell Physiol 302(10):C1421–C1433. doi:10.1152/ajpcell.00085.2012
Burg MB, Green N (1973) Function of thick ascending limb of Henle's loop. Am J Physiol 224:659–668
Cain JE, Islam E, Haxho F, Blake J, Rosenblum ND (2011) GLI3 repressor controls functional development of the mouse ureter. J Clin Invest 121(3):1199–1206. doi:10.1172/JCI45523
Chen J, Edwards A, Layton AT (2009a) A mathematical model of O2 transport in the rat outer medulla. II Impact of outer medullary architecture. Am J Physiol Renal Physiol 297(2):F537–F548. doi:10.1152/ajprenal.90497.2008
Chen J, Layton AT, Edwards A (2009b) A mathematical model of oxygen transport in the rat outer medulla: I. Model formulation and baseline results. Am J Physiol Renal Physiol 297:F537–F548
Chen J, Edwards A, Layton AT (2010) Effects of pH and medullary blood flow on oxygen transport and sodium reabsorption in the rat outer medulla. Am J Physiol Renal Physiol 298 (6):F1369–F1383. doi:10.1152/ajprenal.00572.2009
Chevalier RL (2006) Pathogenesis of renal injury in obstructive uropathy. Curr Opin Pediatr 18 (2):153–160. doi:10.1097/01.mop.0000193287.56528.a4
Chou C-L, Knepper MA (1992) In vitro perfusion of chinchilla thin limb segments: segmentation and osmotic water permeability. Am J Physiol 263(3):F417–F426
Chou C-L, Knepper MA (1993) In vitro perfusion of chinchilla thin limb segments: urea and NaCl permeabilities. Am J Physiol 264:F337–F343
Chou C-L, Nielsen S, Knepper MA (1993) Structural-functional correlation in chinchilla long loop of Henle thin limbs: a novel papillary subsegment. Am J Physiol Renal 265:F863–F874
Christensen EI, Wagner CA, Kaissling B (2012) Uriniferous tubule: structural and functional organization. Compr Physiol 2(2):805–861. doi:10.1002/cphy.c100073
Clapp WL, Madsen KM, Verlander JW, Tisher CC (1989) Morphologic heterogeneity along the rat inner medullary collecting duct. Lab Invest 60(2):219–230
David SG, Cebrian C, Vaughan ED Jr, Herzlinger D (2005) c-kit and ureteral peristalsis. J Urol 173(1):292–295. doi:10.1097/01.ju.0000141594.99139.3d
Dwyer TM, Schmidt-Nielsen B (2003) The renal pelvis: machinery that concentrates urine in the papilla. News Physiol Sci 18:1–6
Espineira M, Pannabecker T (2014) Increase in collecting duct basal: apical AQP2 protein expression ratio with increasing depth along the corticopapillary axis positively correlates with maximum urine concentrating capacity in the kangaroo rat and Sprague-Dawley rat. FASEB J 28(1 Suppl):1137.1132
Feeney MM, Rosenblum ND (2014) Urinary tract pacemaker cells: current knowledge and insights from nonrenal pacemaker cells provide a basis for future discovery. Pediatr Nephrol 29(4):629–635. doi:10.1007/s00467-013-2631-4
Fenton RA, Knepper MA (2007a) Mouse models and the urinary concentrating mechanism in the new millennium. Physiol Rev 87:1083–1112

Fenton RA, Knepper MA (2007b) Urea and renal function in the 21st century: insights from knockout mice. J Am Soc Nephrol 18:679–688
Fenton RA, Chou C-L, Stewart GS, Smith CP, Knepper MA (2004) Urinary concentrating defect in mice with selective deletion of phloretin-sensitive urea transporters in the renal collecting duct. Proc Natl Acad Sci U S A 101:7469–7474
Fry BC, Edwards A, Sgouralis I, Layton AT (2014) Impact of renal medullary three-dimensional architecture on oxygen transport. Am J Physiol Renal Physiol 307(3):F263–F272. doi:10.1152/ajprenal.00149.2014
Garg LC, Knepper MA, Burg MB (1981) Mineralocorticoid effects on Na-K-ATPase in individual nephron segments. Am J Physiol 240:F536–F544
Gilbert RL, Pannabecker TL (2013) Architecture of interstitial nodal spaces in the rodent renal inner medulla. Am J Physio Renal Physiol 305:F745–F752
Gosling JA, Dixon JS (1971) Morphologic evidence that the renal calyx and pelvis control ureteric activity in the rabbit. Am J Anat 130(4):393–408. doi:10.1002/aja.1001300403
Gottschalk CW, Mylle M (1959) Micropuncture study of the mammalian urinary concentrating mechanism: evidence for the countercurrent hypothesis. Am J Physiol 196:927–936
Greenwald L, Stetson D (1988) Urine concentration and the length of the renal papilla. NIPS 3:46–49
Hai MA, Thomas S (1969) The time-course of changes in renal tissue composition during lysine vasopressin infusion in the rat. Pflugers Arch 310:297–319
Hasler U, Vinciguerra M, Vandewalle A, Martin PY, Feraille E (2005) Dual effects of hypertonicity on aquaporin-2 expression in cultured renal collecting duct principal cells. J Am Soc Nephrol 16(6):1571–1582. doi:10.1681/ASN.2004110930
Hasler U, Jeon US, Kim JA, Mordasini D, Kwon HM, Feraille E, Martin PY (2006) Tonicity-responsive enhancer binding protein is an essential regulator of aquaporin-2 expression in renal collecting duct principal cells. J Am Soc Nephrol 17(6):1521–1531. doi:10.1681/ASN.2005121317
Herzlinger D (2011) Upper urinary tract pacemaker cells join the GLI club. J Clin Invest 121(3):836–838. doi:10.1172/JCI46400
Hurtado R, Bub G, Herzlinger D (2010) The pelvis-kidney junction contains HCN3, a hyperpolarization-activated cation channel that triggers ureter peristalsis. Kidney Int 77(6):500–508. doi:10.1038/ki.2009.483
Imai M (1977) Function of the thin ascending limb of Henle of rats and hamsters perfused in vitro. Am J Physiol 232:F201–F209
Issaian T, Urity VB, Dantzler WH, Pannabecker TL (2012) Architecture of vasa recta in the renal inner medulla of the desert rodent *Dipodomys merriami*: potential impact on the urine concentrating mechanism. Am J Physio Regul Integr Comp Physiol 302:R748–R756
Jamison RL, Kriz W (1982) Urinary concentrating mechanism. Oxford University Press, New York
Jeck N, Waldegger S, Lampert A, Boehmer C, Waldegger P, Lang PA, Wissinger B, Friedrich B, Risler T, Moehle R, Lang UE, Zill P, Bondy B, Schaeffeler E, Asante-Poku S, Seyberth H, Schwab M, Lang F (2004) Activating mutation of the renal epithelial chloride channel ClC-Kb predisposing to hypertension. Hypertension 43:1175–1181
Kaissling B, Kriz W (1992) Morphology of the loop of Henle, distal tubule, and collecting duct. In: Windhager EE (ed) Handbook of physiology. Sec 8. Renal physiology. Oxford University Press, New York, pp 109–167
Knepper MA, Roch-Ramel F (1987) Pathways of urea transport in the mammalian kidney. Kidney Int 31(2):629–633
Knepper MA, Saidel GM, Hascall VC, Dwyer T (2003) Concentration of solutes in the renal inner medulla: interstitial hyaluronan as a mechano-osmotic transducer. Am J Physiol Renal Physiol 284:F433–F446
Kokko JP, Rector FC (1972) Countercurrent multiplication system without active transport in inner medulla. Kidney Int 2:214–223

Kone BC, Madsen KM, Tisher CC (1984) Ultrastructure of the thick ascending limb of Henle in the rat kidney. Am J Anat 171(2):217–226. doi:10.1002/aja.1001710207
Kopolovic J, Brezis M, Spokes K, Silva P, Epstein F, Rosen S (1989) The vulnerability of the thin descending limbs of Henle's loop in the isolated perfused rat kidney. Am J Kidney Dis 14:25–30
Koyama S, Yoshitomi K, Imai M (1991) Effect of protamine on ion conductance of ascending thin limb of Henle's loop from hamsters. Am J Physiol 261:F593–F599
Kriz W (1981) Structural organization of the renal medulla: comparative and functional aspects. Am J Physiol 241:R3–R16
Kuhn W, Ryffel K (1942) Herstellung konzentrierter Losüngen aus verdünten durch blosse Membranwirkung: ein Modellversuch zur Funktion der Niere. Hoppe-Seylers Z Physiol Chem 276:145–178
Lacy ER, Schmidt-Nielsen B (1979) Anatomy of the renal pelvis in the hamster. Am J Anat 154:291–320
Lankford SP, Chou C-L, Terada Y, Wall SM, Wade JB, Knepper MA (1991) Regulation of collecting duct water permeability independent of cAMP-mediated AVP response. Am J Physiol 261:F554–F566
Layton AT (2011) A mathematical model of the urine concentrating mechanism in the rat renal medulla: II. Functional implications of three-dimensional architecture. Am J Physiol Renal Physiol 300:F372–F384
Layton AT, Layton HE (2011) Countercurrent multiplication may not explain the axial osmolality gradient in the outer medulla of the rat kidney. Am J Physiol Renal Physiol 301(5):F1047–F1056. doi:10.1152/ajprenal.00620.2010
Layton AT, Pannabecker TL, Dantzler WH, Layton HE (2004) Two modes for concentrating urine in rat inner medulla. Am J Physiol Renal Physiol 287:F816–F839
Layton AT, Layton HE, Dantzler WH, Pannabecker TL (2009) The mammalian urine concentrating mechanism: hypotheses and uncertainties. Physiology 24:250–256
Layton AT, Pannabecker TL, Dantzler WH, Layton HE (2010) Functional implications of the three-dimensional architecture of the rat renal inner medulla. Am J Physiol Renal Physiol 298: F973–F987
Layton AT, Dantzler WH, Pannabecker TL (2012) Urine concentrating mechanism: impact of vascular and tubular architecture and a proposed descending limb urea-Na+ cotransporter. Am J Physiol Renal Physiol 302(5):F591–F605. doi:10.1152/ajprenal.00263.2011
Lemley KV, Kriz W (1987) Cycles and separations: the histotopography of the urinary concentrating process. Kidney Int 31:538–548
Lemley KV, Kriz W (1991) Anatomy of the renal interstitium. Kidney Int 39:370–381
Lopes AG, Amzel LM, Markakis D, Guggino WB (1988) Cell-volume regulation by the thin descending-limb of Henle's loop. Proc Natl Acad Sci U S A 85(8):2873–2877
MacPhee PJ, Michel CC (1995) Fluid uptake from the renal medulla into the ascending vasa recta in anaesthetized rats. J Physiol 487:169–183
Madsen KM, Clapp WL, Verlander JW (1988) Structure and function of the inner medullary collecting duct. Kidney Int 34:441–454
Maeda S, Kuwahara S, Ito H, Tanaka K, Hayakawa T, Seki M (2008) Expression and localization of aquaporins in the kidney of the musk shrew (Suncus murinus). J Histochem Cytochem 56 (1):67–75. doi:10.1369/jhc.7A7281.2007
Marsh DJ (1970) Solute and water flows in thin limbs of Henle's loop in the hamster kidney. Am J Physiol 218:824–831
Michel CC, Curry FE (1999) Microvascular permeability. Physiol Rev 79(3):703–761
Moffat DB, Fourman J (1963) The vascular pattern of the rat kidney. J Anat 97:543–553
Mount DB (2012) Transport of sodium, chloride, and potassium. In: Taal MW, Brenner BM, Rector FC (eds) Brenner and Rector's the kidney, vol 1, 9th edn. Elsevier, Philadelphia, pp 158–201

Nawata CM, Evans KK, Dantzler WH, Pannabecker TL (2014) Transepithelial water and urea permeabilities of isolated perfused Munich-Wistar rat inner medullary thin limbs of Henle's loop. Am J Physiol Renal Physiol 306:F123–F129. doi:10.1152/ajprenal.00491.2013
Nielsen S, Frokiaer J, Marples D, Kwon TH, Agre P, Knepper MA (2002) Aquaporins in the kidney: from molecules to medicine. Physiol Rev 82(1):205–244
Oliver JA (1968) Nephrons and kidneys: a quantitative study of developmental and evolutionary mammalian renal architectonics. Harper and Row, New York
Oliver RE, Roy DR, Jamison RL (1982) Urinary concentration in the papillary collecting duct of the rat—role of the ureter. J Clin Invest 69(1):157–164
Pallone TL, Work J, Myers RL, Jamison RL (1994) Transport of sodium and urea in outer medullary descending vasa recta. J Clin Invest 93:212–222
Pallone TL, Turner MR, Edwards A, Jamison RL (2003) Countercurrent exchange in the renal medulla. Am J Physiol Regul Integr Comp Physiol 284:R1153–R1175
Pannabecker TL (2012) Structure and function of the thin limbs of the loops of Henle. In: Terjung RL (ed) Comprehensive physiology, vol 2. Wiley, Bethesda, pp 2063–2086
Pannabecker TL (2013) Comparative physiology and architecture associated with the mammalian urine concentrating mechanism: role of inner medullary water and urea transport pathways in the rodent medulla. Am J Physiol Regul Integr Comp Physiol 304:R488–R503
Pannabecker TL, Dantzler WH (2006) Three-dimensional architecture of inner medullary vasa recta. Am J Physiol Renal Physiol 290:F1355–F1366
Pannabecker TL, Dantzler WH (2007) Three-dimensional architecture of collecting ducts, loops of Henle, and blood vessels in the renal papilla. Am J Physiol Renal Physiol 293:F696–F704
Pannabecker T, Layton A (2014) Targeted delivery of solutes and oxygen in the renal medulla: role of microvessel architecture. Am J Physio Renal Physiol 307:F649–F655
Pannabecker TL, Dantzler WH, Layton HE, Layton AT (2008a) Role of three-dimensional architecture in the urine concentrating mechanism of the rat renal inner medulla. Am J Physiol Renal Physiol 295:F1271–F1285
Pannabecker TL, Henderson C, Dantzler WH (2008b) Quantitative analysis of functional reconstructions reveals lateral and axial zonation in the renal inner medulla. Am J Physiol Renal Physiol 294:1306–1314
Pennell JP, Lacy FB, Jamison RL (1974) An in vivo study of the concentrating process in the descending limb of Henle's loop. Kidney Int 5:337–347
Pietila I, Ellwanger K, Railo A, Jokela T, Barrantes Idel B, Shan J, Niehrs C, Vainio SJ (2011) Secreted Wnt antagonist Dickkopf-1 controls kidney papilla development coordinated by Wnt-7b signalling. Dev Biol 353(1):50–60. doi:10.1016/j.ydbio.2011.02.019
Pihakaski-Maunsbach K, Vorum H, Honore B, Tokonabe S, Frokiaer J, Garty H, Karlish SJD, Maunsbach AB (2006) Locations, abundances, and possible functions of FXYD ion transport regulators in rat renal medulla. Am J Physiol Renal Physiol 291(5):F1033–F1044
Pruitt MEC, Knepper MA, Graves B, Schmidt-Nielsen B (2006) Effect of peristaltic contractions of the renal pelvic wall on solute concentrations of the renal inner medulla in the hamster. Am J Physiol Renal Physiol 290(4):F892–F896
Reinking LN, Schmidt-Nielsen B (1981) Peristaltic flow of urine in the renal papillary collecting ducts of hamsters. Kidney Int 20:55–60
Rocha AS, Kokko JP (1973) Sodium chloride and water transport in the medullary thick ascending limb of Henle. Evidence for active chloride transport. J Clin Invest 52:612–623
Rollhauser H, Kriz W, Heinke W (1964) Das gefass-system der rattenniere. Zeitschrift fur Zellforschung 64:381–403
Saxen L (1987) Organogenesis of the Kidney. Cambridge University Press, Cambridge, UK
Schmidt-Nielsen B, Graves B (1982) Changes in fluid compartments in hamster renal papilla due to peristalsis in the pelvic wall. Kidney Int 22:613–625
Schmidt-Nielsen B, O'Dell R (1961) Structure and concentrating mechanism in the mammalian kidney. Am J Physiol 200:1119–1124

Schmidt-Nielsen B, Reinking LN (1981) Morphometry and fluid reabsorption during peristaltic flow in hamster renal papillary collecting ducts. Kidney Int 20:789–798
Schmidt-Nielsen B, Churchill M, Reinking LN (1980) Occurrence of renal pelvic refluxes during rising urine flow rate in rats and hamsters. Kidney Int 18:419–431
Schurek HJ, Kriz W (1985) Morphologic and functional evidence for oxygen deficiency in the isolated perfused rat kidney. Lab Invest 53(2):145–155
Schwartz MM, Venkatachalam MA (1974) Structural differences in thin limbs of Henle: physiological implications. Kidney Int 6:193–208
Schwartz MM, Karnovsky MJ, Venkatachalam MA (1979) Regional membrane specialization in the thin limbs of Henle loops as seen by freeze-fracture electron-microscopy. Kidney Int 16 (5):577–589
Schwentner C, Oswald J, Lunacek A, Fritsch H, Deibl M, Bartsch G, Radmayr C (2005) Loss of interstitial cells of Cajal and gap junction protein connexin 43 at the vesicoureteral junction in children with vesicoureteral reflux. J Urol 174(5):1981–1986. doi:10.1097/01.ju.0000176818.71501.93
Sendeski MM (2011) Pathophysiology of renal tissue damage by iodinated contrast media. Clin Exp Pharmacol Physiol 38(5):292–299. doi:10.1111/j.1440-1681.2011.05503.x
Solari V, Piotrowska AP, Puri P (2003) Altered expression of interstitial cells of Cajal in congenital ureteropelvic junction obstruction. J Urol 170(6 Pt 1):2420–2422. doi:10.1097/01.ju.0000097401.03293.f0
Stephenson JL (1972) Concentration of urine in a central core model of the renal counterflow system. Kidney Int 2:85–94
Storm R, Klussmann E, Geelhaar A, Rosenthal W, Maric K (2003) Osmolality and solute composition are strong regulators of AQP2 expression in renal principal cells. Am J Physiol Renal Physiol 284(1):F189–F198. doi:10.1152/ajprenal.00245.2002
Thom RP, Rosenblum ND (2013) A translational approach to congenital non-obstructive hydronephrosis. Pediatr Nephrol 28(9):1757–1761. doi:10.1007/s00467-012-2321-7
Thomas SR (2000) Inner medullary lactate production and accumulation: a vasa recta model. Am J Physiol Renal Physiol 279:F468–F481
Thomas SR (2001) A brief history of theories concerning the mammalian urine concentrating mechanism. Acta Biotheor 49:327–340
Thomas SR, Wexler AS (1995) Inner medullary external osmotic driving force in a 3-D model of the renal concentrating mechanism. Am J Physiol 269:F159–F171
Uchida S, Sasaki S, Furakawa T, Hiraoka M, Imai T, Hirata Y, Marumo F (1993) Molecular cloning of a chloride channel that is regulated by dehydration and expressed predominantly in the kidney medulla. J Biol Chem 268:3821–3824
Uchida S, Sohara E, Rai T, Ikawa M, Okabe M, Sasaki M (2005) Impaired urea accumulation in the inner medulla of mice lacking the urea transporter UT-A2. Mol Cell Biol 25:7357–7363
Urity VB, Issaian T, Braun EJ, Dantzler WH, Pannabecker TL (2012) Architecture of kangaroo rat inner medulla: segmentation of descending thin limb of Henle's loop. Am J Physiol Regul Integr Comp Physiol 302:R720–R726
van Balkom BW, van Raak M, Breton S, Pastor-Soler N, Bouley R, van der Sluijs P, Brown D, Deen PM (2003) Hypertonicity is involved in redirecting the aquaporin-2 water channel into the basolateral, instead of the apical, plasma membrane of renal epithelial cells. J Biol Chem 278(2):1101–1107. doi:10.1074/jbc.M207339200
Vandewalle A, Cluzeaud F, Bens M, Kieferle S, Steinmeyer K, Jentsch TJ (1997) Localization and induction by dehydration of ClC-K chloride channels in the rat kidney. Am J Physiol 272: F678–F688
Vitzthum H, Castrop H, Meier-Meitinger M, Riegger GA, Kurtz A, Kramer BK, Wolf K (2002) Nephron specific regulation of chloride channel CLC-K2 mRNA in the rat. Kidney Int 61 (2):547–554. doi:10.1046/j.1523-1755.2002.00165.x
Wang X, Thomas SR, Wexler AS (1998) Outer medullary anatomy and the urine concentrating mechanism. Am J Physiol Renal Physiol 274:F413–F424

Westrick KY, Serack BJ, Dantzler WH, Pannabecker TL (2013) Axial compartmentation of descending and ascending thin limbs of Henle's loops. Am J Physiol Renal Physiol 304: F308–F316. doi:10.1152/ajprenal.00547.2012

Wexler AS, Kalaba RE, Marsh DJ (1991a) Three-dimensional anatomy and renal concentrating mechanism. I. Modeling results. Am J Physiol 260:F368–F383

Wexler AS, Kalaba RE, Marsh DJ (1991b) Three-dimensional anatomy and renal concentrating mechanism. II. Sensitivity results. Am J Physiol 260:F384–F394

Yu J (2011) Wnt signaling and renal medulla formation. Pediatr Nephrol 26(9):1553–1557. doi:10.1007/s00467-011-1888-8

Yu J, Carroll TJ, Rajagopal J, Kobayashi A, Ren Q, McMahon AP (2009) A Wnt7b-dependent pathway regulates the orientation of epithelial cell division and establishes the cortico-medullary axis of the mammalian kidney. Development 136(1):161–171. doi:10.1242/dev.022087

Yuan J, Pannabecker TL (2010) Architecture of inner medullary descending and ascending vasa recta: pathways for countercurrent exchange. Am J Physiol Renal Physiol 299:F265–F272

Yui N, Lu HA, Chen Y, Nomura N, Bouley R, Brown D (2012) Basolateral targeting and microtubule dependent transcytosis of the aquaporin-2 water channel. Am J Physiol Cell Physiol. doi:10.1152/ajpcell.00109.2012

Zhai X-Y, Thomsen JS, Birn H, Kristoffersen IB, Andreasen A, Christensen EI (2006) Three-dimensional reconstruction of the mouse nephron. J Am Soc Nephrol 17:77–88

Zhai XY, Fenton RA, Andreasen A, Thomsen JS, Christensen EI (2007) Aquaporin-1 is not expressed in descending thin limbs of short-loop nephrons. J Am Soc Nephrol 18 (11):2937–2944

Chapter 12
Non-traditional Models: The Giraffe Kidney from a Comparative and Evolutionary Biology Perspective

Mads Damkjær, Tobias Wang, Kristine H. Østergaard, Emil Brøndum, Ulrik Baandrup, Arne Hørlyck, J.Michael Hasenkam, Niels Marcussen, Carl Christian Danielsen, Mads F. Bertelsen, Carsten Grøndahl, Michael Pedersen, Peter Agger, Geoffrey Candy, John Chemnitz, Christian Aalkjær, and Peter Bie

Abstract Giraffes are the tallest living animals and endowed with the highest arterial blood pressure of any animal on Earth. Here we present novel data on kidney function in this extraordinary animal obtained over the course of two major expeditions in 2010 and 2012. As expected, the anaesthetised giraffes had very high mean arterial blood pressure ranging between 150 and 300 mmHg. However, despite the high filtration pressure, the rate of glomerular filtration (GFR) was only 0.7 ± 0.2 ml/min/kg, which is approximately 40 % below similar-sized mammals. The renal blood flow of 3.1 l/min accounts for approximately 20 % of cardiac output, and the calculated filtration fraction (GFR/ERPF) was approximately 0.3, and hence within typical mammalian values. The normal kidney function of the giraffes appears due to very high interstitial pressures within the kidney, a feature that is possible due to the very thick and strong capsule surrounding the kidney in combination with a vascular valve at the entrance of the renal vein into the abdominal cava. These relatively simple structural modifications normalize the Starling forces driving filtration over the Bowman capsula.

Keywords High blood pressure • Gravity • Glomerular filtration rate (GFR) • Scaling of renal function • Ultrasound, interstial renal pressure • Renal blood flow

M. Damkjær • P. Bie (✉)
Department of Cardiovascular and Renal Research, University of Southern Denmark, Odense, Denmark
e-mail: pbie@health.sdu.dk

T. Wang
Department of Biological Sciences, Institute of Zoophysiology, Aarhus University, Aarhus, Denmark

K.H. Østergaard • U. Baandrup
Department of Clinical Medicine, Aalborg University, Aalborg, Denmark

Centre for Clinical Research, Vendsyssel Hospital, Hjørring, Denmark

K.A. Hyndman, T.L. Pannabecker (eds.), *Sodium and Water Homeostasis*, Physiology in Health and Disease, DOI 10.1007/978-1-4939-3213-9_12

12.1 Introduction

Giraffes are the tallest animals on Earth and unique among all extant animals because of the high arterial blood pressures required to perfuse their brain, which is situated several meters above the heart. The remarkable cardiovascular adaptations required for adequate cerebral perfusion entails tolerance to blood pressure changes at the base of the brain of 3–400 mmHg within seconds when the giraffe lowers it head to drink. The arterial pressure at the base of the giraffe brain probably resembles that of other mammals (apart from during drinking), but a consequence of the high arterial pressure is that all visceral organs are exposed to much higher blood pressures than in other mammals. While much emphasis has been placed on understanding the vascular adaptations that enable and regulate the high arterial pressures, little attention has been directed to understanding how kidney functions are adapted to the high pressures. This is surprising because hypertension in humans, even with blood pressures well below the pressures in the giraffe, invariably causes kidney damage and hence represents a serious clinical concern. On the other hand, even simple classical studies of renal function in the giraffe are difficult to perform for technical reasons, and are also hindered by the challenges of keeping giraffes anesthetised for many hours.

E. Brøndum
Department of Physiology, Institute of Biomedicine, Aarhus University, Aarhus, Denmark

A. Hørlyck
Department of Radiology, Aarhus University Hospital, Aarhus, Denmark

J.M. Hasenkam
Department of Cardiothoracic and Vascular Surgery, Institute of Clinical Medicine, Aarhus University Hospital, Aarhus, Denmark

N. Marcussen
Department of Clinical Pathology, University of Southern Denmark, Odense, Denmark

C.C. Danielsen
Department of Anatomy, Institute of Biomedicine, Aarhus University, Aarhus, Denmark

M.F. Bertelsen • C. Grøndahl
Center for Zoo and Wild Animal Health, Copenhagen Zoo, Copenhagen, Denmark

M. Pedersen • P. Agger
MR Research Centre, Institute of Clinical Medicine, Aarhus University Hospital, Aarhus, Denmark

G. Candy
Department of Surgery, Faculty of Health Sciences, University of the Witwatersrand, Johannesburg, South Africa

J. Chemnitz
Department of Anatomy, University of Southern Denmark, Odense, Denmark

C. Aalkjær
Department of Physiology, Institute of Biomedicine, Aarhus University, Aarhus, Denmark

Department of Biomedicine, University of Copenhagen, Copenhagen, Denmark

The Danish Giraffe Cardiovascular Research Group (DaGiR) was formed in 2007 by a diverse group of scientists with biological, biomedical, clinical and medico-technical competences to generate new insight into the circulatory control in all mammals thereby providing a better background for the analysis also of the pathophysiology of aberrant human blood pressure control. Publications have focused on different topics: cerebral perfusion (Brondum et al. 2009), cardiac function and left ventricular morphology (Ostergaard et al. 2013) the conspicuous absence of edema of the legs (Petersen et al. 2013), and the function of the kidneys (Damkjær et al. manuscript in preparation). Most of the present data were generated during two campaigns to Gauteng Province, South Africa, (2010 and 2012) where a total of 14 giraffes (one female/13 male) were investigated by an array of technologies. Subsequently, fresh giraffe kidneys were obtained on occasions where captured giraffes in Denmark had to be euthanised for other reasons.

12.2 Effects of Giraffe-Sized Arterial Blood Pressure in Other Animals

In most mammals baseline mean arterial blood pressure is close to 100 mmHg (Fig. 12.1) also in species, such as rodents, where the circulation is not influenced by any sizable orthostatic stress. The implication seems to be that arterial pressure

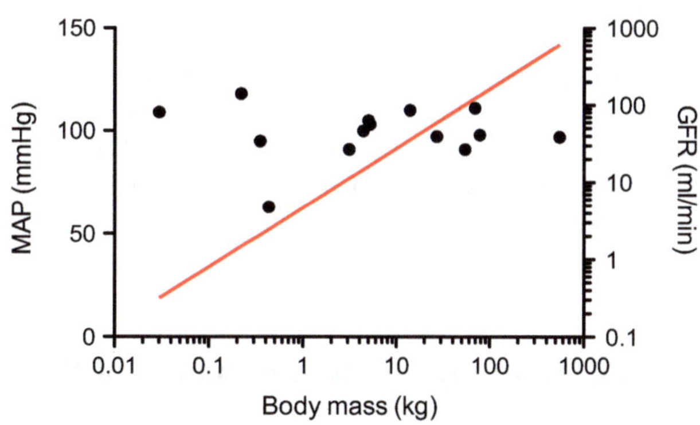

Fig. 12.1 Mean arterial pressure (MAP) plotted as a function of body mass in 14 different species of mammal. To avoid confounders (e.g., anaesthesia, stress etc.) we have chosen only to include data from experiments where MAP have been measured telemetrically in chronically instrumented, conscious and non-stressed animals. Species included are: rat (Sithisarn et al. 2013), rabbit (Malpas et al. 1997), guinea pig (Provan et al. 2005), mouse (Kim et al. 2008), cynomolgus monkey (Haushalter et al. 2008), sheep (King et al. 2007), dog (Mckee et al. 2014) Yucatan miniature swine (Myrie et al. 2012), Göttingen minipig (Markert et al. 2009), Rhesus monkey (Regan et al. 2009), tree toed sloth (Duarte et al. 2003), marmoset (Horii et al. 2002), horse (Hornicke et al. 1977) and humans (Head et al. 2014). Please note that for humans there are no telemetric data, values are from an epidemiological survey. Drawn in full *red line* is the expected glomerular filtration rate calculated from body mass, cf Singer (2001)

is an intrinsic variable unrelated to body size. The developmental background for this remains unidentified, but may, in addition to other aspects of the physics of the cardiovascular system (West et al. 1997), be related to the renal arterial pressure required to drive glomerular filtration rate (GFR) (Nerpin et al. 2014) as well as the post-glomerular blood flows.

Arterial hypertension is the most important risk factor for premature cardiovascular disease, which is highlighted by the fact that 54 % of stroke and 47 % of ischemic heart disease worldwide have been attributed to high blood pressure (Lawes et al. 2008). The kidneys play a central role in both development and maintenance of arterial hypertension. This is seen for instance in patients suffering from Liddle syndrome, a disorder of sodium reabsorption in the distal tubule, who develop severe hypertension (Botero-Velez et al. 1994) and that patients with essential hypertension who undergo renal transplantation become normotensive (Curtis et al. 2000). Although the kidney is the villain in hypertension, it is also a victim, in the sense that arterial hypertension both causes renal damage and increases progression of non-hypertensive renal disease. Hypertensive nephrosclerosis, hypertensive kidney disease and nephroangiosclerosis are clinical terms used to describe the renal damage caused by hypertension (in contrast to hypertension driven by primary renal disease). Histologically this is characterized by hyperplasia of interlobular and afferent arteriolar vessels, hyaline arteriolosclerosis and commonly global glomerulosclerosis. Evidence indicates that nephrosclerosis is an increasing cause of end-stage renal disease (Marin et al. 2005) and an important predictor of cardiovascular disease (Weir 2009). Similar observations on the effect of hypertension on the development of nephrosclerosis and end-stage renal disease have been made in other species such as the spontaneously hypertensive rats (Zhou and Frohlich 2007) and in guinea pigs (Takeda and Grollman 1970). Not only accelerated, but also mild hypertension is known to hasten the progression of primary renal disease (Baldwin and Neugarten 1985; Toto 2005), and it is well-established that clinical interventions to lower blood pressure slow the rate of progression (Toto 2005).

12.3 Functional Anatomy and Biomechanics of the Giraffe Kidney

12.3.1 Macroscopic Features

The renal artery and vein do not show peculiar characteristics (Maluf 2002), but at the junction of the renal vein with the caval vein a semilunar valve partially occludes the lumen, and the surrounding sections of the caudal caval vein show wall characteristics markedly different from more remote parts in caudal as well as cephalic direction. We found that this reno-caval valve has functional importance as demonstrated by ultrasound imaging of renal venous flow and pressure measurements across the reno-caval venous junction (see below: hemodynamics). In addition, we noted a thickening of the caval venous wall at the region of entry of the renal vein, as illustrated by our data on the volume fraction of smooth muscle in the wall of the caval vein, *i.e.*, muscle tissue volume as fraction of total tissue volume. For the first 10 mm of the caval vein, the volume fraction was 0.39 ± 0.10 (see Fig. 12.2). From about 10 mm below to 10 mm above the entrance of the renal vein, the wall of the caval vein had a volume fraction of 0.62 ± 0.05. Near the right atrium (10–30 mm) the wall was again thin with a volume fraction of 0.26 ± 0.04. The functional implications of the marked reinforcement of the caudal caval venous structure at the entry of the renal veins remain difficult to assess.

Maluf (2002) has provided the most extensive description of the structure of the giraffe kidney on the basis of kidneys from two large male Masai giraffes. This study has been extended markedly by our recent work (Damkjær et al. in preparation). The giraffe kidney is a bean-shaped organ without external lobulation. Although the renal cortex in general is similar to that of other species (cf. histological and histochemical analyses below), several conspicuous differences have been found (Maluf 2002) in relation to the kidneys of e.g., pig, cow, and man: Firstly, the renal capsule is a tough, opaque, and almost inextensible collagenous structure without penetrating blood vessels; we confirmed this description and subjected the capsule to biomechanical analysis (see below). Secondly, the cortex does not intrude into the medullary region, *i.e.*, columnae renales (Bertini) are absent and replaced by 'pelvic extensions' and 'vascular processes'. We confirmed these details and have subjected specific tissues to further analysis as described below.

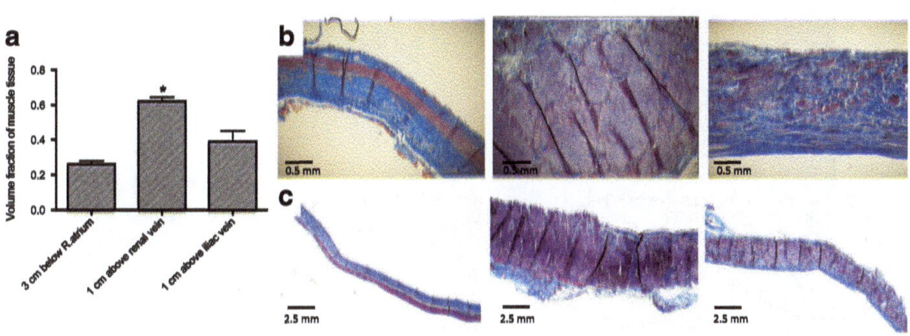

Fig. 12.2 Caval vein histology: Sections taken from one giraffe at three different locations along the caudal vana cava of the giraffe. (**a**) *Bar graph* showing statistically significant differences between the volume fractions of smooth muscle cells at three different locations: (i) 3 cm below the right atrium, (ii) 1 cm above the renal vein and 1 cm above the iliac vein (*$p < 0.001$). (**b**) Masson trichrome staining of vena cava inferior (smooth muscle tissue is *red*, collagen tissue, *blue*). (**b1**) 1 cm above the confluence of the iliac veins, (**b2**) 1 cm above the renal vein and (**b3**) 3 cm below the Right atrium. In b3, a band of slightly stained smooth muscle cells are seen under the intima (↑). (**c**) Sections from the same locations in another giraffe, but in a smaller magnification. Figure reprinted with permission from Damkjaer et al. (2015)

12.3.2 Plastination

Fresh giraffe kidneys were perfused with heparinized saline immediately after necropsy and plastinated according to the method of von Hagen (Dejong and Henry 2007; von Hagen 1986). Before curing, kidneys were hemisected in the coronal plane (Fig. 12.3). The pelvic extensions containing the vascular processes provide regular, deep, complete divisions of the medullary region. Pelvic extensions constitute the final divisions of the pelvic cavity extending all the way to the cortex. Vascular processes are columns of tissue lined with urothelium and containing the interlobar arteries and veins and projecting into the lumen of the pelvis. Some reduction in tissue volume during plastination is unavoidable; it is likely that *in vivo* the pelvic extensions are slit-like cavities containing only small volumes of urine.

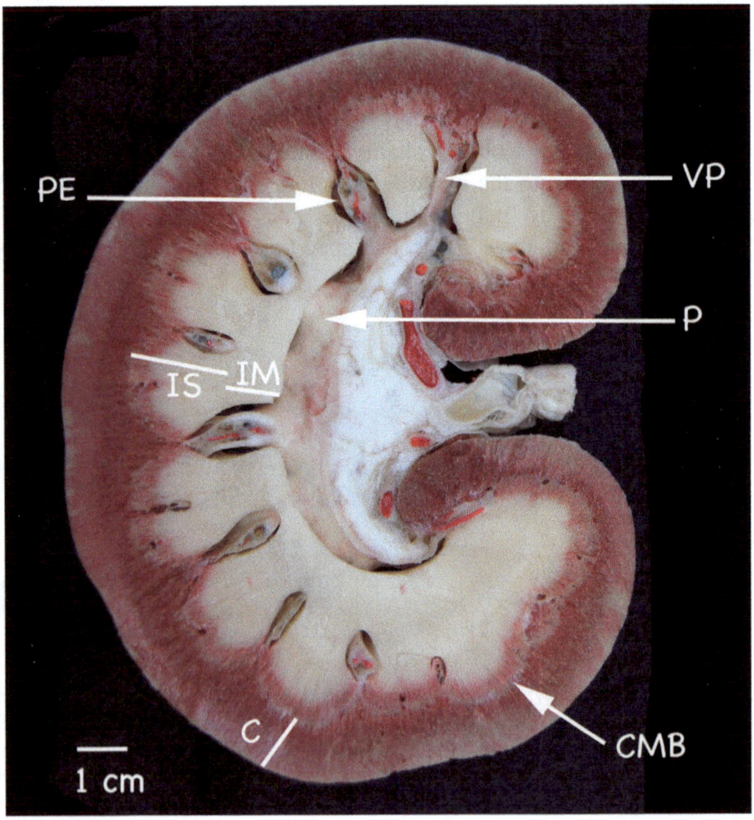

Fig. 12.3 Plastinated giraffe kidney that has been transsected. Abbreviations: *C* cortex, *CMB* corticomedullary border, *IM* inner medulla, *IS* inner stripe, *P* pelvis, *PE* pelvic extension, *VP* vascular process

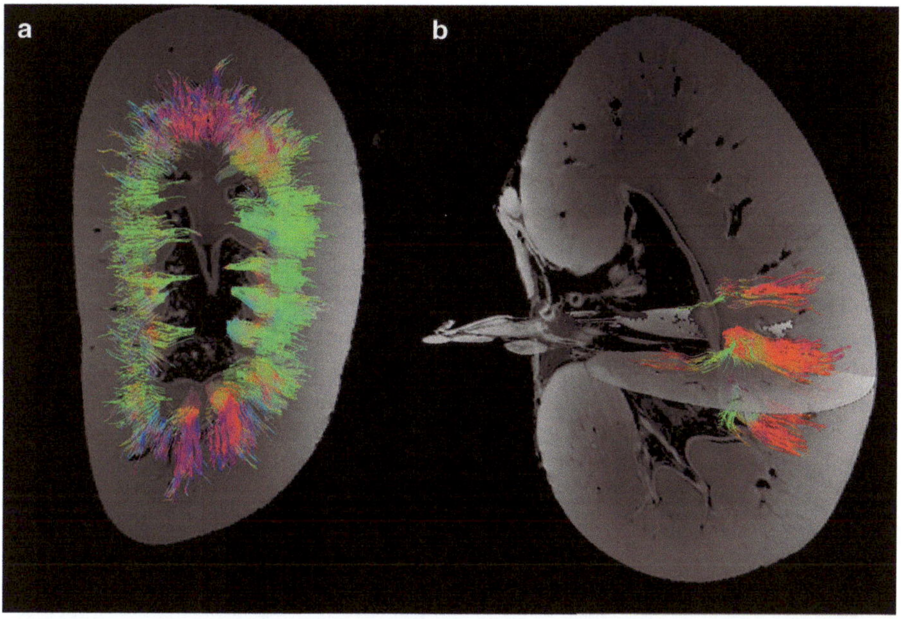

Fig. 12.4 The giraffe kidney imaged by diffusion tensor imaging with magnetic resonance imaging. The fibers in this transverse section show a centripetal alignment of calculated tracks (**a**), representing the medullary tubules, and they follow the minor calyces and join together to form major calyces which in turn drain into the pelvis (**b**). Track colors' are directional color coating derived from the orientation of the diffusion direction relative to the scanner coordinate system. The colors are therefore merely a visual aid and do not represent any anatomical or physiological properties. Reprinted with permission from Damkjaer et al. (2015)

Diffusion tensor imaging (DTI) with magnetic resonance (MR) is a method that allows mapping of diffusion of water molecules in biological tissues. As water movement is restricted by various structures, such as cell membranes, diffusion occurs more easily along a fiber path than across it. We performed DTI on a giraffe kidney with a Philips 1.5 T Achieva system (Fig. 12.4) with a protocol design to visualize the tubular system. The geometric direction of pathways in the kidneys were visualized by directional color coating, and this showed a centripetal alignment of tracks (Fig. 12.4), likely to represent the medullary tubules and collecting ducts, which in turn drain into the pelvis.

12.3.3 Histological Characteristics

Histologically, the giraffe renal cortex resembles the human renal cortex. Occasionally glomeruli were found in close proximity to the wall of the interlobular arteries or even partially embedded in the vessel wall. The medullary rays are broader than the human equivalent and divide the cortex in thin elongated bundles

(Fig. 12.5). The outer stripe of the outer zone of the medulla (OSOMZ) is thin, whereas the inner stripe of the outer zone (ISOMZ) is prominent. *In vivo*, the ISOMZ and the inner medulla are exposed to pelvic urine through the pelvic extensions containing the vascular processes. The surrounding medulla is lined by a single layer of columnar/cuboidal cells covering a layer of α-smooth muscle actin positive cells that also surrounds the adjacent vascular capillary bundles (Fig. 12.5). The vascular processes contain the interlobar blood vessels situated in fibrous tissue (Fig. 12.6). In parts of the vascular processes, the interlobar blood vessels were surrounded by a less cellular connective tissue positive for Alcian Blue at pH 2.7 (Fig. 12.6). This particular connective tissue resembles the mucoid, mesenchymal connective tissue in the human umbilical cord and the vitreous body of the eye. The inner medulla consists of collecting ducts and prominent vascular bundles. Dense connective tissue was located under the urothelium and consisted of 5–6 cell layers. The mesenchymal connective tissue in this part of the vascular processes may be functionally important as it could provide the vascular processes with unusual biomechanical properties.

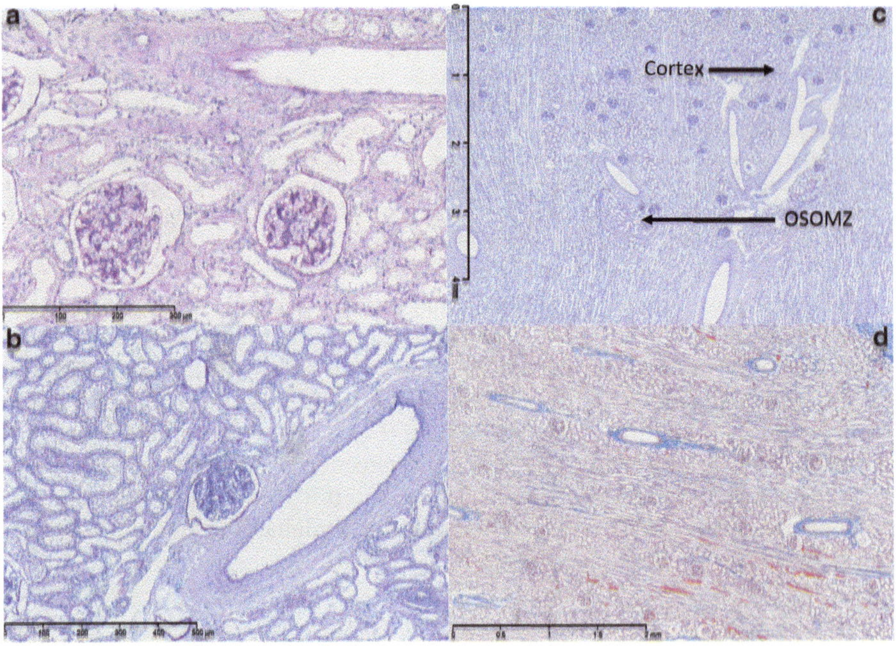

Fig. 12.5 Micrographs of giraffe kidney. (**a**) The two glomerular tufts have a diameter of about 130 mm and Bowman's capsule is lined by a flattened parietal layer. (**b**) In some instances the glomeruli are very close to the wall of the interlobular arteries. (**c**) The proximal tubules are the dominating tubular structures in the cortex. The cortex also contains medullary rays that seem to be broad and split the cortex with proximal tubules in thin elongated bundles. (**d**) The outer stripe of the outer zone of renal medulla (OSOMZ) is very thin and irregular, and it is missing in many parts of the kidney

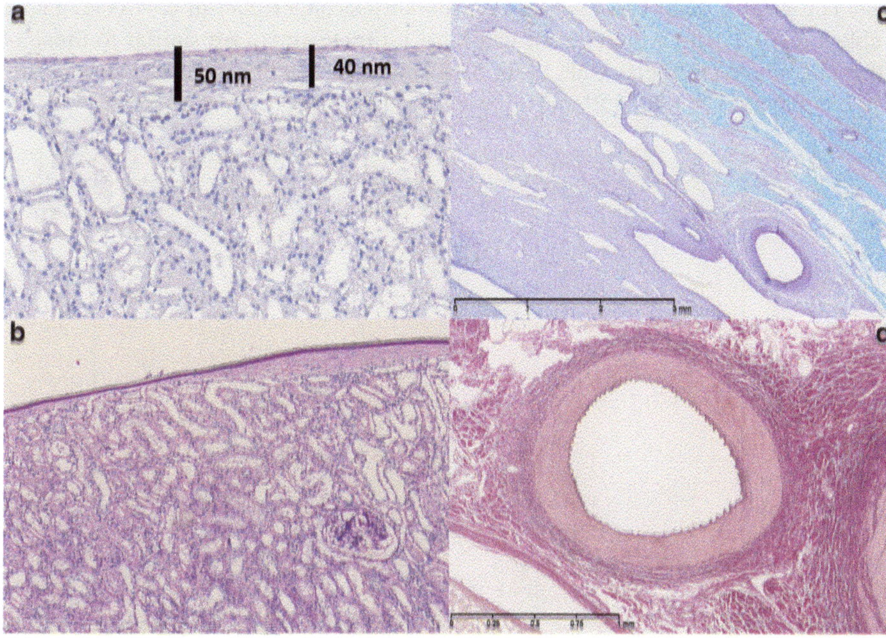

Fig. 12.6 Micrographs of giraffe kidney. (**a**) Superficial part of cortex with inner part of renal capsule. (**b**) Superficial part of cortex with inner part of renal capsule showing slight variation in thickness, stained with PAS. (**c**) Interlobar blood vessels in the vascular processes surrounded by *Alcian Blue* (pH 2.7) positive connective tissue with few cells. This connective tissue is similar to the mucoid, mesenchymal connective tissue present in the umbilical cord and the vitreous body of the eye. (**d**) The interlobar and interlobular arteries show no signs of atherosclerosis. The intima is delicate and thin and the lamina media does not show any signs of hypertrophy

12.3.4 Biomechanical Analysis

Segments of renal vessels and strips of renal capsules were analysed by measurement of (i) cross sectional area, A, (ii) ultimate (*i.e.*, maximal) load, F (in N), (iii) ultimate stress, F/A, (in MPa) and (iv) ultimate strain (fractional increase).

12.3.4.1 Renal Capsule

Longitudinal and transverse strips (5.2 mm × 20 mm) of the lateral surface of renal capsule from seven giraffes were obtained. Strain was calculated as fractional increase in original length determined from the dimensions at minimal load. Five loading cycles were performed before loading until rupture, and the strip was analyzed for collagen content (Danielsen et al. 1988). The ultimate strength was found by dividing maximum load by mg of collagen per mm original strip length (N × mm/mg collagen). Corresponding samples of renal capsule from a young cow was tested. Renal capsular burst pressure was estimated from the Young-Laplace law. The giraffe renal capsule was considerably stronger than in cow (Fig. 12.7a, Table 12.1) for which the present values of ultimate stress are similar to those of other mammals (Herbert et al. 1976; Snedeker et al. 2005). As the burst pressure was estimated to be 600–650 mmHg (compared to some 125 mmHg in the cow) the

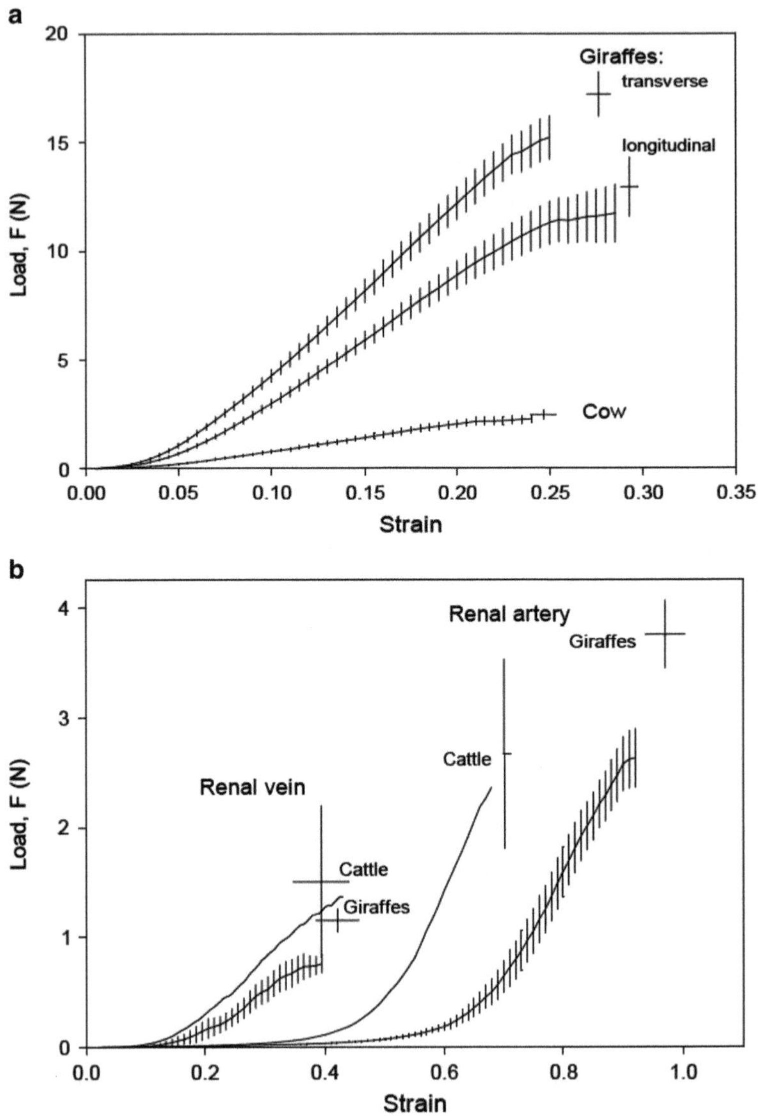

Fig. 12.7 Load-strain curves showing the biomechanical properties of giraffe tissue. (**a**) Biomechanics of renal capsule from giraffes and cow. Mean load-strain curves and the mean parameters (ultimate strain, ultimate load) are shown. Error bars ± SEM. (**b**) Biomechanics of renal vessels from giraffes and cattle (one cow and one young bull). Mean load-strain curves and the mean parameters (ultimate strain, ultimate load) are shown. Error bars ± SEM (for cattle load-strain curves plotted without error bars). Reprinted with permission from Damkjaer et al. (2015)

Table 12.1 Biomechanical characteristics of renal capsules from giraffes (n = 7) and a cow

Renal capsule	Thickness (mm)	Collagen (mg/cm^2)	Ultimate strength (N)	Ultimate stress (MPa)	Normalized ultimate strength (N × mm/mg collagen)
Giraffe					
Combined	0.202 ± 0.011	4.25 ± 0.24	15.1 ± 1.0	14.6 ± 0.6	69.0 ± 2.3
Longitudinal	0.218 ± 0.015	4.41 ± 0.36	13.0 ± 1.4	11.5 ± 1.0	56.8 ± 4.5
Transverse	0.187 ± 0.008	4.09 ± 0.18	17.2 ± 1.0	17.7 ± 0.8	81.2 ± 3.4
Cow	0.081 ± 0.006	0.88 ± 0.11	2.5 ± 0.2	5.9 ± 0.5	58.1 ± 4.6

Data from each giraffe are represented by averages of three longitudinal and three transverse specimens from one kidney. The results for cow are based on 14 specimens of different orientation sampled from both kidneys. Data are mean ± SEM

robust giraffe capsule seems able to resist high intrarenal pressures. Unexpectedly, the ultimate tensile strength of the renal capsule in transverse direction was 43 % above that in longitudinal direction possibly reflecting a prevailing transverse orientation of collagen fibres. The kidney of adult giraffes is unlobulated, but the surprising transverse tensile strength could reside in persistent collagen fibres in the formerly transverse furrows. In a functional perspective, the inflexion point of the load-strain curve is relevant, since straining beyond this point is assumed to result in collagen damage (Herman 2007). For giraffe and cow, our data allow calculation of pressures at inflexion points of 350 and 75 mmHg, respectively.

12.3.4.2 Vessels

Renal artery and vein segments were collected from 10 giraffes (one female) at autopsy. After determination of the cross-sectional area, dynamic analyses were performed at a fixed rate of deformation. Five loading cycles were conducted before loading until rupture. Strain was determined as fractional increase in original luminal perimeter defined as the perimeter at which load increased ≥2 mN. Analogous vessels from a young cow and a young bull were included as reference. Our data did not demonstrate major differences between the properties of renal vessels from giraffes and cows, except that giraffe renal arteries showed higher ultimate strain (Fig. 12.7b, Table 12.2) and lower unstrained luminal radius (r = 2.26 ± 0.05 mm) compared to cows (3.29 ± 0.14 mm).

12.4 Kidney Function in the Giraffe

Body mass exerts fundamental effects on virtually all physiological processes in animals (Schmidt-Nielsen 1984). As the exponent describing how whole body metabolism scales with body mass is somewhere between 0.66–0.75, large animals

Table 12.2 Biomechanical characteristics of renal vessels in giraffe and cattle

Renal vessel	Species	Cross-sectional area (mm^2)	Ultimate load (N)	Ultimate stress (MPa)	Ultimate strain (–)
Vein	Giraffe	2.13 ± 0.12	1.16 ± 0.11	0.56 ± 0.06	0.421 ± 0.036
	Cattle	1.99 ± 0.47	1.51 ± 0.69	0.89 ± 0.55	0.394 ± 0.046
Artery	Giraffe	2.78 ± 0.06	3.76 ± 0.31	1.36 ± 0.12	0.970 ± 0.033
	Cattle	2.49 ± 0.42	2.67 ± 0.86	1.04 ± 0.17	0.702 ± 0.002

Data from each animal are represented by the average of three ring samples from each vessel. Data are mean ± SEM (n = 10 giraffes and two for cattle)

have considerable lower rates of oxygen consumption per unit body mass than smaller mammals. These metabolic differences are evident from the much higher heart rates and cardiac output per unit mass in small mammals, whilst mean arterial pressure (MAP) does not scale appreciably with increased body mass. The relative constancy of MAP over a very large range of body mass is particularly evident when the scaling analysis is carefully restricting to blood pressure measurements from fully conscious animals without the confounding influence of anesthesia or restraint (Fig. 12.8). The lack of association between MAP and body mass implies that total peripheral resistance when expressed per body mass increases with body mass and that the perfusion per tissue mass decreases. This effect includes mass specific renal perfusion, as body mass corrected renal blood flow declines with increased body mass. Perhaps less intuitive from the mass independence of MAP, the mass-specific GFR also declines with body mass in a manner similar to renal blood flow, such that filtration fraction is remarkably similar amongst all mammals, irrespective of range of body mass spanning several orders of magnitude. As a result, it appears that mass corrected GFR and renal blood flow decline similarly to metabolism when body mass increases, thus providing a match between the rate of metabolic waste production and the rate of their filtration in the kidney (Singer 2001).

The giraffe is an interesting animal for elucidation of the interdependency between metabolism and kidney function. With a MAP approximately double that of other mammals, it could be expected that GFR and possibly renal filtration fraction would be considerably higher than in other mammals. However, on the other hand, giraffes are characterized by a lower cardiac output than predicted for similar sized mammals, which can be ascribed to a low cardiac stroke volume at a normal heart rate. The low cardiac output could imply a lower metabolic rate than predicted by mammalian scaling relationship or an unusually high arterio-venous oxygen concentration difference. In any event, if metabolism dictates GFR it would be predicted that GFR of giraffes should be similar or lower than similar-sized mammals. Recently, we conducted the first studies of renal hemodynamic and excretory function in giraffes [for details see Damkjær et al. (in preparation)].

Fig. 12.8 Anaesthetized and instrumented giraffe suspended in upright position on a tilt table. The bladder has been catheterized for continuous urine sampling. Sheaths are placed in the upper and lower part of the left carotid artery and jugular vein. Blood pressure, ECG, saturation and temperature are monitored using the monitor placed on the back of the giraffe. Reprinted with permission from Damkjaer et al. (2015)

12.4.1 Glomerular Filtration, Filtration Fraction

Using standard procedures, we determined GFR on the basis of inulin (sinistrin, Inutest®) clearance while effective renal plasma flow (ERPF) was measured by plasma clearance of p-aminohippurate (PAH). Over the course of these experiments, mean arterial blood pressure ranged from 168 ± 7 to 242 ± 8 mmHg and heart rate from 34 ± 2 to 43 ± 3 bpm. During the renal studies, we also performed numerous determinations of cardiac output using an inert gas rebreathing technique; cardiac output was approximately 15 l/min, which is lower than predicted on basis of body mass relationships in other mammals, but consistent with the low stroke volume (Dawson 2001).

GFR was 342 ± 99 and 348 ± 67 ml/min over the course of two clearance periods. When normalized to body mass, GFR was 0.7 ± 0.2 ml/min/kg, which is considerably lower than the GFR of 1.2 ml/kg BW/min that is predicted for a 493 kg mammal according to mammalian scaling studies (Singer 2001). Thus, we determined a GFR some 40 % below the expected value based on body mass. ERPF averaged 1.2 l/min, based on a hematocrit of 39 % (plasma volume 34 ml/kg by volume of distribution of Evans Blue and blood volume 56 ml/kg); this corresponds to a renal blood flow of 3.1 l/min. Therefore, the kidneys received some 20 % of cardiac output, which is similar to other mammals where renal blood flow constitutes some 22–24 % of cardiac output (Baggot 2001). Filtration fraction was calculated as GFR/ERPF and measured 0.30 ± 0.06 and 0.33 ± 0.04, which is within the range of typical mammalian values.

12.4.2 Renal Interstial Pressure, Urodynamics and the Reno-Caval Valve

Renal interstitial pressure was measured by inserting a 21G fluid-filled needle transrectally or via laparotomy into the kidney parenchyma and connecting it to a pressure transducer placed at heart level. At removal the needle was flushed with 100 μl Evans Blue solution to document the position of the needle tip (Fig. 12.9). For renal pelvic, ureteral and bladder pressures, the urethra was identified through a perineal incision and through a small incision a fiberoptic scope was inserted (MAF, Olympus, Denmark). A side-mounted tip-transducer catheter (SPC350, Millar Instruments, TX, USA) was advanced up each ureter during pressure recording. Subsequently a sheath was placed in the renal vein via a left flank incision and a

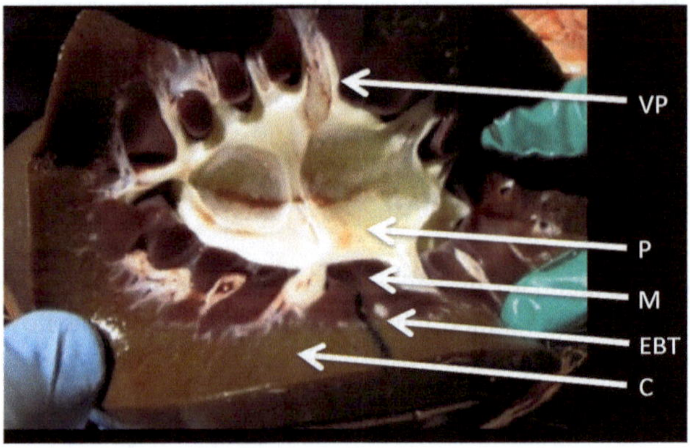

Fig. 12.9 *Post-mortem* hemisected giraffe kidney. *Arrows* indicate: *EBT* Evans blue trace *left* by the needle after recording of renal interstitial hydrostatic pressure, *VP* vascular process, *C* cortex, *M* medulla and *P* pelvis. Reprinted with permission from Damkjaer et al. (2015)

side-mounted tip-transducer catheter was introduced into the vein and pressure recorded in the renal vein. The catheter was then slowly advanced into the caudal caval vein during continuous pressure recording. In the ureter, regular peristaltic pressure waves were observed; however, they disappeared just before the pressure catheter lodged and could be advanced no further; indicating that the pressure catheter was in, or close to, the renal pelvis; pressure at this location averaged 39 mmHg. The renal interstitial pressure averaged 47 mmHg (in the first series of measurements considerably higher values (≈140 mmHg) were obtained, but these results could not be reproduced).

At the site where the renal vein joins the caudal caval vein a semi-lunar valve is present (Figs. 12.10b, 12.11d); catheter pressure fell abruptly by an average of 12 mmHg when the tip was moved from the renal vein to the inferior caval vein (Fig. 12.10c) demonstrating that the valve provided significant resistance to renal venous blood flow. An overview of selected hemodynamic and urodynamic results is presented in Fig. 12.10a.

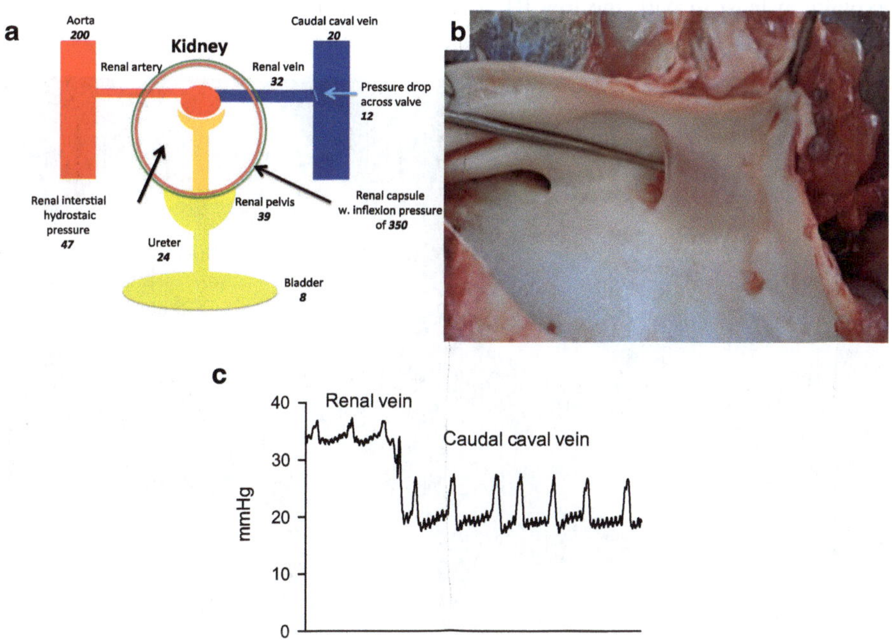

Fig. 12.10 (**a**) Schematic overview of hemodynamic and hydrostatic pressures measured in the giraffe. (**b**) Section through the caudal caval vein showing the fibrous valve structure covering the entrance of the renal vein into the caudal caval vein. (**c**) Continuous pressure recording initially in the renal vein, the catheter is then advanced 5 cm so that it enters the caudal caval vein. This illustrates the pressure drop occurring due to the valve at the junction of the renal vein and the caudal cava

12.4.3 Ultrasound Examinations

The kidneys were localized via transrectal exploration and an ultrasound transducer (curved array 8C-RS, Vivid i, GE Healthcare) was placed on the kidney surface, the aorta, and vena cava. B-mode as well as colour Doppler images and films were obtained from multiple sites in the region (Fig. 12.11). Doppler curves from intrarenal arteries (interlobar/arcuate arteries) were obtained. Intrarenal resistive index (RI) was calculated as (peak systolic velocity—end diastolic velocity) divided by peak systolic velocity; we found a RI of 0.26 ± 0.07 ($n = 8$ giraffes) see Fig. 12.11, compared to a mean of 0.60 in humans (Tublin et al. 2003). The giraffe renal RI seems to be lower than the RI of any other species. The meaning of RI in the kidney is not fully understood (Tublin et al. 2003); an association with vascular resistance appears less direct than previously assumed, especially under conditions of low vascular compliance (Bude and Rubin 1999; Tublin et al. 1999), and RI also correlates with the pulse pressure index (pulse pressure/systolic pressure) (Tublin et al. 1999). Pulse pressure index is low in giraffes and this may explain, at least in part, the low RI. The results of measurements of RI and mean

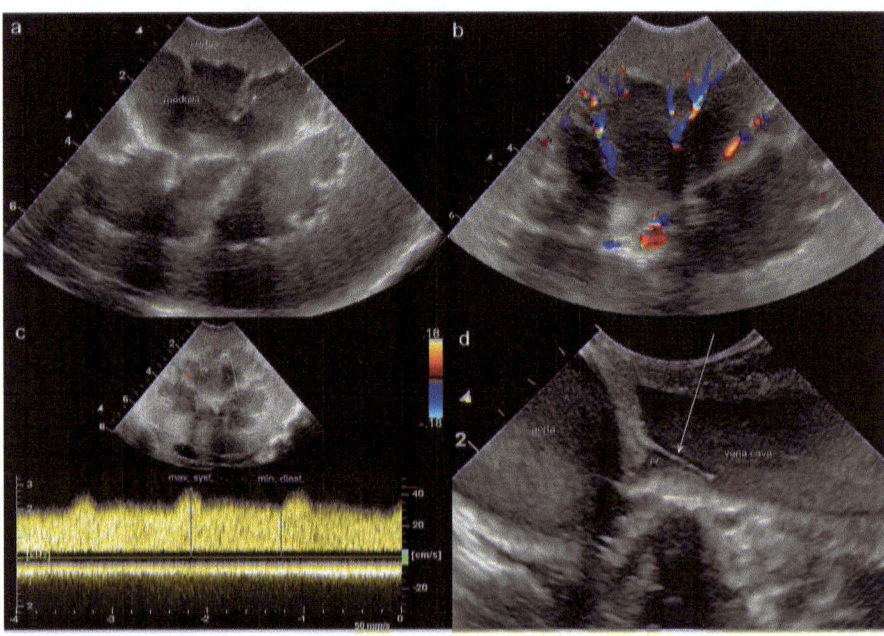

Fig. 12.11 Giraffe kidney imaged by *in-vivo* ultrasound with Doppler. (**a**) B-mode picture. The pelvi-calyceal system not visible (compressed) Vascular process (*arrow*). (**b**) Color Doppler showing flow in the vessels. (**c**) Spectral Doppler with low resistive index (RI). The vertical axis to the *left* shows the Doppler shift in KHz and the vertical axis to the *right* shows the corresponding flow velocity of the blood cells calculated from the Doppler equation. The horizontal axis shows time in seconds. (**d**) Renal vein (RV) junction with vena cava. Note the valve (*arrow*). Reprinted with permission from Damkjaer et al. (2015)

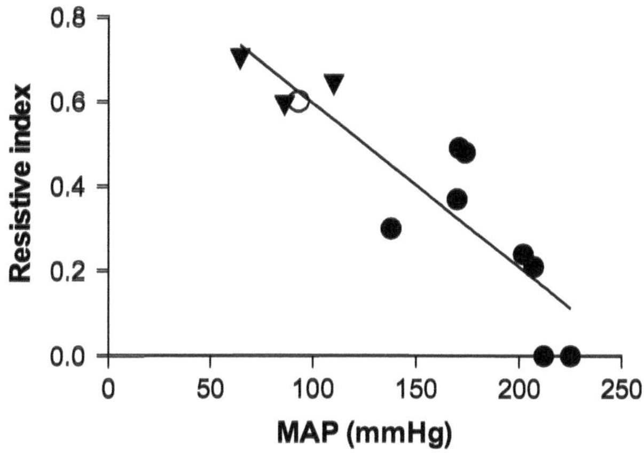

Fig. 12.12 Values for resistive index (RI) as a function of mean arterial blood pressure (MAP) in eight giraffes (*filled circle*), three cows (*filled inverted triangle*), and human normal value (*open circle*). The data seems to suggest that there is an inverse linear correlation (full line, $R^2 = 0.79$) between increasing MAP and a low RI

arterial pressure in different species (giraffes, human, and cows) suggest a linear relationship (Fig. 12.12) possibly indicating that the giraffe is different quantitatively rather than qualitatively. Another possible explanation of the low RI could be related to the structure and composition of the vascular processes (see above) as these might dampen the pressure excursions in the renal vasculature in an organ-specific manner.

12.4.4 Urinary Excretion of Electrolytes

Timed collection of the urine assisted by bladder flushing with water allowed determination of the rates of excretion of electrolytes. In the first period the rate of excretion of Na^+ (NaEx) was 2769 ± 793 µmol/min and in the second 2157 ± 752 µmol/min. The corresponding rates of excretion of K^+ (KEx) were 1900 ± 295 and 1629 ± 88 µmol/min, respectively. The fractional sodium excretion was 5.8 ± 3.4 % and 4.6 ± 2.0 % and the fractional potassium excretion 146.2 ± 33.5 % and 123.1 ± 17.6 %. The fractional excretion of sodium is high compared to baseline values of other mammals under anaesthesia; however, the regimen of anaesthesia included infusion of saline, and the NaEx in these experiments hardly reflect the NaEx of the giraffe prior to induction. KEx was also high. This result is more likely to be a consequence of the existing dietary balances because administration of potassium compounds did not occur in these experiments. A fractional excretion rate of potassium above 1.0 (*i.e.*, 100 %) demonstrates net secretion of potassium, which occurs during high potassium intake in other animals.

Table 12.3 Renal functions and associated plasma concentrations of cardio-renal hormones in two consecutive 30 min periods. Values are mean ± SEM (n = 9 giraffes)

Parameter	Period I	Period II
Glomerular filtration rate (ml/min)	342 ± 99	348 ± 67
Effective renal plasma flow (ml/min)	1252 ± 305	1286 ± 21
Filtration fraction	0.30 ± 0.06	0.33 ± 0.04
Urinary Na + excretion (μmol/min)	2769 ± 793	1130 ± 468
Urinary K+ excretion (μmol/min)	1900 ± 295	980 ± 134
Plasma renin concentration (mIU/l)	2.6 ± 0.5	2.4 ± 0.4
Plasma angiotensin II (pg/ml)	9.1 ± 1.5	10.1 ± 2.1
Plasma arginine vasopressin (pg/ml)	97 ± 62	57 ± 35
Plasma atrial natriuretic peptide (pg/ml)	123 ± 18	95 ± 15
Plasma noradrenaline (pmol/ml)	1.1 ± 0.9	–

Note: Renin was measured as enzyme activity (rate of formation of immunoreactive angiotensin I) at excess of substrate. Angiotensin II, vasopressin, and atrial natriuretic peptide were measured as immunoactivity using highly specific antibodies and conventional radioimmunoassays. Noradrenaline was measured after extraction by HPLC with electrochemical detection

12.4.5 Cardio-Renal Hormones

Blood samples were obtained from an intravenous catheter and stored frozen until assay. The levels of substances binding to highly specific antibodies were determined in standard radioimmunoassays (Table 12.3). Plasma renin concentration was measured by enzyme activity (immunoactivity of generated angiotensin I) and quantified in IU/l. The measurements of plasma immunoactivity indicate that the giraffe has specific peptide hormones (or at least very similar substances) towards which the antibodies are raised. In the future studies more definitive identification could be obtained by mass spectrometry. The plasma levels we found in the giraffe were analogous to the concentrations seen in other species with exception of vasopressin concentrations. These were high and highly variable warranting further studies of the nature of the reactant(s) and—if undisturbed reaction with arginine vasopressin were to be identified—of the reasons for the high and variable plasma concentrations obtained under the present conditions.

12.5 Perspectives

Studies of renal functions in giraffes provide an interesting opportunity to distinguish scaling relationships within mammals. Because giraffes are bestowed with a blood pressure that is roughly twice that of other mammals, yet also endowed with a cardiac output that is 40 % lower than that predicted by body mass, it is interesting to note that GFR matches the low cardiac output and not blood pressure. Thus, if the low cardiac output indeed reflects a lower metabolism than similar-sized mammals and hence a lower rate of metabolic waste production, the giraffe studies provide credence to the notion that mass specific GFR decreases with body mass in accordance to the declining demands on excretion of metabolic waste products. This may appear counterintuitive given the very high blood pressures in the giraffes that would be expected to increase filtration pressure and hence lead to a higher GFR than predicted solely on basis of body mass. However, as demonstrated by the high renal interstitial pressures, part of the driving pressure is abated by the resulting decrease in transmural pressure of the Bowman's capsule. As a result, the giraffe kidney has functional characteristics that resemble other mammals, such that the filtration fraction is similar to other mammals. Our histological studies demonstrate few and remarkable specializations needed for adaptations to the high pressures, where particularly the strengthening of the renal capsule appears to be of paramount importance. From an evolutionary perspective, it seems therefore that relatively few key changes are needed to confer tolerance to high blood pressures while continuing to serve the functional demands on renal excretion.

The physiological solutions that enable animals to cope with extreme environmental situations or allow them to exhibit extreme behaviours, are the result of evolution by natural selection. Although evolution by no means selects for the "perfect solution", the various adaptations in the animal kingdom have proven adequate in securing long-term survival, and hence represent the result of a natural experiment over millions of years. What appear to be relatively simple solutions in the giraffe kidney to tolerate very high blood pressures may therefore serve as an inspiration for future clinical treatments of kidney damage during hypertension in humans and other mammals.

References

Baggot JD (2001) The physiological basis of veterinary clinical pharmacology. Blackwell Science, Oxford

Baldwin DS, Neugarten J (1985) Treatment of hypertension in renal disease. Am J Kidney Dis 5: A57–A70

Botero-Velez M, Curtis JJ, Warnock DG (1994) Brief report: Liddle's syndrome revisited–a disorder of sodium reabsorption in the distal tubule. N Engl J Med 330:178–181

Brondum E, Hasenkam JM, Secher NH, Bertelsen MF, Grondahl C, Petersen KK, Buhl R, Aalkjaer C, Baandrup U, Nygaard H, Smerup M, Stegmann F, Sloth E, Ostergaard KH, Nissen P, Runge M, Pitsillides K, Wang T (2009) Jugular venous pooling during lowering of the head affects blood pressure of the anesthetized giraffe. Am J Physiol Regul Integr Comp Physiol 297:R1058–R1065

Bude RO, Rubin JM (1999) Relationship between the resistive index and vascular compliance and resistance. Radiology 211:411–417

Curtis JJ, Luke RG, Dustan HP, Kashgarian M, Whelchel JD, Jones P, Diethelm AG (2000) Remission of essential hypertension after renal transplantation. 1983. J Am Soc Nephrol 11:2404–2412

Damkjaer M, Wang T, Brøndum E, Østergaard KH, Baandrup U, Hørlyck A, Hasenkam JM, Smerup M, Funder J, Marcussen N, Danielsen CC, Bertelsen MF, Grøndahl C, Pedersen M, Agger P, Candy G, Aalkjaer C, Bie P (2015) The giraffe kidney tolerates high arterial blood pressure by high renal interstitial pressure and low glomerular filtration rate. Acta Physiol (Oxf) 214(4):497–510. doi:10.1111/apha.12531, Epub 2015 June 15

Danielsen CC, Andreassen TT (1988) Mechanical properties of rat tail tendon in relation to proximal-distal sampling position and age. J Biomech 21(3):207–212

Dawson TH (2001) Similitude in the cardiovascular system of mammals. J Exp Biol 204:395–407

Dejong K, Henry RW (2007) Silicone plastination of biological tissue: cold-temperature technique BiodurTM S10/S15 technique and products. J Int Soc Plast 22:2–14

Duarte DP, Silva VL, Jaguaribe AM, Gilmore DP, DA Costa CP (2003) Circadian rhythms in blood pressure in free-ranging three-toed sloths (Bradypus variegatus). Braz J Med Biol Res 36:273–278

Haushalter TM, Friedrichs GS, Reynolds DL, Barecki-Roach M, Pastino G, Hayes R, Bass AS (2008) The cardiovascular and pharmacokinetic profile of dofetilide in conscious telemetered beagle dogs and cynomolgus monkeys. Br J Pharmacol 154:1457–1464

Head GA, Andrianopoulos N, Mcgrath BP, Martin CA, Carrington MJ, Lukoshkova EV, Davern PJ, Jennings GL, Reid CM (2014) Predictors of mean arterial pressure morning rate of rise and power function in subjects undergoing ambulatory blood pressure recording. PLoS One 9, e93186

Herbert LA, Chen WC, Hartmann A, Garancis JC (1976) Mechanical properties of the dog renal capsule. J Appl Physiol 40:164–170

Herman IP (2007) Physics of the human body Biological and Medical Physics, Biomedical Engineering . Springer, Heidelberg

Horii I, Kito G, Hamada T, Jikuzono T, Kobayashi K, Hashimoto K (2002) Development of telemetry system in the common marmoset–cardiovascular effects of astemizole and nicardipine. J Toxicol Sci 27:123–130

Hornicke H, Von Engelhardt W, Ehrlein HJ (1977) Effect of exercise on systemic blood pressure and heart rate in horses. Pflugers Arch 372:95–99

Kim SM, Eisner C, Faulhaber-Walter R, Mizel D, Wall SM, Briggs JP, Schnermann J (2008) Salt sensitivity of blood pressure in NKCC1-deficient mice. Am J Physiol Renal Physiol 295: F1230–F1238

King AJ, Olivier NB, Mohankumar PS, Lee JS, Padmanabhan V, Fink GD (2007) Hypertension caused by prenatal testosterone excess in female sheep. Am J Physiol Endocrinol Metab 292: E1837–E1841

Lawes CM, vander hoorn S, Rodgers A, International Society of Hypertension (2008) Global burden of blood-pressure-related disease, 2001. Lancet 371:1513–1518

Malpas SC, Groom AS, Head GA (1997) Baroreflex control of heart rate and cardiac hypertrophy in angiotensin II-induced hypertension in rabbits. Hypertension 29:1284–1290

Maluf NS (2002) Kidney of giraffes. Anat Rec 267:94–111

Marin R, Gorostidi M, Fernandez-Vega F, Alvarez-Navascues R (2005) Systemic and glomerular hypertension and progression of chronic renal disease: the dilemma of nephrosclerosis. Kidney Int Suppl 99:S52–S56

Markert M, Stubhan M, Mayer K, Trautmann T, Klumpp A, Schuler-Metz A, Schumacher K, Guth B (2009) Validation of the normal, freely moving Gottingen minipig for pharmacological safety testing. J Pharmacol Toxicol Methods 60:79–87

Mckee J, Daller J, Baumgartner B, Pettinger S (2014) Evaluation of the sensitivity of a new fully implantable telemetry device and the importance of simultaneously measuring cardiac output and left ventricular pressure. J Pharmacol Toxicol Methods 69:229–236

Myrie SB, Mcknight LL, King JC, Mcguire JJ, Van Vliet BN, Bertolo RF (2012) Effects of a diet high in salt, fat, and sugar on telemetric blood pressure measurements in conscious, unrestrained adult Yucatan miniature swine (Sus scrofa). Comp Med 62:282–290

Nerpin E, Ingelsson E, Riserus U, Sundstrom J, Andren B, Jobs E, Larsson A, Lars L, Arnlov J (2014) The association between glomerular filtration rate and left ventricular function in two independent community-based cohorts of elderly. Nephrol Dial Transplant 29:2069–2074
Ostergaard KH, Baandrup UT, Wang T, Bertelsen MF, Andersen JB, Smerup M, Nyengaard JR (2013) Left ventricular morphology of the giraffe heart examined by stereological methods. Anat Rec (Hoboken) 296:611–621
Petersen KK, Horlyck A, Ostergaard KH, Andresen J, Broegger T, Skovgaard N, Telinius N, Laher I, Bertelsen MF, Grondahl C, Smerup M, Secher NH, Brondum E, Hasenkam JM, Wang T, Baandrup U, Aalkjaer C (2013) Protection against high intravascular pressure in giraffe legs. Am J Physiol Regul Integr Comp Physiol 305:R1021–R1030
Provan G, Stanton A, Sutton A, Rankin-Burkart A, Laycock SK (2005) Development of a surgical approach for telemetering guinea pigs as a model for screening QT interval effects. J Pharmacol Toxicol Methods 52:223–228
Regan HK, Lynch JJ, Regan CP (2009) Long term assessment of blood pressure transducer drift in rhesus monkeys chronically instrumented with telemetry implants. J Pharmacol Toxicol Methods 59:35–38
Schmidt-Nielsen K (1984) Scaling: why is animal size so important. Cambridge University Press, NewYork
Singer MA (2001) Of mice and men and elephants: metabolic rate sets glomerular filtration rate. Am J Kidney Dis 37:164–178
Sithisarn T, Bada HS, Charnigo RJ, Legan SJ, Randall DC (2013) Effects of perinatal oxycodone exposure on the cardiovascular response to acute stress in male rats at weaning and in young adulthood. Front Physiol 4:85
Snedeker JG, Niederer P, Schmidlin FR, Farshad M, Demetropoulos CK, Lee JB, Yang KH (2005) Strain-rate dependent material properties of the porcine and human kidney capsule. J Biomech 38:1011–1021
Takeda T, Grollman A (1970) Spontaneously occurring renal disease in the guinea pig. Am J Pathol 60:103–118
Toto RD (2005) Treatment of hypertension in chronic kidney disease. Semin Nephrol 25:435–439
Tublin ME, Bude RO, Platt JF (2003) Review. The resistive index in renal Doppler sonography: where do we stand? AJR Am J Roentgenol 180:885–892
Tublin ME, Tessler FN, Murphy ME (1999) Correlation between renal vascular resistance, pulse pressure, and the resistive index in isolated perfused rabbit kidneys. Radiology 213:258–264
von Hagen G (1986) Heidelberg plastination folder: collection of technical leaflets for plastination. Biodur Products, Heidelberg, Germany
Weir MR (2009) Hypertension and the kidney: perspectives on the relationship of kidney disease and cardiovascular disease. Clin J Am Soc Nephrol 4:2045–2050
West GB, Brown JH, Enquist BJ (1997) A general model for the origin of allometric scaling laws in biology. Science 276:122–126
Zhou X, Frohlich ED (2007) Analogy of cardiac and renal complications in essential hypertension and aged SHR or L-NAME/SHR. Med Chem 3:61–65

Chapter 13
Non-traditional Models: The Molecular Physiology of Sodium and Water Transport in Mosquito Malpighian Tubules

Peter M. Piermarini and Christopher M. Gillen

Abstract The Malpighian (renal) tubules of adult female mosquitoes are a valuable comparative model system for elucidating the mechanisms of transepithelial fluid secretion. Here we summarize the current state of knowledge on the cellular and molecular mechanisms of sodium and water transport in the tubule epithelium, and highlight key areas in which further research is needed. In brief, the epithelium contains several mechanisms for the transport of sodium, including cation proton antiporters, cation chloride cotransporters, sodium channels, sodium-coupled bicarbonate cotransporters, and a Na,K-ATPase. Moreover, the epithelium contains mechanisms for the paracellular and transcellular transport of water via septate junctions and aquaporin water channels, respectively. Collectively, these transport systems contribute to the renal maintenance of extracellular fluid homeostasis in mosquitoes, provide an accessible comparative model system for understanding the evolution of Na^+ and water transport in animal epithelia, and present potential physiological targets to exploit in the control of disease vectors.

Keywords Epithelial transport • Principal cell • Stellate cell • Mosquito • Malpighian tubules • Diuresis

13.1 Introduction

The Malpighian (renal) tubules are the kidneys of insects. These excretory structures are ubiquitous in nearly all groups of insects examined with the exception of the aphids, which have mysteriously lost them in their evolution, despite facing obvious needs for kidneys while gorging on plant fluids. In the insects that possess

P.M. Piermarini (✉)
Department of Entomology, Ohio Agricultural Research and Development Center, The Ohio State University, 1680 Madison Ave/OARDC, 224 Thorne Hall, Wooster, OH 44691, USA
e-mail: piermarini.1@osu.edu

C.M. Gillen (✉)
Department of Biology, Kenyon College, Higley Hall 310, Gambier, OH 43022, USA
e-mail: gillenc@kenyon.edu

K.A. Hyndman, T.L. Pannabecker (eds.), *Sodium and Water Homeostasis*,
Physiology in Health and Disease, DOI 10.1007/978-1-4939-3213-9_13

Malpighian tubules, the total number of tubules can vary greatly among species, from as few as 1 to over 200 (Pacheco et al. 2014). As expected for a kidney, the central physiological function typically attributed to Malpighian tubules is salt and water balance. However, it is also important to appreciate that like the kidneys of other animals the Malpighian tubules are multifunctional and can contribute to nitrogenous waste excretion, xenobiotic/metabolite detoxification and excretion, and immunity (Beyenbach et al. 2010). And, in some species the tubules take on additional specialized roles, such as light and silk production (Pacheco et al. 2014).

In the spirit of the present book, the goal of our chapter is to discuss recent advances made in the Malpighian tubules of mosquitoes in regards to how the epithelium transports sodium and water from the extracellular fluid (hemolymph) to the tubule lumen. In particular, we focus on Malpighian tubules in adult female mosquitoes, which are a well-studied 'non-traditional' model system for elucidating the mechanisms of transepithelial sodium and water transport, because of the vital physiological need of female mosquitoes to excrete copious amounts of sodium and water after they engorge on the blood of a vertebrate host (Beyenbach 2003; Benoit and Denlinger 2010). That is, although a blood meal provides the female mosquito with the nutrients she needs to develop her eggs, it also provides her with a glut of ions and water that need to be excreted to maintain salt and water balance.

13.2 Transepithelial Fluid Secretion in Mosquito Malpighian Tubules

Malpighian tubules are functionally analogous to the kidneys of mammals, but their mechanisms for producing urine are quite different. Mammalian kidneys produce urine via ultrafiltration of the blood at glomeruli, whereas insect Malpighian tubules produce urine via transepithelial fluid secretion. In other words, urine production in insects is more akin to the way the salivary, tear, and sweat glands of mammals produce their secretions. Below, we briefly review the main cellular and molecular players in this process to provide context for our focused discussion on the mechanisms of sodium and water transport in mosquito Malpighian tubules.

13.2.1 Anatomy of Mosquito Malpighian Tubules

To our knowledge, all mosquitoes examined to date possess five Malpighian tubules. In the yellow fever mosquito *Aedes aegypti*, the tubules are morphologically and functionally identical (Beyenbach et al. 1993), which is presumably the case for other mosquitoes as well. Each tubule has two general anatomical segments: the blind-ended distal (relative to the gut) segment, which composes about

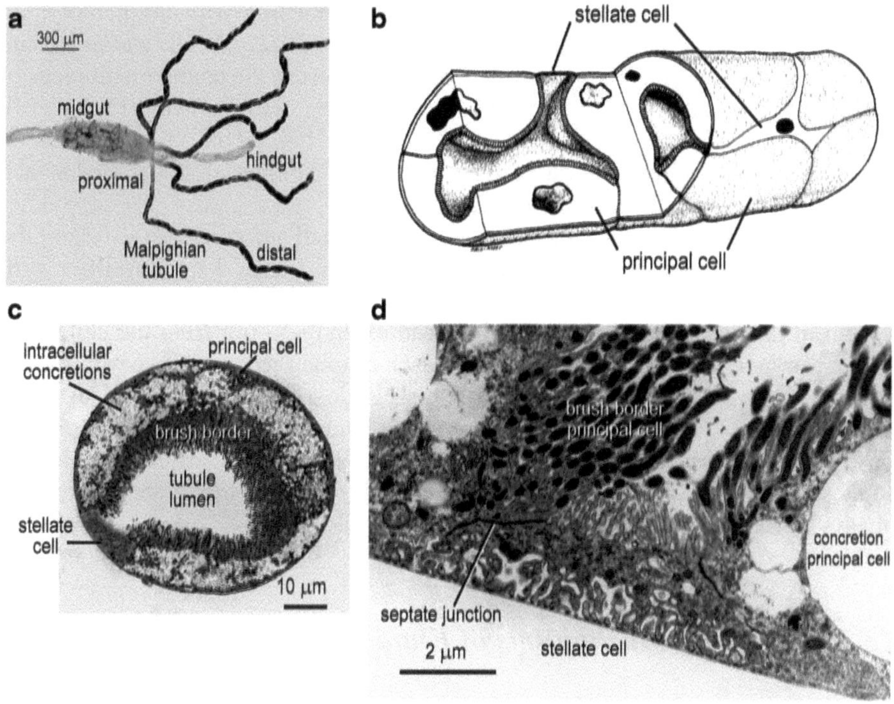

Fig. 13.1 Anatomy of mosquito Malpighian tubules. (**a**) Isolated alimentary canal of an adult female yellow fever mosquito *A. aegypti* showing the five Malpighian tubules (*black structures*) in relation to the midgut and hindgut. From Beyenbach and Piermarini (2011) with permission. (**b**) Drawing of the distal, blind end of a Malpighian tubule showing the cellular composition of the epithelium. From Bradley et al. (1982) with permission. (**c**) Transmission electron micrograph (TEM) of a cross-section through an isolated Malpighian tubule showing the ultrastructure of a principal cell and arm of a stellate cell. From Beyenbach and Piermarini (2011) with permission. (**d**) TEM of a cross-section through an arm of a stellate cell flanked by principal cells. From Beyenbach and Piermarini (2011) with permission

2/3 of the total tubule length, and the open-ended proximal segment, which composes the remaining 1/3 of the tubule length and attaches to the alimentary canal between the midgut and hindgut (Fig. 13.1a). The distal segment is the primary site of active transepithelial fluid secretion and is the main focus of this chapter; direct functional studies on the proximal segment are lacking, but it is hypothesized to serve a reabsorptive role (Patrick et al. 2006).

The distal segment consists of principal cells that are intercalated by stellate cells (Fig. 13.1b). Principal cells are large (~100 μm long), thick (~30 μm thick), and fusiform, with a centrally located nucleus (Fig. 13.1b). On the ultrastructural level, these cells have an elaborate, luminal brush border where each microvillus is filled with a mitochondrion (Fig. 13.1c, d), suggestive of a region of energetically expensive processes. The cytoplasm is packed with membrane-bound, mineralized concretions of divalent cations and/or uric acid, and the basal membrane possesses short in-foldings that increase the surface area interacting with the hemolymph (Fig. 13.1c, d). Principal cells are neighbored by other principal cells and/or stellate cells; a ladder-like septate junction keeps separate most of the lateral membranes of the neighboring cells to form a paracellular (between cell) pathway (Fig. 13.1d). Gap junctions are also present, which allow for direct contact of the lateral membranes and communication between the cells (Weng et al. 2008; Beyenbach and Piermarini 2011; Piermarini and Calkins 2014).

Relative to principal cells, the stellate cells are small and comprise ~10 % of the tubule mass. They have a nucleated cell body from which 3 or 4 arm-like projections radiate, giving them a star-like appearance (Fig. 13.1b). The cells are only ~5 μm in thickness, but their projections can extend ~50 μm from the cell body, which can give them an impressive axial reach despite their small mass. In further contrast to principal cells, the stellate cells (1) do not possess a luminal brush border, (2) do not have cytoplasmic concretions, and (3) have elaborate in-foldings on their basolateral membrane (Fig. 13.1c, d). We are not aware of evidence suggesting that stellate cells can connect to each other; thus they are assumed to always intercalate between principal cells.

13.2.2 General Mechanisms of Fluid Secretion

Early in vitro physiological studies of isolated mosquito Malpighian tubules by the Beyenbach laboratory established that the epithelium transports Na^+ and K^+ against their respective electrochemical potentials from the hemolymph to the tubule lumen (Williams and Beyenbach 1984). This active transport of cations generates a lumen-positive transepithelial potential, which drives the passive transport of Cl^- into the lumen. The accumulation of these ions provides an osmotic gradient for water to follow into the lumen from the hemolymph, thereby forming urine that is isosmotic to the hemolymph.

In general, the principal cells are considered the primary sites of active transcellular cation (Na^+, K^+) secretion, whereas the paracellular pathway is considered the major route of passive Cl^- movements from the hemolymph to the lumen (Beyenbach and Piermarini 2011). The stellate cells also contribute a transcellular pathway for Cl^- secretion, and a potential role of stellate cells in cation transport cannot be ruled out. Stellate cells may also contribute to the regulation of principal cell metabolism during periods of intense metabolic activity (e.g., the post-prandial diuresis) by serving as a sink for intracellular bicarbonate (Piermarini et al. 2010; Beyenbach and Piermarini 2011; Piermarini and Calkins 2014). Movements of water across the epithelium can occur through paracellular and transcellular pathways (Beyenbach and Piermarini 2011). Figure 13.2 provides a detailed model of the specific molecular mechanisms of ion and fluid transport in the tubule epithelium. We briefly introduce these mechanisms below, and then provide a detailed review of the evidence for the Na^+ and H_2O transporters in this system.

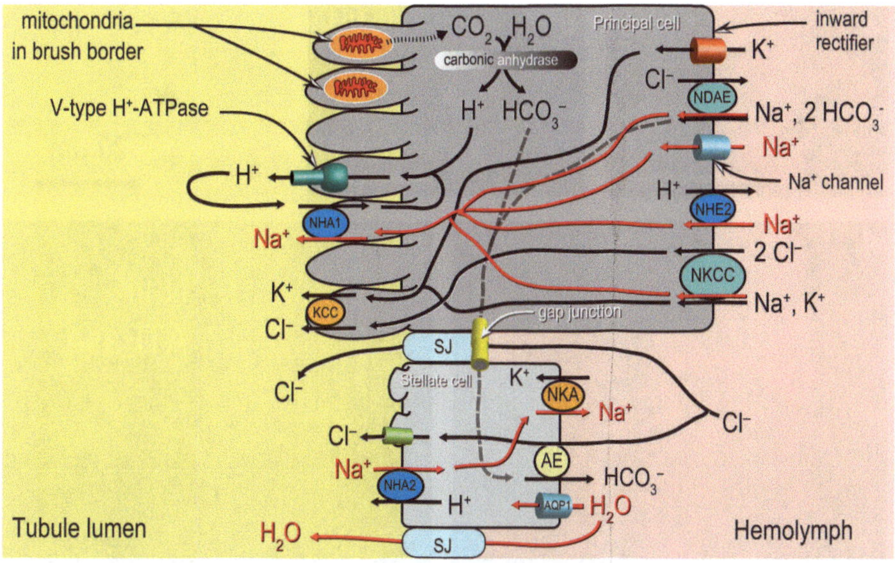

Fig. 13.2 Generalized transport model of transepithelial fluid secretion in mosquito Malpighian tubules. The model is derived from physiological, pharmacological, immunochemical, and molecular studies on Malpighian tubules from larval and adult mosquitoes (*A. aegypti* and *A. gambiae*; see text for details). *Red arrows* highlight movements of sodium and water. Abbreviations: *AE* anion exchanger, *AQP* aquaporin, *KCC* potassium chloride cotransporter, *NDAE* sodium-driven anion exchanger, *NHA* sodium hydrogen antiporter, *NHE* sodium hydrogen exchanger, *NKCC* sodium potassium chloride cotransporter, *SJ* septate junction

13.2.3 Molecular Mechanisms of Fluid Secretion

The brush border of principal cells is the site of expression of the V-type H^+-ATPase (Figs. 13.2 and 13.3b), which is the dominant ATPase in the epithelium that drives transepithelial movements of ions and water (Beyenbach 2001). The active secretion of protons by this pump (fueled by the nearby mitochondria) establishes the membrane voltages of both the apical and basolateral membranes (Beyenbach 2001). On the basolateral membrane, the voltage helps drive the uptake of Na^+ and K^+ from the hemolymph via ion-selective channels (Fig. 13.2). The evidence for Na^+ channels is discussed in Sect. 13.3.3; the major K^+ channels are thought to be barium-sensitive inward rectifiers (Piermarini et al. 2013). On the apical membrane, the voltage is hypothesized to drive the uptake of protons in exchange for the secretion of Na^+ and/or K^+ via an electrogenic cation-proton antiporter (Fig. 13.2; see Sect. 13.3.1). In addition to these conductive pathways, there is physiological and/or molecular evidence for electroneutral transport systems, such as sodium proton exchangers (NHEs, Sect. 13.3.1), Na,K,2Cl cotransporters (NKCCs; Sect. 13.3.2), K,Cl cotransporters (Piermarini et al. 2011), and sodium-driven anion exchangers (NDAEs, Sect. 13.3.4) in the apical and/or basolateral membranes that contribute to the transepithelial movement of ions (Fig. 13.2).

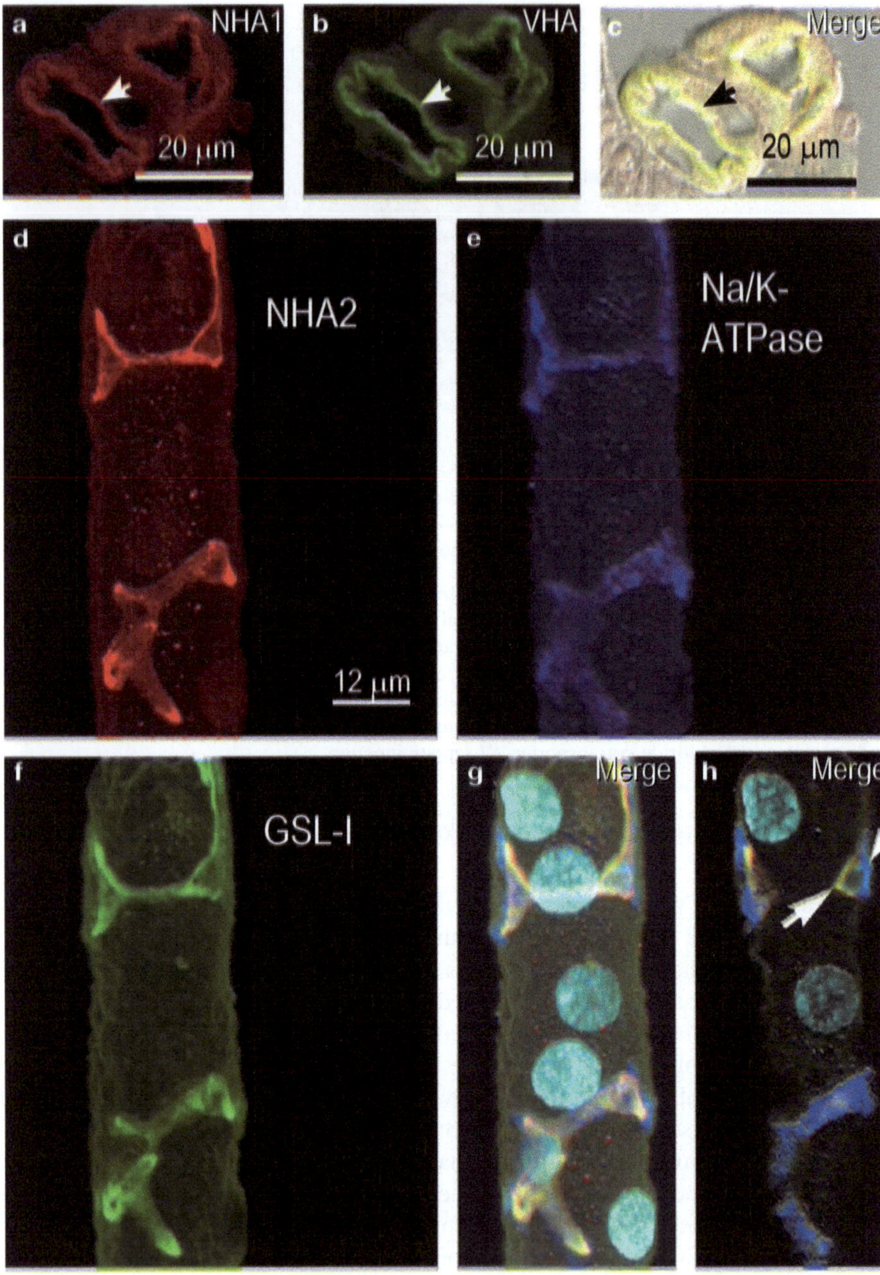

Fig. 13.3 Immunolabeling of NHA1 and NHA2 in Malpighian tubules of larval *A. gambiae*. Panels (**a–c**) respectively show labeling of NHA1 (*red*), the V-type H^+-ATPase, VHA (*green*), and their colocalization (*yellow*) in the single z-plane of a tubule cross section. *Arrows* indicate the luminal brush border. Panels (**d–g**) respectively show labeling of NHA2 (*red*), the Na,K-ATPase (*blue*), a lectin apical membrane marker, GSL1 (*green*), and their colocalization (*yellow*) with nuclear counterstain (*cyan*) in a z-stack of a tubule. Panel (**h**) shows the colocalization in a single z-plane. *Arrow* indicates colocalization of NHA2 and GSL1 in apical membrane of a stellate cell;

The stellate cells are typically considered a site of chloride transport. They are the exclusive site of expression for a basolateral Cl/HCO_3 anion exchanger (AE) and apical chloride channels (Fig. 13.2) (O'Connor and Beyenbach 2001; Piermarini et al. 2010). Moreover, the stellate cells are the exclusive site of expression for a Na/H antiporter (see Sect. 13.3.1) and the alpha subunit of the Na,K-ATPase (see Sect. 13.3.5) in the distal segment, suggesting possible roles in cation transport (Fig. 13.2). Stellate cells are also the only known site of aquaporin water channel (AQP) expression in the distal segment of mosquito Malpighian tubules (see Sect. 13.4.2), suggesting an important role in transcellular water movements (Fig. 13.2). In addition to stellate cells, transepithelial chloride and water movements can take place through the paracellular pathway via septate junctions (Fig. 13.2; see Sect. 13.4.1). This pathway likely plays a key role when the tubule is secreting at high rates in vivo during the post-prandial diuresis or in vitro under stimulation by kinin diuretic peptides (Beyenbach and Piermarini 2011; Beyenbach 2012).

With that general overview, we now revisit the specific evidence for the sodium and water transport mechanisms. In some cases, evidence for particular molecular mechanisms in adult female mosquitoes is lacking in which case we defer to evidence in larval mosquitoes and/or the more 'traditional' model system of the fruit fly *Drosophila melanogaster* for guidance and the generation of hypotheses.

13.3 Sodium Transporters, Channels, and Pumps in Mosquito Malpighian Tubules

As shown in Fig. 13.2, Na^+ can enter or exit the Malpighian tubules through a variety of carriers, channels, and pumps. Below, we review the pharmacological, physiological, and molecular evidence for these mechanisms.

13.3.1 Cation Proton Antiporters

The cation proton antiporters (CPAs) are a large superfamily of genes (also known as the solute carrier 9, SLC9, superfamily) encoding proteins that mediate the exchange transport of Na^+ and/or K^+ for protons across plasma and organellar membranes (Brett et al. 2005; Donowitz et al. 2013). The superfamily consists of two sub-families referred to as CPA1 (SLC9A) and CPA2 (SLC9B). The CPA1 members are the classical amiloride-sensitive, electroneutral Na,H exchangers

Fig. 13.3 (continued) arrowhead indicates localization of the Na,K-ATPase in the basolateral membrane of a stellate cell. From Xiang et al. (2012) with permission

(NHEs) that have been the subject of much investigation in a variety of eukaryotic systems, whereas the CPA2 members are Na,H antiporters (NHAs) that have a closer phylogenetic relation to the prokaryotic NHAs than to eukaryotic CPA1 members (Brett et al. 2005; Rheault et al. 2007). CPA2 members are known to have the capacity for electrogenic exchange (e.g., $2Na^+/H^+$ or $Na^+/2H^+$ exchange) (Brett et al. 2005; Rheault et al. 2007).

13.3.1.1 Basolateral CPAs

In Malpighian tubules, there is strong pharmacological and physiological evidence for an electroneutral CPA in the basolateral membrane of principal cells. In vitro studies of isolated Malpighian tubules in *A. aegypti* have shown that peritubular amiloride inhibits the transepithelial secretion of fluid and Na^+ by ~60 % and ~70 %, respectively, without major effects on the electrophysiology of the tubule epithelium (Hegarty et al. 1992). Moreover, an analog of amiloride with superior selectivity for NHEs [5-(N-Ethyl-N-isopropyl)amiloride; EIPA] is the most potent amiloride-based inhibitor of fluid secretion in isolated tubules (Petzel 2000). Furthermore, the peritubular addition of EIPA blocks Na^+-dependent mechanisms of (1) steady-state pH_i regulation in principal cells, and (2) pH_i recovery in principal cells after an acid load (Petzel 2000). Thus, an amiloride/EIPA-sensitive NHE appears to contribute to the uptake of Na^+ and the extrusion of H^+ across the basolateral membrane of principal cells (Fig. 13.2).

The specific NHE responsible for the aforementioned transport activity has proven elusive to identify. The most promising candidate is *Aedes* (*Ae*) NHE2,[1] which is an ortholog of mammalian NHE3 cloned from the Malpighian tubules of *A. aegypti* (Hart et al. 2002; Pullikuth et al. 2006). The *Ae*NHE2 cDNA is expressed as two splice variants ('a' and 'b'), which are identical except for the cytosolic COOH-terminal domain; the 'b' variant is truncated by ~500 amino acids relative to the 'a' variant (Hart et al. 2002; Pullikuth et al. 2006). Both variants mediate Na/H exchange when expressed heterologously in fibroblasts (Pullikuth et al. 2006). Furthermore, immunolabeling of the 'a' variant shows expression along the basolateral membrane of principal cells (Pullikuth et al. 2006), consistent with the location of physiological NHE activity in the tubule. However, neither *Ae*NHE2 variant is highly sensitive to EIPA (Pullikuth et al. 2006).

Thus, until proven otherwise, it should be assumed that *Ae*NHE2 along with another unidentified amiloride/EIPA-sensitive NHE populates the basolateral membrane of principal cells. The employment of RNA interference (RNAi) experiments to selectively knock down the expression of *Ae*NHE2 would be useful to determine

[1] The nomenclature of insect CPA members is confusing, because different naming systems have been used for mosquitoes vs. fruit flies. For clarity, we defer to the system used by Rheault et al. (2007) where the mosquito CPAs are named after their closest orthologs in *Drosophila melanogaster* rather than in vertebrates.

if this transporter is indeed distinct from the amiloride/EIPA-sensitive NHE activity of the basolateral membrane.

13.3.1.2 Apical CPAs

An area of enthusiastic pursuit in the Malpighian tubules of mosquitoes and other insects is the identification of a hypothesized CPA in the luminal membrane of principal cells. The hypothesis is an extension of the Wieczorek model for K^+ secretion in the midgut of the tobacco hornworm *Manduca sexta* in which the activity of a V-type H^+-ATPase establishes a voltage and/or proton gradient for energizing the secretion of K^+ via an electrogenic K^+/nH^+ exchanger (Wieczorek et al. 1991), presumably of the CPA2 clade.

The genomes of both mosquitoes and fruit flies possess genes for two CPA2 members, referred to as NHA1 and NHA2, which phylogenetically cluster independently from vertebrate NHAs (Brett et al. 2005; Rheault et al. 2007). An extensive functional genetics study by the laboratory of Dow in *Drosophila melanogaster* demonstrated that both CPA2 members (NHA1, NHA2) exhibit similar patterns of transcript enrichment as the V-type H^+-ATPase, suggesting a potential functional relationship (Day et al. 2008). Furthermore, immunolabeling experiments showed that NHA1 and NHA2 colocalize with the V-type H^+-ATPase in the apical membrane of *D. melanogaster* principal cells (Day et al. 2008). Interestingly, when heterologously expressed in NHA-deficient yeast, NHA1 and NHA2 appear to mediate protection against a K^+ and Na^+ stress, respectively, suggesting different potential cation preferences of these transporters (Day et al. 2008).

Harvey and colleagues have provided the most convincing evidence that the NHAs are the best candidates for the Wieczorek exchanger in mosquito Malpighian tubules (Xiang et al. 2012). Although their study was conducted in larvae of *Anopheles gambiae* and not adults, the authors demonstrated the colocalization of *Anopheles* (*An*) NHA1 with the V-type H^+-ATPase in the brush border of principal cells (Fig. 13.3a–c) (Xiang et al. 2012). Surprisingly, *An*NHA2 localized exclusively to the apical membrane of stellate cells (Fig. 13.3d–h), suggesting a potential novel role for these cells in luminal cation movements (Xiang et al. 2012). In particular, Harvey and colleagues proposed a role for *An*NHA2 in the reabsorption of luminal Na^+ by stellate cells (together with a basolateral Na,K-ATPase, Fig. 13.3e) of larval tubules to minimize Na^+ wasting in an ion-poor freshwater environment (Fig. 13.2) (Xiang et al. 2012). Amazingly, without the help of molecular or immunochemical tools, Berridge and Oschman (1969) proposed a very similar role for stellate cells in the reabsorption of Na^+ in the Malpighian tubules of blow flies *Calliphora erythrocephala* over 40 years prior. Whether the localization of *An*NHA2 is similar in the Malpighian tubules of adult female mosquitoes remains to be determined.

In summary, NHA1 appears to be the best candidate for the apical secretion of Na^+ and/or K^+ in principal cells, whereas a role for NHA2 in the transport of Na^+

and/or K^+ in stellate cells cannot be ruled out. Important remaining experiments to conduct with the NHAs of mosquitoes (and fruit flies) are their functional characterization in a heterologous expression system, such as *Xenopus* oocytes, which will be essential to (1) demonstrate the putative electrogenic properties of these exchangers, (2) determine their proposed cation preferences, and (3) characterize their pharmacology. Although this has proven difficult to date (Xiang et al. 2012), the recent heterologous expression and functional characterization of human NHA2 by Rao and colleagues in MDCK cells offers a potential model to follow (Kondapalli et al. 2012). Such a characterization will allow for a more comprehensive understanding of how the respective NHAs contribute to the transepithelial secretion of Na^+ and/or K^+ in Malpighian tubules.

13.3.1.3 Intracellular NHEs

Mosquitoes also possess two other CPA1 members: NHE1 and NHE3. *Ae*NHE1 is an ortholog of mammalian NHE8, and was originally considered an apical CPA in mosquito Malpighian tubules by the laboratory of Gill based on immunolabeling experiments (Kang'ethe et al. 2007). However, the laboratory of Beyenbach later demonstrated that the expression of *Ae*NHE1 occurs in intracellular vesicles basal to the brush border (Fig. 13.4) (Piermarini et al. 2009), which is consistent with the localization of NHE1 in *Drosophila* Malpighian tubules (Day et al. 2008). Nevertheless, both the Gill and Beyenbach laboratories have cloned *Ae*NHE1 and demonstrated that it mediates EIPA-sensitive Na/H exchange in heterologous expression systems (Kang'ethe et al. 2007; Piermarini et al. 2009). It is unclear what physiological role the intracellular *Ae*NHE1 plays in Malpighian tubules, but its sensitivity to EIPA suggests that it may in part be responsible for some of the aforementioned inhibitory effects of amiloride and EIPA on fluid secretion and regulation of pH_i in isolated tubules.

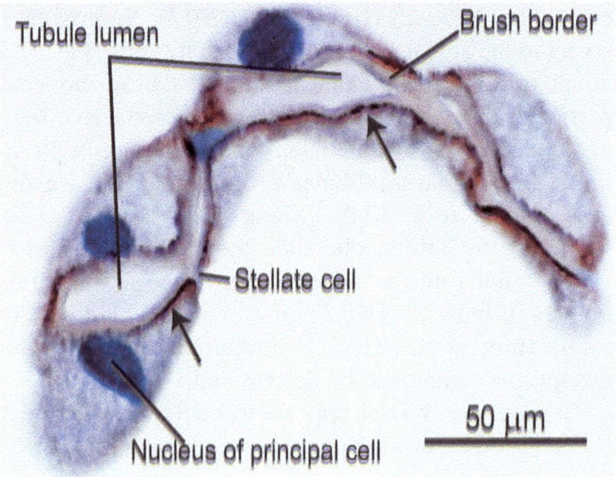

Fig. 13.4 Immunolabeling of *Ae*NHE1 in a paraffin-embedded section of a Malpighian tubule isolated from an adult female *A. aegypti*. *Red* staining (*arrows*) indicates labeling of the anti-*Ae*NHE1 antibody. *Blue* staining is from hematoxylin. Note the distinct labeling of the sub-apical region of principal cells basal to the brush border and not in the brush border. From Piermarini et al. (2009) with permission

Mosquito NHE3 has not received much investigative attention in Malpighian tubules. This NHE is an ortholog of the intracellular NHEs of vertebrates (i.e., NHE6, NHE7, NHE9) that play important roles in the regulation of intraorganellar pH (Donowitz et al. 2013). In *D. melanogaster* Malpighian tubules, NHE3 mRNAs and immunoreactivity are expressed weakly (Day et al. 2008). Thus, NHE1 is likely the predominant intracellular NHE in Malpighian tubules.

13.3.2 Cation-Coupled Chloride Cotransporters

The cation-coupled chloride cotransporters (CCCs), which are also known as the solute carrier 12 superfamily (SLC12), perform electroneutral transport of Cl^- coupled to Na^+ and/or K^+ (Payne 2012; Arroyo et al. 2013; Gagnon and Delpire 2013). CCCs fall into three classes with differing ion requirements and inhibitor sensitivities. The Na,K,2Cl cotransporters (NKCCs) are inhibited by the loop diuretics bumetanide and furosemide, K^+-independent Na,Cl cotransporters (NCCs) are inhibited by thiazide diuretics, and Na^+-independent K,Cl cotransporters (KCCs) are selectively inhibited by (Dihydroindenyl) oxyalkanoic acid (DIOA) (Gamba 2005). NCCs localize to luminal membranes of absorptive epithelia, whereas NKCCs and KCCs can participate in either ion secretion or absorption, depending on their localization in polarized epithelia (Russell 2000; Adragna et al. 2004). In *A. aegypti*, a KCC (*Ae*KCC1) localizes to the apical membrane of principal cells of adult female Malpighian tubules (Piermarini et al. 2011). Mosquito genomes contain three additional genes that cluster with Na^+-dependent CCCs (Pullikuth et al. 2003; Sun et al. 2010; Piermarini et al. 2011).

Studies conducted in the Beyenbach laboratory strongly implicate a basolateral NKCC in ion secretion by the distal Malpighian tubule of adult female mosquitoes. Early studies in *A. aegypti* provided evidence that tubular transport of Na^+, K^+, and Cl^- are coupled (Williams and Beyenbach 1984). Exposure of tubules to 100 μM bumetanide blocked the cAMP-stimulated increase in fluid and Na^+ secretion (Hegarty et al. 1991). Intriguingly, in the same study, bumetanide moderately increased Na^+ secretion with a proportional decrease in K^+ secretion in unstimulated tubules. However, combined treatment of unstimulated tubules with bumetanide and barium, which blocks the basolateral K^+ conductance, almost completely abolished epithelial fluid, K^+, and Na^+ secretion (Scott et al. 2004). Electrophysiological measurements conducted in conjunction with the above experiments indicated that bumetanide inhibits an electroneutral transport process (Hegarty et al. 1991). Together, these studies point toward participation of an NKCC in tubular ion secretion, but suggest that the consequences of NKCC activity may vary depending on which other transport proteins are active.

Studies in other insects demonstrate a role for basolateral NKCCs in ion secretion (Ianowski et al. 2004; Ianowski and O'Donnell 2004; Coast 2012; Gamez et al. 2012). In *D. melanogaster* Malpighian tubules, 100 μM bumetanide inhibits fluid secretion, K^+ flux, and Na^+ flux after treatment with ouabain (Ianowski and O'Donnell 2004). Also, calculations of electrochemical gradients based upon membrane potential and intracellular ion activities strongly support NKCC as a key mediator of basolateral K^+ and Cl^- flux in *D. melanogaster* tubules (Ianowski and O'Donnell 2004). The *D. melanogaster* ncc69 gene (also called cg4357) encodes a bumetanide-sensitive NKCC, as demonstrated by studies in which its transport properties have been characterized following expression in tissue culture cells (Sun et al. 2010; Leiserson et al. 2011). Tubules of flies null for ncc69 showed reduced epithelial fluid and K^+ secretion (Rodan et al. 2012). However, Na^+ flux was unaltered in ncc69-null tubules, probably because Na^+ that flows through ncc69 is recycled across the basolateral membrane through a Na,K-ATPase (Rodan et al. 2012). This work convincingly demonstrates a role for ncc69 in K^+ and fluid secretion but not in Na^+ secretion by *Drosophila* tubules.

Functions of the mosquito orthologs of ncc69 have not been characterized. However, preliminary experiments demonstrate that the mRNA of an *A. aegypti* ortholog of ncc69 (AAEL006180) is strongly expressed in adult female Malpighian tubules, consistent with a potential role in ion and fluid secretion (Fig. 13.5a). Moreover, an antibody (M6) against a putative NKCC from *Manduca sexta* binds to the basolateral membrane of adult female Malpighian tubules (Fig. 13.5b). Finally, the ortholog of ncc69 and AAEL006180 in the Asian tiger mosquito *Aedes albopictus* is down-regulated in Malpighian tubules of female mosquitoes at 12 and 24 h following a blood meal (Esquivel et al. 2014). This result supports the involvement of a ncc69/AAEL006180 ortholog in secretion, because the demands for ion and fluid secretion decline 2 h after a blood meal (Williams et al. 1983; Esquivel et al. 2014). If the mosquito orthologs of ncc69 are indeed the basolateral NKCCs of principal cells, then their activity may have different consequences for epithelial transport than ncc69 in *D. melanogaster* tubules, because principal cells of mosquito distal Malpighian tubules lack the Na,K-ATPase (see Sect. 13.3.5). Thus, less Na^+ may be recycled across the basolateral membrane of mosquito principal cells compared to those of *D. melanogaster*, which may allow the mosquito orthologs of ncc69 to participate in Na^+ as well as K^+ secretion (Hegarty et al. 1991; Scott et al. 2004).

Mosquito genomes contain two other genes with sequence similarity to NKCCs (Sun et al. 2010; Piermarini et al. 2011). In preliminary experiments with *A. aegypti*, transcripts of one of these genes (AAEL009886) were predominantly expressed in larval mosquitoes. Transcripts of the other gene (AAEL009888) were found in Malpighian tubules of adult females, but were more abundant in the hindgut (Fig. 13.5a), which may suggest a more prominent role of this transporter in absorption vs. secretion. Consistent with this notion, an ortholog of AAEL009888 in *Manduca sexta* has been localized to the apical membrane of midgut epithelial cells (Gillen et al. 2006). Clearly, further functional and molecular studies are needed to elucidate the roles of NKCC-like genes in mosquito Malpighian tubules.

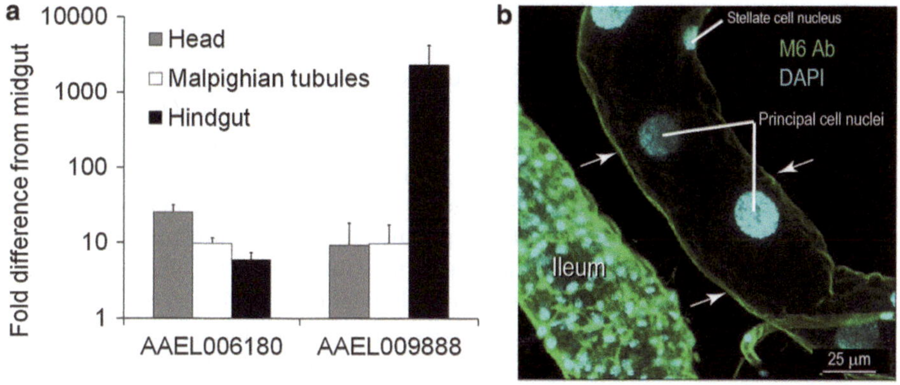

Fig. 13.5 (**a**) Quantification of *A. aegypti* cation-coupled chloride cotransporter (AAEL006180, AAEL009888) transcripts by qPCR in adult female tissues. Expression in head, Malpighian tubules, and hindgut are expressed as fold difference from midgut. Values are means ± SEM (N = 3 cDNA samples per tissue). Data were previously unpublished. (**b**) Na,K,2Cl cotransporter (NKCC)-like immunoreactivity (*green*) in an isolated Malpighian tubule and hindgut (*ileum*) using an antibody (M6) raised against a putative NKCC in *M. sexta* (Gillen et al. 2006). The M6 antibody was obtained from the Gillen laboratory (Kenyon College). *Arrows* indicate labeling of the basolateral aspect of principal cells. Nuclei are counterstained with DAPI (*cyan*). Data were previously unpublished

13.3.3 Na^+-Channels

Under control conditions, only ~15 % of the basolateral membrane conductance of principal cells is associated with Na^+ (Beyenbach and Masia 2002). This Na^+ conductance is presumably mediated by amiloride-insensitive Na^+ channels and/or electrogenic cotransporters and exchangers (Beyenbach and Masia 2002). On the other hand, nearly 65 % of the basolateral membrane conductance of principal cells is attributed to K^+ (Beyenbach and Masia 2002). This K^+ conductance is blocked by barium (Beyenbach and Masia 2002) and is likely mediated by inward-rectifier K^+ (Kir) channels (Piermarini et al. 2013).

When isolated Malpighian tubules are stimulated to secrete at diuretic rates by the calcitonin-like peptide (also known as the mosquito natriuretic peptide or diuretic hormone 31) or cyclic AMP (cAMP, the second message of the calcitonin-like peptide), the Na^+ conductance of the basolateral membrane significantly increases (Williams and Beyenbach 1984; Petzel et al. 1985; Beyenbach and Masia 2002; Coast et al. 2005). The opening of Na^+ channels enhances the uptake of Na^+ across the basolateral membrane, thereby promoting the secretion of Na^+ across the apical membrane and excretion of Na^+ by the whole animal (Williams and Beyenbach 1984; Petzel et al. 1985; Beyenbach and Masia 2002; Coast et al. 2005). This natriuresis is of particular importance to adult female mosquitoes during the early stages of processing a blood meal (<30–60 min after engorgement) when they need to excrete large amounts of Na^+ and water that they absorb into their hemolymph from the ingested blood plasma (Williams et al. 1983; Coast 2009).

Notably, the cAMP-stimulated Na^+ conductance of the basolateral membrane is blocked nearly completely by amiloride (Beyenbach and Masia 2002), suggesting the potential involvement of degenerin/epithelial Na^+ channels (DEG/ENaC), which are encoded by '*pickpocket*' (*ppk*) genes in *D. melanogaster* (Zelle et al. 2013). Although the DEG/ENaC genes have not been specifically studied in mosquito Malpighian tubules, a transcriptomic study of *A. gambiae* has revealed the significant enrichment of a transcript for one DEG/ENaC member (AGAP007945) in Malpighian tubules of adult females (Baker et al. 2011); this gene is an ortholog of *ppk13* in *D. melanogaster*. Furthermore, in the Malpighian tubules of the Asian tiger mosquito *A. albopictus*, an ortholog of AGAP007945 and *ppk13* is significantly down-regulated at 12 and 24 h after a blood meal, which are times when a natriuresis is no longer necessary (Esquivel et al. 2014). Thus, the *ppk13* ortholog of mosquitoes appears to be the best candidate for mediating the cAMP/CLP-stimulated, amiloride-sensitive Na^+-conductance on the basolateral membrane of principal cells. However, additional functional characterization and molecular studies are clearly needed to test this hypothesis.

13.3.4 Na^+-Coupled Bicarbonate Transporters

The Na^+-coupled bicarbonate transporters (NCBTs) belong to a large superfamily of genes (also known as the solute carrier 4, SLC4, superfamily) encoding proteins that mediate the transport of bicarbonate across cell membranes (Romero et al. 2013). The NCBTs form one subfamily of SLC4 members, whereas the Na^+-independent, Cl/HCO_3 anion exchangers (AEs) form the other subfamily. The AEs include the classical electroneutral 4,4′-Diisothiocyanatostilbene-2,2′-disulfonic acid (DIDS)-sensitive 'band 3' exchanger first described in mammalian red blood cells. Mosquitoes have one or two genes encoding predicted AEs that have been examined in Malpighian tubules (Piermarini et al. 2010; Linser et al. 2012).

The NCBTs include members that can mediate electroneutral Na, HCO_3 cotransport (e.g., 1:1 cotransport of $Na^+:HCO_3^-$), electroneutral Na^+-driven anion exchange (e.g., 1:2 cotransport of $Na^+:HCO_3^-$ in exchange for Cl), or electrogenic Na, HCO_3 cotransport (e.g., 1:2 or 1:3 cotransport of $Na^+:HCO_3^-$). To date, genomic evidence suggests that insects only possess electroneutral Na^+-driven anion exchangers (NDAEs); in mosquitoes there is only one gene encoding a putative NDAE, referred to as NDAE1.

In brief, there is no direct functional evidence for a role of NDAEs in transepithelial Na^+ or fluid secretion in Malpighian tubules of adult female mosquitoes. Only molecular evidence suggests that an NDAE is expressed in the tubule epithelium. Linser and colleagues cloned four full-length NDAE1 splice variants from larvae of *A. gambiae* and demonstrated that at least one of the variants mediates the expected electroneutral Na^+-driven Cl/HCO_3 exchange when expressed in *Xenopus* oocytes (Linser et al. 2012), consistent with the functional properties of NDAE1 in *D. melanogaster* (Romero et al. 2000). Furthermore, the Beyenbach laboratory has cloned three NDAE splice variants from the Malpighian tubules of adult female *A. aegypti* (Yamahiro et al. 2008). In contrast to the mosquito AEs, the deduced mosquito NDAEs do not possess the conserved extracellular binding motifs for DIDS, which likely makes them weakly sensitive to this pharmacological agent (Yamahiro et al. 2008; Piermarini et al. 2010).

Nevertheless, the Linser laboratory has demonstrated that *An*NDAE1 immunoreactivity localizes to the basolateral membrane of principal cells in Malpighian tubules of larval *A. gambiae* (Fig. 13.6a) (Linser et al. 2012). This localization is consistent with that found for the NDAE of *D. melanogaster* (Sciortino et al. 2001), and suggests that *An*NDAE1 has the potential to contribute to the uptake of Na^+ and HCO_3^- by principal cells of larval tubules. On the other hand, AE immunoreactivity localizes to the basolateral membrane of stellate cells in larval *A. gambiae* and *A. aegypti* and adult female *A. aegypti* and *A. albopictus* (Fig. 13.6b) (Piermarini et al. 2010; Linser et al. 2012). Thus, each cell type within the tubule epithelium appears to express its own SLC4-like protein, suggesting putative novel roles of each cell type in hemolymph acid–base regulation.

Whether the aforementioned expression of NDAE1 in larval tubules persists in adult female mosquitoes is unknown, but preliminary studies of relative NDAE1 and AE immunoreactivity in the midgut and Malpighian tubules of adult female *A. aegypti* suggest that NDAE1 expression dominates in the midgut, whereas AE expression dominates in the Malpighian tubules (Fig. 13.6c). Thus, the AE may be the most important SLC4-like transporter in the tubules of adult females, whereas the NDAE may be the most important SLC4-like transporter in the midgut of adult females. The use of RNAi to selectively knockdown the NDAE vs. AE would be illuminating to help resolve the relative contributions of these two transporters in the tubule epithelium.

13.3.5 Na,K-ATPase

The role of the Na,K-ATPase in the transport of Na^+ and water in mosquito Malpighian tubules has not received much investigative attention, because the V-type H^+-ATPase in the apical membrane of principal cells is considered the primary energizer of transepithelial fluid secretion. Indeed, over 50 % of the ATPase activity in the epithelium is sensitive to bafilomycin (Weng et al. 2003; Tiburcy et al. 2013), and in isolated Malpighian tubules peritubular bafilomycin inhibits fluid secretion and the voltages across both the apical and basolateral membranes (Beyenbach et al. 2000). However, if the results of physiological, biochemical, immunochemical, and molecular studies are integrated, a putative role of the ouabain-sensitive Na,K-ATPase in Malpighian tubule function starts to emerge.

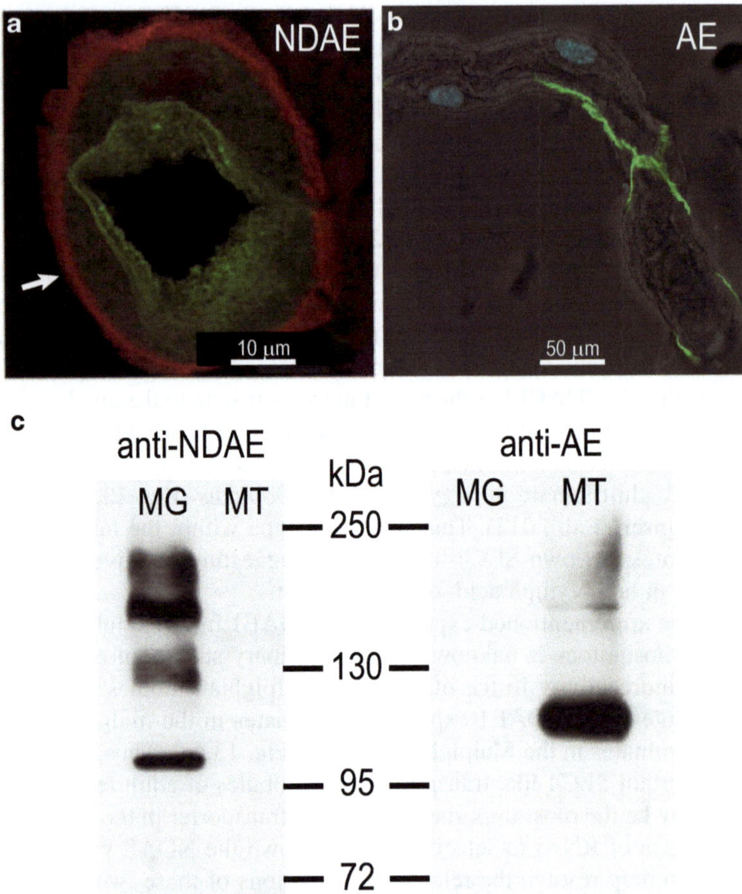

Fig. 13.6 Immunoreactivities of the Na^+-driven anion exchanger (NDAE) and Cl/HCO_3 anion exchanger (AE) in mosquito Malpighian tubules. (**a**) Immunolabeling of NDAE1 (*red*) and the V-type H^+-ATPase (*green*) in a cross section of a Malpighian tubule isolated from a larval *A. gambiae*. Only principal cells express NDAE immunoreactivity on the basolateral membrane. From Linser et al. (2012) with permission. (**b**) Immunolabeling of the AE (*green*) in a whole mount of a Malpighian tubule isolated from an adult female *A. albopictus* using an antibody from Piermarini et al. (2010). Nuclei are counterstained with DAPI (*cyan*). Only stellate cells express AE immunoreactivity on the basolateral membrane. Data were previously unpublished. (**c**) Comparison of NDAE and AE immunoreactivity in membrane enrichments isolated from midguts (MG) and Malpighian tubules (MT) of adult female *A. aegypti*. Each lane was loaded with 25 μg of total protein. Note that NDAE immunoreactivity is prominent in the MG, but not detectable in MT. Reciprocally, AE immunoreactivity is not detectable in the MG, but prominent in the MT. Antibodies raised against the NH_2-terminal domain of *D. melanogaster* NDAE (Sciortino et al. 2001) and a NH_2-terminal peptide of *A. aegypti* AE (Piermarini et al. 2010) were used to detect NDAE and AE immunoreactivity, respectively. Data were previously unpublished

The Beyenbach laboratory first demonstrated that peritubular ouabain had no significant acute (<5 min) effects on the electrophysiology of isolated mosquito Malpighian tubules (Williams and Beyenbach 1984), but chronically (30 min) led to an inhibition of fluid secretion, especially at high concentrations (0.1 and 1 mM) (Hegarty et al. 1991). The high concentrations were likely required due to powerful, endogenous detoxification mechanisms for ouabain in the tubules, which at least in *D. melanogaster* mask the inhibitory effects of ouabain on fluid secretion (Torrie et al. 2004).

Furthermore, biochemical experiments by the Wieczorek laboratory using membrane-enriched fractions of tubules revealed that 28 % of the total membrane-associated ATPase activity is ouabain sensitive; the majority, 60 %, is still bafilomycin sensitive (Tiburcy et al. 2013). Apparently, the enrichment of membrane fractions is essential to detect ouabain-sensitive activity, because in crude lysates of Malpighian tubules this activity is nominal (Weng et al. 2003). The lower fractional activity of the Na,K-ATPase vs. the V-type H^+-ATPase is consistent with their respective localizations in mosquito Malpighian tubules. That is, in the distal segment, an antibody to the alpha-subunit of the Na,K-ATPase selectively labels the basolateral membrane of the minority stellate cells (Patrick et al. 2006), whereas antibodies to the V-type H^+-ATPase selectively label the brush border of the majority principal cells (Weng et al. 2003; Patrick et al. 2006). In *D. melanogaster* Malpighian tubules, the localization of the V-type H^+-ATPase is similar to mosquitoes, but the alpha-subunit of the Na,K-ATPase is localized exclusively to the basolateral membrane of principal cells and not stellate cells (Torrie et al. 2004), suggesting a more prominent role of the Na,K-ATPase in tubules of flies vs. mosquitoes.

Thus, in the distal segment of mosquito Malpighian tubules, the Na,K-ATPase appears positioned to contribute to the uptake of peritubular K^+ and the extrusion of intracellular Na^+ across the basolateral membrane of stellate cells (Fig. 13.2). Indeed, a recent physiological study by Beyenbach and colleagues indicates that ouabain inhibits fluid secretion in Malpighian tubules of adult female mosquitoes by inhibiting the transepithelial secretion of K^+, suggesting a role of the Na,K-ATPase and stellate cells in K^+ secretion (Hine et al. 2014). Whether the transport of Na^+ by this pump contributes to the reabsorption of luminal Na^+ in conjunction with an apical NHA in stellate cells as proposed by Harvey's group (Xiang et al. 2012) and/or to the recycling of Na^+ across the basolateral membrane of stellate cells to drive other Na^+-coupled transporters (e.g., NKCCs, NHEs, NDAEs) remains to be determined. In conclusion, principal cells are the predominant site of transcellular cation transport powered by the V-type H^+-ATPase, but a secondary, significant role of the stellate cells in cation transport powered by the Na,K-ATPase cannot be excluded.

13.4 Water Transport in Mosquito Malpighian Tubules

As described in Sect. 13.2.2, the transport of water across the Malpighian tubule epithelium is driven by the transport of Na^+, K^+, and Cl^- into the tubule lumen, which creates a favorable osmotic gradient for water to follow. The movement of water can occur either through a paracellular pathway via septate junctions or a transcellular pathway via aquaporin water channels in principal and/or stellate cells. Below, we discuss the evidence for both pathways.

13.4.1 Paracellular Water Transport via Septate Junctions

The Beyenbach laboratory has measured the permeability of *A. aegypti* Malpighian tubules to sucrose and inulin for estimating the contributions of the paracellular pathway to water transport. They found that the permeability of the tubule epithelium to both sucrose and inulin—which are assumed to cross via paracellular routes—increase when isolated tubules are stimulated to secrete at diuretic rates with a kinin peptide (i.e., leucokinin), but not when they are stimulated with cAMP (Wang et al. 1996). The selective effect of the kinin is important, because this peptide is also known to make the paracellular pathway more permeable to Cl^- (Beyenbach 2012), whereas cAMP primarily enhances transcellular Na^+ transport (Beyenbach and Piermarini 2011). Moreover, the net secretory flux of sucrose, but not inulin, was found to correlate well with the rates of fluid secretion, which indicates that in addition to diffusion, the passage of sucrose through the paracellular pathway occurs with water—i.e., solvent drag (Wang et al. 1996). Taken together, these observations indicate that a component of water transport in mosquito Malpighian tubules indeed occurs via the paracellular pathway (Fig. 13.2).

The molecular basis for paracellular water transport across any epithelium is not well understood, but a recent study in a mammalian cell line indicates that expression of the tight junction protein claudin-2 enhances transepithelial water fluxes (Rosenthal et al. 2010). Septate junctions are the insect analogs of vertebrate tight junctions, and in some insect epithelia, the orthologs of claudins (e.g., sinuous, megatrachea, snakeskin) are known to contribute to the barrier functions of septate junctions (Beyenbach and Piermarini 2011; Beyenbach 2012). It will be interesting to determine if any of these claudin-like proteins in mosquitoes influence the permeability of the paracellular pathway to water in Malpighian tubules.

13.4.2 Transcellular Water Transport via Aquaporins

The movement of water through a transcellular pathway requires the presence of aquaporins (AQPs), which are a family of channels that are best known to mediate the passive transport of water across cell membranes (Campbell et al. 2008; Rojek et al. 2008). The AQPs must be present on both the apical and basolateral membranes of a given epithelium to form a complete transcellular pathway for water. As explained below, our understanding of AQP-mediated movements of water across mosquito Malpighian tubules is far from complete, but recent studies are beginning to reveal their contributions.

The laboratory of Pietrantonio was the first to clone an ortholog of mammalian AQP4 from mosquito Malpighian tubules (Pietrantonio et al. 2000) and demonstrated that this AQP (*Ae*AQP1) indeed mediates the transport of H_2O and not glycerol when expressed heterologously in *Xenopus* oocytes (Duchesne et al. 2003). The ortholog of *Ae*AQP1 in *A. gambiae* (*An*AQP1) exhibits similar functional characteristics (Liu et al. 2011). However, in situ hybridization and immunolabeling experiments revealed that the *Ae*AQP1 mRNA and immunoreactivity localize to the tracheolar tubes that are tethered to the Malpighian tubules and not to the tubule epithelium proper (Pietrantonio et al. 2000; Duchesne et al. 2003), whereas *An*AQP1 immunoreactivity localizes to the basolateral membrane of stellate cells (Fig. 13.7a) (Liu et al. 2011; Tsujimoto et al. 2013). The localization of *An*AQP1 is consistent with that of AQP1 mRNA and AQP-immunoreactivity in the Malpighian tubules of *D. melanogaster* (Dow et al. 1995; Dow and Davies 2003; Kaufmann et al. 2005). Thus, there are either species-specific differences in the localization of AQP1 in mosquito Malpighian tubules or the expression of AQP1 in stellate cells of *A. aegypti* was below the detectable limits of the nucleotide and antibody probes that were utilized.

The laboratory of Hansen has conducted the most comprehensive functional-genetic study of the mosquito AQPs (Drake et al. 2010). They identified six and seven genes encoding putative AQPs in the genomes of *A. aegypti* and *A. gambiae*, respectively. In adult female *A. aegypti*, the Malpighian tubules primarily express mRNAs for AQP1, AQP2, AQP4, and AQP5 (Drake et al. 2010); in *A. gambiae* the Malpighian tubules at least express AQP1 (Liu et al. 2011). In *A. aegypti*, the

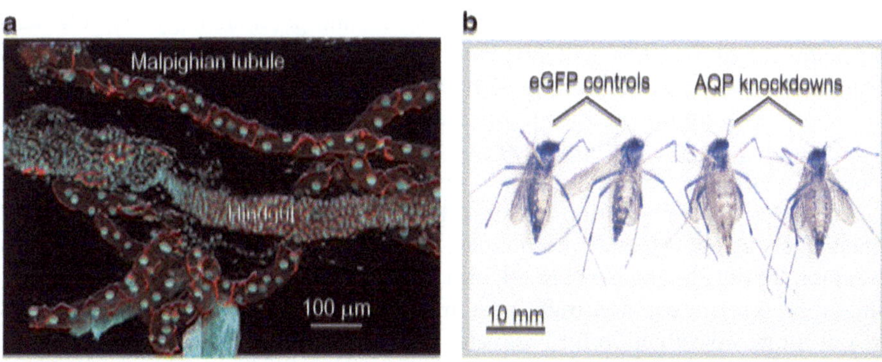

Fig. 13.7 (**a**) Immunolabeling of AQP1 in whole mounts of Malpighian tubules isolated from an adult female *A. gambiae*. In the distal segment, only stellate cells express immunoreactivity (*red*) on the basolateral membrane. Nuclei are counterstained with DAPI (*cyan*). From Tsujimoto et al. (2013) with permission. (**b**) Photograph of adult female *A. aegypti* 3 h after injection with 2.5 μl of phosphate-buffered saline. The mosquitoes on the *left* are controls that were previously treated with double-stranded RNA (dsRNA) against a non-mosquito gene (enhanced green-fluorescent protein, eGFP). The mosquitoes on the *right* were previously treated with dsRNA against AQP1, 2, 4, and 5 to knockdown AQP expression. Note the bloated abdomens in the AQP knockdown mosquitoes compared to the eGFP controls, suggesting the retention of injected fluid. From Drake et al. (2010) with permission

hemolymph injection of $HgCl_2$ or collective knockdown of the four renal AQPs by RNAi significantly reduced the excretory capacity of adult female mosquitoes (Fig. 13.7b) (Drake et al. 2010), consistent with AQPs forming a transcellular pathway for water movements in the Malpighian tubules. In *A. gambiae*, Rasgon and colleagues knocked down the expression of *An*AQP1 (Liu et al. 2011), but did not examine the effects on excretory capacity. However, the knockdown enhanced the survival of mosquitoes during a desiccation stress (Liu et al. 2011), which may in part be explained by decreased excretory water losses (i.e., knockdown of *An*AQP1 decreases urine production and thereby urinary loss of water).

In summary, the molecular precursors for a transcellular pathway of water transport are present in mosquito Malpighian tubules. However, it is still necessary to develop antibodies against AQP2, AQP4, and AQP5 and confirm their localizations to the basolateral and apical membranes of Malpighian tubules to determine if they form an intact pathway for transcellular water transport.

13.5 Perspectives

The preceding pages have summarized the current state of knowledge on the mechanisms of Na^+ and water transport in mosquito Malpighian tubules. Remarkably, this 'simple' epithelium has been the subject of physiological investigation for over 60 years (Ramsay 1951), and the cellular and molecular mechanisms of Na^+ and water transport are still far from being resolved. The integration of modern transcriptomics, proteomics, functional genetics, and pharmacology with classical physiological approaches will no doubt help us elucidate the contributions of known and new molecular players. Further motivation for discovering these mechanisms comes from the prominent role that mosquitoes play in vectoring emerging and remerging diseases of importance to global health, such as chikungunya fever, dengue fever, malaria, and West Nile fever. Given the importance of the Malpighian tubules in the regulation of extracellular fluid homeostasis and processing of blood meals by adult female mosquitoes, they provide a novel physiological target to exploit for developing new insecticides for vector control (Raphemot et al. 2013). Thus, this comparative 'non-traditional' model system promises to remain a fruitful area of future investigation for basic and applied scientists alike.

References

Adragna NC, Fulvio MD, Lauf PK (2004) Regulation of K-Cl cotransport: from function to genes. J Membr Biol 201:109–137

Arroyo JP, Kahle KT, Gamba G (2013) The SLC12 family of electroneutral cation-coupled chloride cotransporters. Mol Aspects Med 34:288–298

Baker DA, Nolan T, Fischer B, Pinder A, Crisanti A, Russell S (2011) A comprehensive gene expression atlas of sex- and tissue-specificity in the malaria vector, *Anopheles gambiae*. BMC Genomics 12:296

Benoit JB, Denlinger DL (2010) Meeting the challenges of on-host and off-host water balance in blood-feeding arthropods. J Insect Physiol 56:1366–1376

Berridge MJ, Oschman JL (1969) A structural basis for fluid secretion by Malpighian tubules. Tissue Cell 1:247–272

Beyenbach KW (2001) Energizing epithelial transport with the vacuolar H^+-ATPase. News Physiol Sci 16:145–151

Beyenbach KW (2003) Transport mechanisms of diuresis in Malpighian tubules of insects. J Exp Biol 206:3845–3856

Beyenbach KW (2012) A dynamic paracellular pathway serves diuresis in mosquito Malpighian tubules. Ann N Y Acad Sci 1258:166–176

Beyenbach KW, Masia R (2002) Membrane conductances of principal cells in Malpighian tubules of *Aedes aegypti*. J Insect Physiol 48:375–386

Beyenbach KW, Piermarini PM (2011) Transcellular and paracellular pathways of transepithelial fluid secretion in Malpighian (renal) tubules of the yellow fever mosquito *Aedes aegypti*. Acta Physiol 202:387–407

Beyenbach KW, Oviedo A, Aneshansley DJ (1993) Malpighian tubules of *Aedes aegypti*: five tubules, one function. J Insect Physiol 39:639–648

Beyenbach KW, Pannabecker TL, Nagel W (2000) Central role of the apical membrane H^+-ATPase in electrogenesis and epithelial transport in Malpighian tubules. J Exp Biol 203:1459–1468

Beyenbach KW, Skaer H, Dow JA (2010) The developmental, molecular, and transport biology of Malpighian tubules. Annu Rev Entomol 55:351–374

Bradley TJ, Stuart AM, Satir P (1982) The ultrastructure of the larval Malpighian tubules of a saline-water mosquito. Tissue Cell 14:759–773

Brett CL, Donowitz M, Rao R (2005) Evolutionary origins of eukaryotic sodium/proton exchangers. Am J Physiol Cell Physiol 288:C223–C239

Campbell EM, Ball A, Hoppler S, Bowman AS (2008) Invertebrate aquaporins: a review. J Comp Physiol B 178:935–955

Coast GM (2009) Neuroendocrine control of ionic homeostasis in blood-sucking insects. J Exp Biol 212:378–386

Coast GM (2012) Intracellular Na^+, K^+ and Cl^- activities in *Acheta domesticus* Malpighian tubules and the response to a diuretic kinin neuropeptide. J Exp Biol 215:2774–2785

Coast GM, Garside CS, Webster SG, Schegg KM, Schooley DA (2005) Mosquito natriuretic peptide identified as a calcitonin-like diuretic hormone in *Anopheles gambiae* (Giles). J Exp Biol 208:3281–3291

Day JP, Wan S, Allan AK, Kean L, Davies SA, Gray JV et al (2008) Identification of two partners from the bacterial Kef exchanger family for the apical plasma membrane V-ATPase of Metazoa. J Cell Sci 121:2612–2619

Donowitz M, Ming Tse C, Fuster D (2013) SLC9/NHE gene family, a plasma membrane and organellar family of Na^+/H^+ exchangers. Mol Aspects Med 34:236–251

Dow JT, Davies SA (2003) Integrative physiology and functional genomics of epithelial function in a genetic model organism. Physiol Rev 83:687–729

Dow JAT, Kelly DC, Davies SA, Maddrell SHP, and Brown D (1995) A novel member of the intrinsic protein family in *Drosophila*: are aquaporins involved in insect Malpighian (renal) tubule fluid secretion? J Physiol 489, 110P

Drake LL, Boudko DY, Marinotti O, Carpenter VK, Dawe AL, Hansen IA (2010) The Aquaporin gene family of the yellow fever mosquito, *Aedes aegypti*. PLoS One 5, e15578

Duchesne L, Hubert JF, Verbavatz JM, Thomas D, Pietrantonio PV (2003) Mosquito (*Aedes aegypti*) aquaporin, present in tracheolar cells, transports water, not glycerol, and forms orthogonal arrays in *Xenopus* oocyte membranes. Eur J Biochem 270:422–429

Esquivel CJ, Cassone BJ, Piermarini PM (2014) Transcriptomic evidence for a dramatic functional transition of the malpighian tubules after a blood meal in the asian tiger mosquito *Aedes albopictus*. PLoS Negl Trop Dis 8, e2929

Gagnon KB, Delpire E (2013) Physiology of SLC12 transporters: lessons from inherited human genetic mutations and genetically engineered mouse knockouts. Am J Physiol Cell Physiol 304:C693–C714

Gamba G (2005) Molecular physiology and pathophysiology of electroneutral cation-chloride cotransporters. Physiol Rev 85:423–493

Gamez AD, Gutierrez AM, Garcia R, Whittembury G (2012) Recent experiments towards a model for fluid secretion in *Rhodnius* Upper Malpighian Tubules (UMT). J Insect Physiol 58:543–550

Gillen CM, Blair CR, Heilman NR, Somple M, Stulberg M, Thombre R et al (2006) The cation-chloride cotransporter, masBSC, is widely expressed in *Manduca sexta* tissues. J Insect Physiol 52:661–668

Hart SJ, Knezetic JA, Petzel DH (2002) Cloning and tissue distribution of two Na^+/H^+ exchangers from the Malpighian tubules of *Aedes aegypti*. Arch Insect Biochem Physiol 51:121–135

Hegarty JL, Zhang B, Pannabecker TL, Petzel DH, Baustian MD, Beyenbach KW (1991) Dibutyryl cAMP activates bumetanide-sensitive electrolyte transport in Malpighian tubules. Am J Physiol Cell Physiol 261:C521–C529

Hegarty JL, Zhang B, Carroll MC, Cragoe EJJ, Beyenbach KW (1992) Effects of amiloride on isolated Malpighian tubules of the yellow fever mosquito (*Aedes aegypti*). J Insect Physiol 38:329–337

Hine RM, Rouhier MF, Park ST, Qi Z, Piermarini PM, Beyenbach KW (2014) The excretion of NaCl and KCl loads in mosquitoes: 1. Control data. Am J Physiol Regul Integr Comp Physiol 307:R837–R849

Ianowski JP, O'Donnell MJ (2004) Basolateral ion transport mechanisms during fluid secretion by *Drosophila* Malpighian tubules: Na^+ recycling, Na^+:K^+:$2Cl^-$ cotransport and Cl^- conductance. J Exp Biol 207:2599–2609

Ianowski JP, Christensen RJ, O'Donnell MJ (2004) Na^+ competes with K^+ in bumetanide-sensitive transport by Malpighian tubules of *Rhodnius prolixus*. J Exp Biol 207:3707–3716

Kang'ethe W, Aimanova KG, Pullikuth AK, Gill SS (2007) NHE8 mediates amiloride-sensitive Na^+/H^+ exchange across mosquito Malpighian tubules and catalyzes Na^+ and K^+ transport in reconstituted proteoliposomes. Am J Physiol Renal Physiol 292:F1501–F1512

Kaufmann N, Mathai JC, Hill WG, Dow JA, Zeidel ML, Brodsky JL (2005) Developmental expression and biophysical characterization of a *Drosophila melanogaster* aquaporin. Am J Physiol Cell Physiol 289:C397–C407

Kondapalli KC, Kallay LM, Muszelik M, Rao R (2012) Unconventional chemiosmotic coupling of NHA2, a mammalian Na^+/H^+ antiporter, to a plasma membrane H^+ gradient. J Biol Chem 287:36239–36250

Leiserson WM, Forbush B, Keshishian H (2011) *Drosophila* glia use a conserved cotransporter mechanism to regulate extracellular volume. Glia 59:320–332

Linser PJ, Neira Oviedo M, Hirata T, Seron TJ, Smith KE, Piermarini PM et al (2012) Slc4-like anion transporters of the larval mosquito alimentary canal. J Insect Physiol 58:551–562

Liu K, Tsujimoto H, Cha SJ, Agre P, Rasgon JL (2011) Aquaporin water channel AgAQP1 in the malaria vector mosquito *Anopheles gambiae* during blood feeding and humidity adaptation. Proc Natl Acad Sci U S A 108:6062–6066

O'Connor KR, Beyenbach KW (2001) Chloride channels in apical membrane patches of stellate cells of Malpighian tubules of *Aedes aegypti*. J Exp Biol 204:367–378

Pacheco CA, Alevi KCC, Ravazi A, de Azeredo Oliveria MTV (2014) Review: Malpighian tubule, an essential organ for insects. Entomol Ornithol Herpetol 3:1000122

Patrick ML, Aimanova K, Sanders HR, Gill SS (2006) P-type Na^+/K^+-ATPase and V-type H^+-ATPase expression patterns in the osmoregulatory organs of larval and adult mosquito *Aedes aegypti*. J Exp Biol 209:4638–4651

Payne JA (2012) Molecular operation of the cation chloride cotransporters: ion binding and inhibitor interaction. Curr Top Membr 70:215–237

Petzel DH (2000) Na^+/H^+ exchange in mosquito Malpighian tubules. Am J Physiol Regul Integr Comp Physiol 279:R1996–R2003

Petzel DH, Hagedorn HH, Beyenbach KW (1985) Preliminary isolation of mosquito natriuretic factor. Am J Physiol 249:R379–R386

Piermarini PM, Calkins TL (2014) Evidence for intercellular communication in mosquito renal tubules: a putative role of gap junctions in coordinating and regulating the rapid diuretic effects of neuropeptides. Gen Comp Endocrinol 203:43–48

Piermarini PM, Weihrauch D, Meyer H, Huss M, Beyenbach KW (2009) NHE8 is an intracellular cation/H^+ exchanger in renal tubules of the yellow fever mosquito *Aedes aegypti*. Am J Physiol Renal Physiol 296:F730–F750

Piermarini PM, Grogan LF, Lau K, Wang L, Beyenbach KW (2010) A SLC4-like anion exchanger from renal tubules of the mosquito (*Aedes aegypti*): evidence for a novel role of stellate cells in diuretic fluid secretion. Am J Physiol Regul Integr Comp Physiol 298:R642–R660

Piermarini PM, Hine RM, Schepel M, Miyauchi JT, Beyenbach KW (2011) Role of an apical K, Cl cotransporter in urine formation by renal tubules of the yellow fever mosquito (Aedes aegypti). Am J Physiol Regul Integr Comp Physiol 301:R1318–R1337

Piermarini PM, Rouhier MF, Schepel M, Kosse C, Beyenbach KW (2013) Cloning and functional characterization of inward-rectifying potassium (Kir) channels from Malpighian tubules of the mosquito *Aedes aegypti*. Insect Biochem Mol Biol 43:75–90

Pietrantonio PV, Jagge C, Keeley LL, Ross LS (2000) Cloning of an aquaporin-like cDNA and in situ hybridization in adults of the mosquito *Aedes aegypti* (Diptera: Culicidae). Insect Mol Biol 9:407–418

Pullikuth AK, Filippov V, Gill SS (2003) Phylogeny and cloning of ion transporters in mosquitoes. J Exp Biol 206:3857–3868

Pullikuth AK, Aimanova K, Kang'ethe W, Sanders HR, Gill SS (2006) Molecular characterization of sodium/proton exchanger 3 (NHE3) from the yellow fever vector, *Aedes aegypti*. J Exp Biol 209:3529–3544

Ramsay JA (1951) Osmotic regulation in mosquito larvae; the role of the Malpighian tubules. J Exp Biol 28:62–73

Raphemot R, Rouhier MF, Hopkins CR, Gogliotti RD, Lovell KM, Hine RM et al (2013) Eliciting renal failure in mosquitoes with a small-molecule inhibitor of inward-rectifying potassium channels. PLoS One 8, e64905

Rheault MR, Okech BA, Keen SB, Miller MM, Meleshkevitch EA, Linser PJ et al (2007) Molecular cloning, phylogeny and localization of AgNHA1: the first Na^+/H^+ antiporter (NHA) from a metazoan, *Anopheles gambiae*. J Exp Biol 210:3848–3861

Rodan AR, Baum M, Huang CL (2012) The *Drosophila* NKCC Ncc69 is required for normal renal tubule function. Am J Physiol Cell Physiol 303:C883–C894

Rojek A, Praetorius J, Frokiaer J, Nielsen S, Fenton RA (2008) A current view of the mammalian aquaglyceroporins. Annu Rev Physiol 70:301–327

Romero MF, Henry D, Nelson S, Harte PJ, Dillon AK, Sciortino CM (2000) Cloning and characterization of a Na^+-driven anion exchanger (NDAE1). A new bicarbonate transporter. J Biol Chem 275:24552–24559

Romero MF, Chen AP, Parker MD, Boron WF (2013) The SLC4 family of bicarbonate (HCO_3^-) transporters. Mol Aspects Med 34:159–182

Rosenthal R, Milatz S, Krug SM, Oelrich B, Schulzke JD, Amasheh S et al (2010) Claudin-2, a component of the tight junction, forms a paracellular water channel. J Cell Sci 123:1913–1921

Russell JM (2000) Sodium-potassium-chloride cotransport. Physiol Rev 80:211–276

Sciortino CM, Shrode LD, Fletcher BR, Harte PJ, Romero MF (2001) Localization of endogenous and recombinant Na^+-driven anion exchanger protein NDAE1 from *Drosophila melanogaster*. Am J Physiol Cell Physiol 281:C449–C463

Scott BN, Yu MJ, Lee LW, Beyenbach KW (2004) Mechanisms of K^+ transport across basolateral membranes of principal cells in Malpighian tubules of the yellow fever mosquito, *Aedes aegypti*. J Exp Biol 207:1655–1663

Sun Q, Tian E, Turner RJ, Ten Hagen KG (2010) Developmental and functional studies of the SLC12 gene family members from *Drosophila melanogaster*. Am J Physiol Cell Physiol 298: C26–C37

Tiburcy F, Beyenbach KW, Wieczorek H (2013) Protein kinase A-dependent and -independent activation of the V-ATPase in Malpighian tubules of *Aedes aegypti*. J Exp Biol 216:881–891

Torrie LS, Radford JC, Southall TD, Kean L, Dinsmore AJ, Davies SA et al (2004) Resolution of the insect ouabain paradox. Proc Natl Acad Sci U S A 101:13689–13693

Tsujimoto H, Liu K, Linser PJ, Agre P, Rasgon JL (2013) Organ-specific splice variants of aquaporin water channel AgAQP1 in the malaria vector *Anopheles gambiae*. PLoS One 8, e75888

Wang S, Rubenfeld A, Hayes T, Beyenbach K (1996) Leucokinin increases paracellular permeability in insect Malpighian tubules. J Exp Biol 199:2537–2542

Weng XH, Huss M, Wieczorek H, Beyenbach KW (2003) The V-type H^+-ATPase in Malpighian tubules of *Aedes aegypti*: localization and activity. J Exp Biol 206:2211–2219

Weng XH, Piermarini PM, Yamahiro A, Yu MJ, Aneshansley DJ, Beyenbach KW (2008) Gap junctions in Malpighian tubules of *Aedes aegypti*. J Exp Biol 211:409–422

Wieczorek H, Putzenlechner M, Zeiske W, Klein U (1991) A vacuolar-type proton pump energizes K^+/H^+ antiport in an animal plasma membrane. J Biol Chem 266:15340–15347

Williams JC, Beyenbach KW (1984) Differential effects of secretagogues on the electrophysiology of the Malpighian tubules of the yellow fever mosquito. J Comp Physiol B 154:301–309

Williams JC, Hagedorn HH, Beyenbach KW (1983) Dynamic changes in flow rate and composition of urine during the post blood meal diuresis in *Aedes aegypti*. J Comp Physiol B 153:257–266

Xiang MA, Linser PJ, Price DA, Harvey WR (2012) Localization of two Na^+- or K^+-H^+ antiporters, AgNHA1 and AgNHA2, in *Anopheles gambiae* larval Malpighian tubules and the functional expression of AgNHA2 in yeast. J Insect Physiol 58:570–579

Yamahiro A, Piermarini PM, Beyenbach KW (2008) Identification of Na-driven anion exchanger (NDAE) splice variants from Malpighian (renal) tubules of the adult yellow-fever mosquito. FASEB J 22:757.22

Zelle KM, Lu B, Pyfrom SC, Ben-Shahar Y (2013) The genetic architecture of degenerin/epithelial sodium channels in *Drosophila*. G3(Bethesda) 3:441–450

Chapter 14
Circadian Rhythms of Ion Transporters in the Visual System of Insects

Jolanta Górska-Andrzejak, Milena Damulewicz, and Elżbieta Pyza

Abstract Active ion transporters that maintain gradients of ion concentration across cell membranes take part in generating daily structural changes in neurons and glial cells of the visual system of Diptera. In the optic lobe of two fly species, *Drosophila melanogaster* and *Musca domestica*, significant changes in the abundance of two major types of membrane pumps, the sodium pump (Na^+/K^+-ATPase) and the vacuolar proton pump (V-ATPase) have been observed in the 24 h day/night cycle. These oscillations have been shown to depend on circadian inputs from the pacemaker neurons as well as from photoreceptors and glial cells that express the clock genes. Therefore, these two types of ion pumps seem to be the target elements of the circadian output regulating daily structural changes observed in the presynaptic terminals of the compound eye photoreceptors, their postsynaptic partners—the so-called L1 and L2 large monopolar cells (LMCs), and the glial cells of the optic lobe.

Keywords Circadian clock • *Drosophila melanogaster* • Neuronal plasticity • Proton pump • Sodium pump

14.1 Introduction

Ion transporters are membrane proteins that maintain different ion concentrations inside and outside the cells by transporting ions, using ATP, against ion gradients. They are involved in a great variety of cellular processes. In neurons they maintain the membrane resting potential, the pH homeostasis and the cation-coupled uptake of glucose, neurotransmitters and other molecules that are substantial for neuronal functions (Sweadner 1989; Lingrel et al. 1990). Due to these numerous functions, it is not surprising that the level of expression of at least certain ion transporter proteins in the brain changes in the course of day reflecting the alterations in brain activity during the day/night or the sleep/wakefulness cycle. So far such

J. Górska-Andrzejak • M. Damulewicz • E. Pyza (✉)
Department of Cell Biology and Imaging, Institute of Zoology, Jagiellonian University, Gronostajowa 9, Kraków 30-387, Poland
e-mail: elzbieta.pyza@uj.edu.pl

K.A. Hyndman, T.L. Pannabecker (eds.), *Sodium and Water Homeostasis*, Physiology in Health and Disease, DOI 10.1007/978-1-4939-3213-9_14

phenomenon has been observed in the visual system of flies; however, the daily activity of Na^+/K^+-ATPase has also been reported in the suprachiasmatic nucleus (SCN), a site of the mammalian circadian clock, and in other tissues (Morise et al. 1989; Wang and Huang 2004).

Circadian rhythmicity of behavioral, physiological and biochemical processes is generated by the endogenous pacemakers—the circadian clocks, that generate oscillations with a period of approximately 24 h to maintain temporal homeostasis of an organism and its synchrony with the external (environmental) cycle of day and night. The clock-controlled rhythms are observed in both day/night (Light/ Dark; LD) and constant darkness (DD) conditions. They are abolished in DD in clock mutants.

In flies, in addition to the above mentioned circadian rhythms, a new type of circadian rhythmicity has been discovered—in volume and morphological changes of neurons and glial cells (Pyza and Meinertzhagen 1995, 1997, 1999; Pyza and Górska-Andrzejak 2004). The mechanism of this volume differences is unknown, however, we have found that injections of some neurotransmitters can mimic these changes (Pyza and Meinertzhagen 1996; Meinertzhagen and Pyza 1999). We have also observed that they involve protein synthesis when neurons swell but not when they shrink (Kula and Pyza 2007).

Although not yet entirely understood, these rhythmic cell volume alterations strongly suggest ion shifts and substantial changes in the properties of plasma membranes, as well as the involvement of ion transporters in their generation. Looking for mechanisms of the circadian rhythms in structure of cells, the so-called circadian plasticity, we have monitored the abundance of several subunits of two types of membrane pumps in the optic lobe of *Musca domestica* and/or *Drosophila melanogaster*. We have found that the abundance of major pumps that regulate the ionic homeostasis in various cell types (Lingrel et al. 1990), the Na^+/K^+-ATPase or sodium pump, as well as the vacuolar and plasma membrane proton-adenosine triphosphatase—the proton pump (V-ATPase), changes in a circadian manner (Pyza et al. 2004; Górska-Andrzejak et al. 2009; Damulewicz et al. 2013). Due to these oscillations, the above mentioned ion transporters are important elements of the output of the circadian system.

In the following chapter we describe the pattern of expression of these transporters in the visual system of flies and changes in the level of their expression in the course of day and night. We discuss these rhythms in connection with the observed circadian structural rhythmicity of neurons and glial cells of the optic lobe.

14.2 Circadian Plasticity of Neurons and Glial Cells of the Visual System of Diptera

In the visual system of Diptera (Fig. 14.1a), the circadian plasticity has been observed not only in the retina of the compound eye (Chen et al. 1992), but also in the structure of the visual pathway beneath the retina—in the underlying optic lobe (Pyza and Górska-Andrzejak 2008; Pyza 2010; Górska-Andrzejak 2013).

The optic lobe contains three optic neuropils (synaptic regions) (Strausfeld 1976; Strausfeld and Nässel 1980): the lamina, the medulla and the lobula (Fig. 14.1a). Owing to its relatively simple, modular, and already well described cellular organization, the most distal neuropil (lamina) has become a model for studying circadian rhythms at the cellular level in the central nervous system. The lamina neuropil is composed of synaptic units of columnar shape—the cartridges (Strausfeld and Nässel 1980) (Fig. 14.1b). Each cartridge has an invariable cellular composition. The cylindrical terminals of photoreceptors (R1–R6), which provide the visual input to the entire cartridge, and the axons of their primary postsynaptic partners—the L1 and L2 interneurons (the so-called large monopolar cells—LMCs) are the main elements that constitute the cartridge (Fig. 14.1b–d). Each cartridge is also enwrapped by three cells of the lamina neuropil glia, known as the epithelial glia (Fig. 14.1b, e) (Boschek 1971; Meinertzhagen and O'Neil 1991). These cells not only surround the entire cartridge, but also send their thin projections between the prominent profiles of R1–R6 and L1, L2 (Fig. 14.1b, e), penetrating the whole inside of the cartridge. In addition, they make intimate connections with the terminals of R1–R6. These are the characteristic close invaginating appositions called capitate projections (Fig. 14.1b). They seem to be involved in vesicle endocytosis (Fabian-Fine et al. 2003; Edwards and Meinertzhagen 2010). It has been shown that the aforementioned chief elements of the cartridge, including the epithelial glia (Fig. 14.1c–e), exhibit robust structural changes during the cycle of day and night (Pyza and Meinertzhagen 1995, 1997, 1999; Pyza and Górska-Andrzejak 2004; Górska-Andrzejak et al. 2005; Weber et al. 2009; Górska-Andrzejak 2013).

In the lamina of the housefly (*Musca domestica*), the photoreceptor terminals display circadian changes in the number of screening pigment granules and interreceptor invaginations of the terminals of neighboring photoreceptors (Pyza and Meinertzhagen 1997). On the other hand, their postsynaptic partners, the L1 and L2 monopolar cells, show circadian changes in the volume of their axons (Pyza and Meinertzhagen 1993), which was also revealed by studies conducted on another two fly species: *Drosophila melanogaster* (Pyza and Meinertzhagen 1999) and *Calliphora vicina* (the blowfly) (Pyza and Cymborowski 2001). In each of these species the daily pattern of changes of the L1, L2 axon volume is different. In *M. domestica* the axons of L1 and L2 swell during the day and shrink by night, whereas in *D. melanogaster* they are larger at the beginning of the day and the night, bearing a close resemblance to the daily pattern of locomotor activity of a given species. These rhythmic changes are caused partially by light exposure but

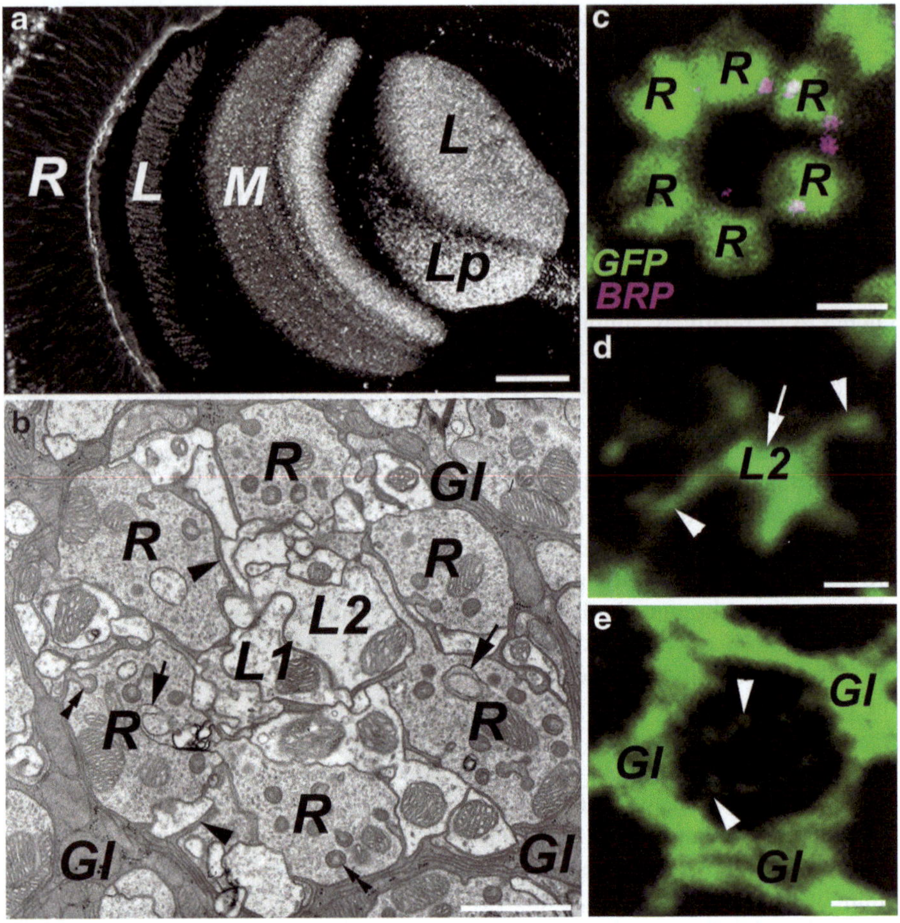

Fig. 14.1 (**a**) The visual system of flies is composed of the retina of the compound eye (*R*) and three optic neuropils, the lamina (*L*), the medulla (*M*) and the lobula, which in flies is divided into the lobula (*L*) and the lobula plate (*Lp*). Scale bar: 20 μm. (**b**) The Electron Microscope (EM) micrograph of the lamina cartridge in a cross-section. The photoreceptor terminals (*R*) make inter-receptor invaginations to the terminals of neighboring photoreceptors (*arrows*). The epithelial glia that surround each lamina cartridge (*Gl*) direct small processes into the cartridge (*arrow head*) and deep invaginations with a characteristic spherical head (the capitate projections) into the photoreceptor terminals. *L1*, *L2*—the axons of L1 and L2 large monopolar cells. Scale bar: 1 μm. (**c–e**) Cross sections of the lamina cartridges of transgenic lines of *Drosophila melanogaster* showing targeted expression of green fluorescent protein (*GFP*) in cylindrical photoreceptor terminals (**c**), axon (*arrow*) and dendrites (*arrow head*) of L2 monopolar cell (**d**) and processes of epithelial glial cells (**e**). (**c**) The cartridge with GFP-labeled photoreceptors shows also the distribution of presynaptic elements (*magenta*) visualized by anti–BRP immunostaining using Mab nc82 (Hybridoma). Scale bar for (**c–e**): 1 μm

predominantly by clock input from circadian oscillators. Mechanism of these morphological changes is not fully understood but it was shown that it depends on neurotransmitters (PDF, histamine, 5HT, GABA), V-ATPase activity as well as function of the epithelial glia surrounding every single cartridge (Pyza and Meinertzhagen 1996; Pyza et al. 2004; Pyza and Górska-Andrzejak 2004). Moreover, in the housefly, daily and circadian rhythms are also observed in the number of tetrad synapses formed between the photoreceptor terminals (R1–R6) and four postsynaptic cells (L1, L2, amacrine and glial cells), and feedback synapses that are formed back onto the R1–R6 terminals in the proximal lamina (Meinertzhagen and Sorra 2001; Górska-Andrzejak et al. 2013; Pyza and Meinertzhagen 1993). The number of tetrads and feedbacks increases at the beginning of the day and at the beginning of the night, respectively. In case of the feedback synapses this rhythm is also maintained in DD. Thus it is generated by the circadian clock and not by daily changes of light intensity. In spite of the fact that the precise functional significance of feedback transmission at their infrequent sites (Meinertzhagen and Sorra 2001; Górska-Andrzejak et al. 2013) is still not entirely clear, it has been suggested that this transmission nevertheless shapes the photoreceptor's response to some extent (presumably also in a circadian manner) by a negative feedback loop (Zheng et al. 2006). Circadian remodeling of synaptic contacts in the first optic neuropil of *D. melanogaster* has also been confirmed by recent findings on circadian expression of synaptic proteins: the presynaptic structural protein Bruchpilot (BRP) (Górska-Andrzejak et al. 2013), the protein involved in synaptic vesicle release—Synapsin (SYN) and the postsynaptic protein—Disc Large (DLG) (Krzeptowski et al. 2014). The amount of BRP and DLG increases two times during the day/night cycle—in the morning and in the evening. In the case of BRP, the evening peak of expression is controlled primarily by the circadian clock and the morning peak is light-dependent (Górska-Andrzejak et al. 2013), whereas in the case of DLG both peaks appear to be clock controlled (Krzeptowski et al. 2014).

In addition to the circadian changes of axon size and the number of feedback synapses, also the cell nucleus and the dendritic tree of L2 (Fig. 14.1d) undergo circadian remodeling, as revealed by the studies using transgenic lines of *D. melanogaster* (*21D-Gal 4* $\times$ *UAS-GFP*) expressing either cytoplasmic or membranous GFP reporter targeted to L2 (Górska-Andrzejak et al. 2005; Weber et al. 2009). The length of L2 dendrites (Fig. 14.1d) that carry postsynaptic elements of tetrad synapses is the most prominent at the beginning of the day (Weber et al. 2009), which coincides with the increase in the amount of BRP (Fig. 14.1c) in the lamina neuropil (Górska-Andrzejak et al. 2013).

Because of the circadian structural changes of photoreceptor terminals and the axons of L1 and L2, it is not surprising that the epithelial glial cells of *M. domestica* have also been found to display daily alterations in cell size (Pyza and Górska-Andrzejak 2004). Interestingly, they expand at night and shrink during the day, which is exactly in antiphase to the circadian axonal changes of L1 and L2, and which therefore implies a compensatory nature of glial structural oscillations (Pyza and Górska-Andrzejak 2004). In fact, glial oscillators must constitute an important component of this phenomenon, as they clearly influence the rhythmic structural changes of L1 and L2 (Pyza and Górska-Andrzejak 2004). Disrupting either glial metabolism (by injecting glial metabolic toxins, e.g., fluorocitrate or iodoacetate, Kitano et al. 2003), or glial communication via the gap junction channels (by injecting their closing agent—octanol, Pappas et al. 1996), affects the pattern and/or amplitude of the neuronal (L1, L2) rhythms (Pyza and Górska-Andrzejak 2004). Such results strongly support a hypothesis that the lamina epithelial glial cells regulate or modulate the circadian plasticity of the L1 and L2 target interneurons. In contrast to L1, L2 monopolar cells, glial cells cyclically express the clock gene *period* (*per*)—the core gene of the molecular mechanism for keeping the circadian time (Fig. 14.2a), and therefore contain the PER-based circadian oscillator (Siwicki et al. 1988; Jackson 2011). Their connection with the circadian system is additionally supported by the fact that they belong to a subtype of glia that express EBONY, an enzyme conjugating β-alanine to biogenic amines (such as histamine, dopamine and serotonin). Since histamine is a neurotransmitter in tetrad synapses in the lamina (Hardie 1989), and dopamine and serotonin are important in regulating behavior, β-alanine has been proved to be the first exclusively glial factor required for fly's behavioral rhythmicity (Suh and Jackson 2007; Jackson 2011).

The precise mechanism of generation of the circadian rhythms in the visual system of Diptera is very complex. The circadian input to the lamina, particularly to the L1 and L2 interneurons, seems to originate from the retina photoreceptors (Cheng and Hardin 1998), the circadian pacemaker neurons of the brain (Bałys and Pyza 2001; Damulewicz and Pyza 2011) and from the glial cells of the optic lobe (Pyza and Górska-Andrzejak 2004; Górska-Andrzejak et al. 2009; Damulewicz et al. 2013; Górska-Andrzejak 2013) (Fig. 14.2b). Our studies indicate that ion transporters such as Na^+/K^+-ATPase and V-ATPase may play an important role in each of these pathways.

14.3 Circadian Expression of the Na^+/K^+-ATPase in the Visual System of Diptera

The Na^+/K^+-ATPase is an integral plasma membrane protein that uses the energy from the hydrolysis of one ATP molecule to transport three Na^+ out of the cell and two K^+ into the cell to maintain ion balance and membrane potential. As the major pump regulating the ionic homeostasis in different cell types (Lingrel et al. 1990), the Na^+/K^+-ATPase might have been expected to be engaged in the diurnal modulation of morphology of neurons and glial cells (especially their volumetric changes) that have been detected in the optic lobe of Diptera (Pyza and Górska-Andrzejak 2008; Pyza 2010; Jackson 2011; Górska-Andrzejak 2013).

In *D. melanogaster*, the functional enzyme is composed of at least two subunits that are non-covalently linked to each other to form a heterodimer composed of the catalytic α subunit (Na^+/K^+-ATPase α; ATPα) and the regulatory β subunit (ATPβ) (Chow and Forte 1995; Blanco and Mercer 1998). The relative amount of the

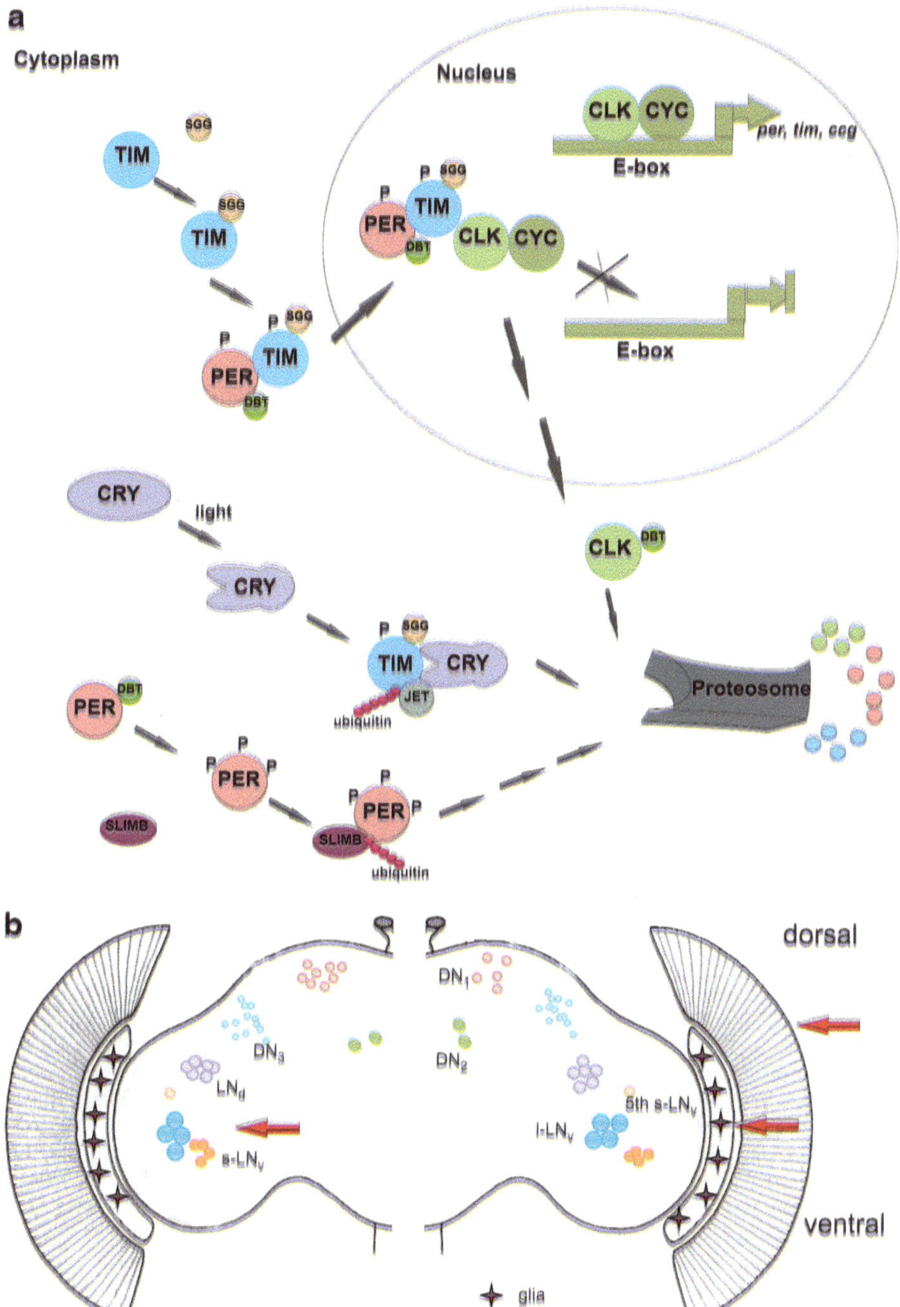

Fig. 14.2 (**a**) The molecular mechanism of the circadian clock of *Drosophila melanogaster*. The main feedback loop of the molecular clock is composed of PERIOD (PER), TIMELESS (TIM) proteins and transcription factors of their genes: CLOCK (CLK) and CYCLE (CYC). In the nucleus CLK and CYC form heterodimers that bind to the E-box—specific sequences of the promoter of clock genes *per* and *tim* and clock-controlled genes (*ccg*), and activate their transcription. During the night, when TIM and PER levels are high in the cytoplasm, the SHAGGY

catalytic ATPα, which mirrors the enzymatic activity of the complete pump, shows a robust circadian rhythmicity at both mRNA and protein levels under LD and DD conditions. The level of Na^+/K^+-ATPase α–subunit is the highest at the beginning of the night (1 h after the beginning of the dark phase) or the subjective night in DD (Górska-Andrzejak et al. 2009) and also at the onset of the day (or the subjective day in DD) (Damulewicz et al. 2013). The α subunit of the ubiquitous Na^+/K^+-ATPase can be detected, using specific antibodies, in the membranes of all cells in the optic lobe; however, its level is especially high in the cells of the first (lamina) and the second (medulla) optic neuropils. Moreover, in the lamina, its expression is particularly robust at the plasma membrane of the epithelial glial cells (Fig. 14.3) (Górska-Andrzejak et al. 2009; Damulewicz et al. 2013; Górska-Andrzejak 2013), suggesting that these cells may chiefly contribute to the rhythm of the sodium pump concentration in the lamina neuropil. By displaying circadian changes in the abundance of ATPα, they once again prove to be the likely modulators of L1 and L2 structural rhythmicity, which coincides with the circadian profile of ATPα immunoreactivity. Presumably, at certain times of the day glial cells need to provide the enhanced ion buffering capacity for the axons of highly active neurons and they do so by recruitment into their membranes of more ATPase. A failure to supply sufficient activity of this pump can result in a loss of the transmembrane gradients (for Na^+ and K^+) and eventually membrane excitability (Green 2004).

The circadian oscillations of the amount of ATPα subunit in the membrane of the epithelial glial cells could be regulated by their own circadian oscillators. It has

---

Fig. 14.2 (continued) (SGG) kinase phosphorylates TIM protein, so TIM can bind PER, forming heterodimers. In turn PER is phosphorylated by the DOUBLETIME (DBT) kinase that is necessary to transport the PER/TIM/DBT complex to the nucleus. Hyper-phosphorylation of PER by DBT results in binding of SLIMB (Supernumerary limbs) and targeting PER, by ubiquitination, to degradation in proteasome. The PER/TIM/DBT complex enters the nucleus, binds to CLK-CYC heterodimer and inhibits *per* and *tim* transcription. So PER and TIM inhibit transcription of their own genes. In meantime DBT binds to CLK causing its exit from the nucleus and degradation in proteasome. During the day, in response to light CRYPTOCHROME (CRY), a blue-light photoreceptor, changes its conformation and binds to TIM phosphorylated by SGG. This complex is recognized by JETLAG protein. That leads to ubiquitination of TIM and CRY and their degradation in proteasome (reviewed by Tataroglu and Emery, 2015). (**b**) Different types of circadian oscillators in the brain of *Drosophila melanogaster*. The main clock (pacemaker) in the brain of *Drosophila melanogaster* is composed of about 150 cells, the so-called clock neurons. The most important clock neurons for maintaining the circadian rhythms are small ventral lateral (s-LN_v) and large ventral lateral neurons (l-LN_v). All of them, with the exception of the fifth s-LN_v, express the neuropeptide Pigment-Dispersing Factor (PDF) and they constitute the Morning Oscillators (M Cells). The fifth s-LN_v expresses the second clock neuropeptide—Ion Transport Peptide (ITP) and with a group of six dorsal lateral neurons (LN_d) (one of them also expresses ITP) forms the Evening Oscillators (E cells). Dorsal clock neurons (DN) are divided into three subgroups: 15 DN_1 (E cells), 2 DN_2 (involved in synchronization of the clock to the temperature cycles) and about 40 DN_3 (reviewed by Hermann-Luibl and Helfrich-Förster, 2015). In addition to the main clock, there are also peripheral oscillators located in the retinal photoreceptors and glial cells. In all cellular oscillators the clock genes are cyclically expressed. The circadian expression of the α subunit of Na^+/K^+-ATPase depends on three inputs, from the pacemaker neurons, the eye photoreceptors and the glial cells (*red arrows*)

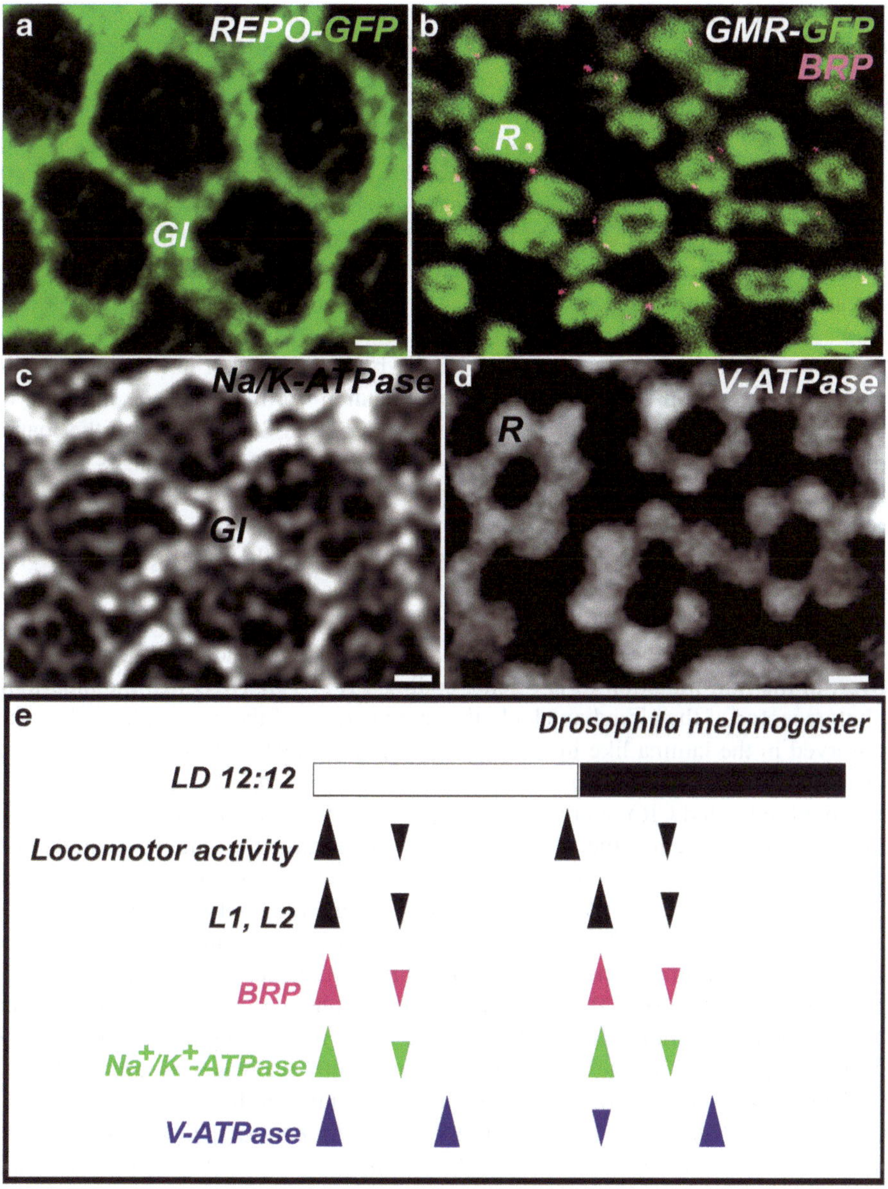

Fig. 14.3 Confocal images of cartridges of the transgenic flies with GFP expression in the epithelial glial cells (**a**) or photoreceptors (**b**) and immunolabeled with antibodies against the α-subunit of Na^+/K^+-ATPase (**c**) or anti-β subunit of V-ATPase (**d**) reveals the localization of the first transporter predominantly in glia, whereas the second one in photoreceptors. *R*—photoreceptors, *Gl*—epithelial glial cells. Scale bar: 1 μm. (**e**) The schematic representation of the timing of maximum (*arrow head* pointing up) or minimum (*arrow head* pointing down) of locomotor activity, L1 and L2 axon volumes (*L1, L2*), level of expression of the presynaptic protein Bruchpilot (*BRP*), the amount of sodium-potassium pump (*Na^+/K^+-ATPase*) and proton pump (*V-ATPase*) in the lamina of *D. melanogaster* during the 24 h cycle of day (*white bar*) and night (*black bar*)

been shown that they express the clock genes *per* and *timeless* (*tim*) (Siwicki et al. 1988; Zerr et al. 1990; Cheng and Hardin 1998) and for that reason must possess the PER-based circadian oscillator (Jackson 2011; Górska-Andrzejak 2013). Consequently, it is not surprising that the circadian oscillations in the abundance of the ATPα in the lamina of *D. melanogaster* proved to be *per* and *tim*-dependent (Górska-Andrzejak et al. 2009; Damulewicz et al. 2013). They are not maintained in the optic lobe of per^0 and tim^0 null mutants. Nevertheless, these oscillations result from the lack of PER and TIM proteins not only in glia but also in photoreceptors and clock neurons, which are also important components of the circadian regulation of ATPα expression, as it has been revealed in the most recent studies (Damulewicz et al. 2013).

This rhythm is strongly controlled by the brain pacemaker since after genetic ablation of the main pacemaker cells, the circadian oscillation of the intensity of ATPα immunostaining is abolished (Damulewicz et al. 2013). On the other hand, disruption of the circadian peripheral clock located in the retina changes the usual pattern of ATPα oscillations, causing suppression of ATPα expression in the middle of the night (Damulewicz et al. 2013).

ATPα expression also seems to be partially controlled by light. In DD the oscillations of ATPα are maintained but their pattern is different than in LD conditions. In DD ATPα level is higher during the subjective night than during the subjective day (Górska-Andrzejak et al. 2009; Damulewicz et al. 2013). Moreover, in the null mutation of *cry* gene, coding CRYPTOCHROME, the blue-light photoreceptor of the circadian clock, the same pattern of the α subunit rhythm was observed in the lamina like in wild-type flies kept in DD. In turn, inserting *cry* by using the UAS/GAL4 system can rescue the bimodality in LD conditions. It means that light-activated CRY increases ATPα protein level in the morning. CRY is also involved in reducing the ATPα expression during the night (Damulewicz et al. 2013). Our results, obtained after silencing of *cry*, provide strong evidence that CRY plays a cell-autonomous function in regulating ATPα expression. CRY expressed in the pacemaker cells plays a crucial role in maintaining phase of the rhythm, with maximum and minimum of ATPα level at specific time points during the 24 h period. In turn, CRY expressed in the peripheral oscillators located in glial cells influences the amplitude of this rhythm. Silencing *cry* in glia does not shift the rhythm in phase or change its period but significantly decreases its amplitude.

The circadian pattern of ATPα expression is controlled by the brain pacemaker neurons, the LN_vs, by at least two neuropeptides; Pigment Dispersing Factor (PDF) and Ion Transport Peptide (ITP). PDF is expressed in the LN_vs, in four small (s-LN_vs) and four large (l-LN_vs) neurons, and released in a circadian manner in the medulla and in the dorsal protocerebrum while PDF receptors (PDFR) are expressed in some clock neurons and in glial cells in the lamina (Sheeba 2008; Im and Taghert 2010). It indicates that PDF synchronizes the pacemaker neurons within their network and transmits circadian information to the visual system. Flies with Pdf^0 mutation show a characteristic unimodal pattern of the daily rhythm of ATPα expression, with a peak at the beginning of the night. This unimodal pattern in Pdf^0 null mutant is observed in both LD and DD conditions (Damulewicz et al. 2013). It suggests that PDF regulates expression of ATPα in the morning. It is in accordance with the morning/evening oscillator model (MO-EO), which implies that the LN_vs regulate morning peak of locomotor activity and morning cell functions by releasing PDF (Picot et al. 2007; Helfrich-Förster 2004).

The second neuropeptide ITP is also expressed in clock neurons, in one of the dorsal lateral neurons (LN_ds) and in the fifth s-LN_v (Johard et al. 2009). The latter neuron innervates the lamina (Damulewicz and Pyza 2011). The role of ITP in the brain and the circadian system is unknown. Silencing of *itp* in the clock neurons, however, causes changes in ATPα expression in the lamina. This procedure increases ATPα level during the day in day/night conditions and advances the phase of the rhythm by several hours in DD. It suggests that ITP plays a role in regulating ATPα rhythm in the lamina epithelial glia, probably by regulating the evening peak of this rhythm.

Another subunit of the sodium pump, the β-subunit (ATPβ), encoded by *Nervana* gene is necessary for the pump's maturation, localization to the membrane and trafficking (Fambrough et al. 1994). This 40 kDa glycosylated transmembrane protein works as an obligatory component of the pump. In *D. melanogaster* there are three *Nervana* genes, which produce the splicing forms of *Nrv1*, *Nrv2.1*, *Nrv2.2* and *Nrv3* mRNAs (Sun and Salvaterra 1995a, b). While *Nrv1* is expressed especially in muscles, the *Nrv2.1*, *Nrv2.2* and *Nrv3* code isoforms that are expressed in the nervous system (Sun et al. 1998; Xu et al. 1999). *Nrv2.2*, the major β subunit isoform, is expressed not only in the nervous system but also in ectodermal epithelia, where it takes part in the formation of septate junctions (Paul et al. 2003, 2007).

While the developmental studies document the prominent changes of composition of ATPβ as well as changes in the level of expression of its different isoforms during the development of *D. melanogaster* eye (Baumann et al. 2010), the studies on daily expression of *Nrv2* in the mature visual system were less conclusive, except the retina, revealing only small day/night differences. This might have been related to the fact that these studies were based on the expression of the long lasting form of GFP reporter in transgenic flies (*Nrv2-Gal4* × *UAS-S65T-GFP*) rather than on labeling using Nrv2-specific antibodies (Górska-Andrzejak et al. 2009). Nevertheless, they revealed that the GFP signal was strong in the neuropil glia of the second visual neuropil, the medulla (Górska-Andrzejak et al. 2009). This is a particularly interesting type of glia because its long processes penetrate the medulla neuropil, run close to both L1 and L2 terminals and neighbor dense varicose arborization of the circadian clock neuron terminals releasing PDF (Górska-Andrzejak et al. 2009). Due to such localization, these type of glia should have a considerable capacity to modulate the circadian rhythmicity in the optic lobe, based on the instructive circadian signals (PDF) from the neighboring clock neurons. However, further studies are needed to establish their precise role in this process. It is also possible that ATPβ might be additionally involved in the circadian regulation of neuron—glia communication. Apart from the pump function it is well known that the β-subunit takes part in the formation of cell junctions (Genova and Fehon 2003; Paul et al. 2007). In this case, the circadian changes in its expression in the optic lobe of *D. melanogaster* might indicate the circadian modulation of neuron-glia communication as well. It is also interesting to mention that mRNA of *Nrv2* cycles only in the retina but not in the lamina of *D. melanogaster* (Damulewicz and Pyza, unpublished results).

14.4 Circadian Expression of the V-ATPase in the Visual System of Diptera

V-ATPases, the proton pumps, are the common enzymes that are essential for energy conversion and maintaining the ionic homeostasis in eukaryotic cells (Finbow and Harrison 1997). In neurons, they are also important components of synaptic vesicle membrane (Nelson 1992; Yamagata and Parsons 1989; Moriyama and Futai 1990; Xiao et al. 2008). They produce an electrochemical gradient necessary for transporting neurotransmitters into vesicles (Liu and Edwards 1997). Apart from this, they seem to modulate the synaptic activity of neurons by regulating the pH of their cytoplasm and extracellular space (Takahashi and Copenhagen 1996; Palmer et al. 2003). The alterations of pH have been shown to induce changes in the size of the extracellular space as well as the size of neurons and glial cells by influencing the architecture of their cytoskeleton (Faff and Nolte 2000). The involvement of V-ATPase in regulation of size and shape of the cell has also been reported in other cell types: erythrocytes (Nishiguchi et al. 1995), osteoclasts (Lees and Heersche 2000), and the stomatal guard cells of plants (Kinoshita and Shimazaki 1999).

Based on the above characteristics, the V-ATPase ought to be considered a very likely participant in the circadian regulation of cell volume in the visual system of Diptera. It could ultimately be an element of the mechanism underlying the structural circadian plasticity of this system, provided that it is expressed in this part of the nervous system and that its expression is controlled by the circadian clock. Both of these conditions appear to be fulfilled.

V-ATPase was reported to localize in endomembranes and plasma membranes of insect tissues (Wieczorek et al. 2000). For example, it occurs in large amounts in the salivary glands of the blowfly *Calliphora vicina* (Zimmermann et al. 2003). The visual system of *D. melanogaster* and *M. domestica* was also found to be immunoreactive to V-ATPase, when it was examined immunohistochemically using different antibodies against the β-subunit, one of the eight subunits that build this pump (Pyza et al. 2004).

In comparison with the expression of the Na^+/K^+-ATPase, however, the pattern of the V-ATPase expression in the optic lobe of Diptera differs considerably. While the sodium pump subunits are ubiquitously expressed in the membranes of many different cell types, the expression of the V-ATPase subunits is restricted to the retina photoreceptor cells (R1–R8). Even distribution of immunostaining throughout the terminals of R1–R6 (Fig. 14.3d) (Pyza et al. 2004) suggests the presence of

V-ATPase both at the plasma membrane and in the cytoplasm of photoreceptors. The latter is presumably related to its presence at the membranes of synaptic vesicles (Yamagata and Parsons 1989; Moriyama and Futai 1990; Moriyama et al. 1992). It has been shown that in the brain of the rat, the V-ATPase represents no less than 20 % of synaptic vesicle proteins, which explains why it has been considered the primary pump essential for accumulation of neurotransmitters (Moriyama and Futai 1990). In the terminals of *D. melanogaster* photoreceptors, the V-ATPase is engaged in loading synaptic vesicles with histamine (Borycz and Pyza, unpublished results), which is released by these cells in tetrad synapses (Hardie 1989) to act as a fast neurotransmitter (Stuart 1999) at postsynaptic ligand-gated ion channels on L1 and L2 monopolar cells (Hardie 1989; Gisselmann et al. 2001). It is then taken up by the terminals for recycling, in which a fast metabolic pathway, regulated by products of two genes, *ebony* and *tan,* has been reported to be crucial (Borycz et al. 2002). Taken up by the terminal, histamine is pumped into synaptic vesicles by V-ATPase for subsequent release (Borycz and Pyza, unpublished results).

When the ratios of vesicular and non-vesicular histamine were studied by HPLC (Borycz et al. 2000) it occurred that normally most of the histamine of *D. melanogaster* head (~70 %) is loaded into synaptic vesicles (70 % of histamine stays in the pellet). However, after feeding flies with bafilomycin, the high-affinity blocker of V-ATPase activity (Bowman et al. 1988; Wang and Telfer 1998), the relative content of histamine in the vesicles decreases to ~30 %, while the non-vesicular pool increases to more than 60 %. Bafilomycin-induced redistribution of histamine from vesicular to non-vesicular compartments implies that pumping this neurotransmitter into vesicles must indeed depend on functioning of the V-ATPase (Borycz and Pyza, unpublished results).

Examination of the level of expression of the V-ATPase subunits in the optic lobes of *M. domestica* and *D. melanogaster* during the 24 h day/night cycle confirmed an involvement of the V-ATPase in the phenomenon of circadian plasticity (Pyza et al. 2004). An intensity of the V-ATPase-specific immunolabeling showed clear circadian oscillations, when examined at different time points of LD cycle and in DD (Pyza et al. 2004). The pattern of this circadian rhythm was quite different, however, than the pattern of the Na^+/K^+-ATPase rhythm (Fig. 14.3a): the V-ATPase level is the lowest at the beginning of the night (when the level of ATPα and its activity are the highest) and the highest later during the night (when the ATPα subunit level and the pump activity are the lowest). In flies kept in DD (in both species), the level of V-ATPase β subunit was higher during the subjective night than during the subjective day.

In the housefly, the V-ATPase concentration is the highest during the night when the fly's locomotor activity is low and the L1 and L2's axons shrink (Pyza and Meinertzhagen 1995) (Fig. 14.3b). This may be connected with the action of the pump in maintaining a proper ionic and/or pH balance between the photoreceptors and the other cellular compartments of the cartridge: the monopolar cells, the glial cells and the extracellular space (Pyza et al. 2004). Note that the energy that is required to service this pump appears to be relatively small in comparison with that required for phototransduction (Laughlin et al. 1998). Therefore, one can assume that the level of its expression can be high even when the metabolic activity of photoreceptors is low. Interestingly, the L1 and L2, unlike the surrounding glia, show their lower activity and the lower uptake of deoxyglucose at that phase of the circadian cycle (Bausenwein 1994).

When daily histamine level was examined using HPLC in flies fed with bafilomycin (Borycz and Pyza, unpublished results), the largest increase of histamine level was detected at the beginning of the night, whereas its rapid decrease was detected in the middle of the night. This presumably was the consequence of the high activity of V-ATPase in the middle of the day and its low activity at the beginning of the night, respectively (Borycz and Pyza, unpublished results).

On the other hand, the role of the proton pump in maintaining a proper ionic and/or pH balance inside a cartridge is confirmed by the obvious influence of the circadian changes in V-ATPase concentration on the circadian rhythmicity of the L1 and L2 axons, which could have been observed after blocking the activity of the V-ATPase in the optic lobe of *M. domestica* with bafilomycin (Pyza et al. 2004). The night injections of bafilomycin into the second optic neuropil (medulla) resulted in the large increase of L1 and L2 axon sizes (by 57 % and 69 %, respectively) and consequently a disappearance of day/night differences—the axons had the same volume during the day as during the night. It should also be mentioned that such effect was not visible when the injection was applied during the day. Therefore the V-ATPase blocker might be responsible for the inhibition of the night axonal shrinking. The exact action of V-ATPase inside the lamina cartridge at night is yet unknown. However, it suggests that V-ATPase regulates the proton shifts between the photoreceptors and the extracellular space. Blocking the shifts that the pump normally creates may have an indirect influence on the volume of L1 and L2 axons.

The robust daily changes in the amount of V-ATPase in the visual system of Diptera in both LD and DD indicate that they must have the circadian nature and stay under control of the circadian clock. Circadian control of V-ATPase has also been reported in the male reproductive system of *Spodoptera littoralis* (Bebas et al. 2002).

14.5 Conclusions

Circadian changes in the activity of ion transporters have been shown in the brain (Senatorov and Hu 1997; Wang and Huang 2004) and in other tissues (Morise et al. 1989; Wang et al. 1993), so the circadian activity of ion pumps seems to be crucial for generation of cyclic physiological processes in cells of most if not of all tissues. We have found that circadian regulation of Na^+/K^+-ATPase, by circadian control of ATPα, may be an important mechanism of clock-controlled plasticity in the brain. The lowest ATPα expression corresponds to the time when the L1 and L2 interneurons in the lamina are shrunk (Pyza and Meinertzhagen 1999) and when the level of the presynaptic protein BRP is minimal in the retina photoreceptors (Górska-Andrzejak et al. 2013). These daily changes in Na^{+}/K^{+}-ATPase activity may also provide energy savings during the LD cycle because low ATPα level and the sodium pump activity are correlated with the resting time of flies, their nap in the middle of day and sleep in the middle of night. Na^{+}/K^{+}-ATPase activity seems to be cyclic in the brain and other tissues to maintain rhythmic processes but it may also adapt cells to changing external and internal environments.

In turn V-ATPase cyclic activity may be important in specific regions of the nervous system, where pH or cell and extracellular space sizes have to be regulated because of intense synaptic transmission.

14.6 Perspectives

Knowing as much as we do now, we still do not understand the exact mechanism by which the circadian clock modulates the output expression of different ion transporters in different cell types, e.g., cells L1 and L2, which do not express the clock genes themselves, and epithelial glia, which are known to express *per*. It is also crucial for our understanding to define the specific role of different transporters and their isoforms in the circadian plasticity. Taking into account that ion channels also participate in volume regulation of tumor cells (as well as in their survival, death and motility), studying the mechanism of their circadian control and the consequences of its disruption should be of great importance.

References

Bałys M, Pyza E (2001) Localization of the clock controlling circadian rhythms in the first neuropile of the optic lobe in the housefly. J Exp Biol 204:3303–3310

Baumann O, Salvaterra PM, Takeyasu K (2010) Developmental changes in beta-subunit composition of Na+, K+-ATPase in the *Drosophila* eye. Cell Tissue Res 340:215–228. doi:10.1007/s004410100948

Bausenwein B (1994) Struktur-Funktionskartierung im Nerven system von *Drosophila*. Habilitationschrift der Fakultät für Biologie der Albert-Ludwigs-Universität Freiburg

Bebas P, Maksimiuk E, Gvakharia B et al (2002) Circadian rhythm of glycoprotein secretion in the vas deferens of the moth, *Spodoptera littoralis*. BMC Physiol 2:15

Blanco G, Mercer RW (1998) Isozymes of the Na^{+}-K^{+}-ATPase: heterogeneity in structure, diversity in function. Am J Physiol 275:F633–F650

Borycz J, Vohra M, Tokarczyk G et al (2000) The determination of histamine in the *Drosophila* head. J Neurosci Methods 101:141–148

Borycz J, Borycz JA, Loubani M et al (2002) Tan and ebony genes regulate a novel pathway for transmitter metabolism at fly photoreceptor terminals. J Neurosci 22:10549–10557

Boschek CB (1971) On the fine structure of the peripheral retina and lamina ganglionaris of the fly, *Musca domestica*. Z Zellforsch Mikrosk Anat 118:369–409

Bowman EJ, Siebers A, Altendorf K (1988) Bafilomycins: a class of inhibitors of membrane ATPases from microorganisms, animal cells, and plant cells. Proc Natl Acad Sci U S A 85:7972–7976
Chen DM, Christianso JS, Sapp RJ et al (1992) Visual receptor cycle in normal and period mutant *Drosophila*: microspectrophotometry, electrophysiology, and ultrastructural morphometry. Vis Neurosci 9:125–135
Cheng Y, Hardin PE (1998) *Drosophila* photoreceptors contain an autonomous circadian oscillator that can function without period mRNA cycling. J Neurosci 18:741–750
Chow DC, Forte JG (1995) Functional significance of the beta-subunit for heterodimeric P-type ATPases. J Exp Biol 198:1–17
Damulewicz M, Pyza E (2011) The clock input to the first optic neuropil of *Drosophila melanogaster* expressing neuronal circadian plasticity. PLoS One 6, e21258. doi:10.1371/journal.pone.0021258
Damulewicz M, Rosato E, Pyza E (2013) Circadian regulation of the Na^+/K^+-ATPase alpha subunit in the visual system is mediated by the pacemaker and by retina photoreceptors in *Drosophila melanogaster*. PLoS One 8(9), e73690. doi:10.1371/journal.pone.0073690
Edwards TN, Meinertzhagen IA (2010) The functional organisation of glia in the adult brain of *Drosophila* and other insects. Prog Neurobiol 90:471–497
Fabian-Fine R, Verstreken P, Hiesinger PR et al (2003) Endophilin promotes a late step in endocytosis at glial invaginations in *Drosophila* photoreceptor terminals. J Neurosci 23:10732–10744
Faff L, Nolte C (2000) Extracellular acidification decreases the basal motility of cultured mouse microglia via the rearrangement of the actin cytoskeleton. Brain Res 853:22–31
Fambrough DM, Lemas MV, Hamrick M et al (1994) Analysis of subunit assembly of the Na-K-ATPase. Am J Physiol 266:C579–C589
Finbow ME, Harrison MA (1997) The vacuolar H^+-ATPase: a universal proton pump of eukaryotes. Biochem J 324:697–712
Genova JL, Fehon RG (2003) Neuroglian, Gliotactin, and the Na^+/K^+ ATPase are essential for septate junction function in *Drosophila*. J Cell Biol 161:979–989
Gisselmann G, Pusch H, Hovemann BT et al (2001) Two cDNAs coding for histamine-gated ion channels in *D. melanogaster*. Nat Neurosci 5:11–12
Górska-Andrzejak J (2013) Glia-related circadian plasticity in the visual system of Diptera. Front Physiol 4:36. doi:10.3389/fphys.2013. 00036
Górska-Andrzejak J, Keller A, Raabe T et al (2005) Structural daily rhythms in GFP-labelled neurons in the visual system of *Drosophila melanogaster*. Photochem Photobiol Sci 4:721–726
Górska-Andrzejak J, Salvaterra PM, Meinertzhagen IA et al (2009) Cyclical expression of Na^+/K^+-ATPase in the visual system of *Drosophila melanogaster*. J Insect Physiol 55:459–468. doi:10.1016/j.jinsphys.2009.02.003
Górska-Andrzejak J, Makuch R, Stefan J et al (2013) Circadian expression of the presynaptic active zone protein bruchpilot in the lamina of *Drosophila melanogaster*. Dev Neurobiol 73:14–26. doi:10.1002/dneu.22032
Green HJ (2004) Membrane excitability, weakness, and fatigue. Can J Appl Physiol 29 (3):291–307
Hardie RC (1989) A histamine-activated chloride channel involved in neurotransmission at a photoreceptor synapse. Nature 339:704–706
Helfrich-Förster C (2004) The circadian clock in the brain: a structural and functional comparison between mammals and insects. J Comp Physiol A Neuroethol Sens Neural Behav Physiol 190:601–613
Hermann-Luibl C, Helfrich-Förster C (2015) Clock network in Drosophila. Curr Opin Insect Sci 7:65–70
Im SH, Taghert PH (2010) PDF receptor expression reveals direct interactions between circadian oscillators in Drosophila. J Comp Neurol 518:1925–1945
Jackson FR (2011) Glial cell modulation of circadian rhythms. Glia 59:1341–1350

Johard HA, Yoishii T, Dircksen H, Cusumano P, Rouyer F, Helfrich-Förster C, Nässel DR (2009) Peptidergic clock neurons in Drosophila: ion transport peptide and short neuropeptide F in subsets of dorsal and ventral lateral neurons. J Comp Neurol 516:59–73

Kinoshita T, Shimazaki K (1999) Blue light activates the plasma membrane H(+)-ATPase by phosphorylation of the C-terminus in stomatal guard cells. EMBO J 18:5548–5558

Kitano T, Nisimaru N, Shibata E et al (2003) Monocarboxylates and glucose utilization as energy substrates in rat brain slices under selective glial poisoning—a 31P NMR study. Mol Cell Biochem 244:77–81

Krzeptowski W, Górska-Andrzejak J, Kijak E et al (2014) External and circadian inputs modulate synaptic protein expression in the visual system of *Drosophila melanogaster*. Front Physiol 5:102. doi:10.3389/fphys.2014.00102

Kula E, Pyza E (2007) Effects of locomotor stimulation and protein synthesis inhibition on circadian rhythms in size changes of L1 and L2 interneurons in the fly's visual system. Dev Neurobiol 67:1433–1442

Laughlin SB, de Ruyter, van Steveninck RR et al (1998) The metabolic cost of neural information. Nat Neurosci 1:36–41

Lees RL, Heersche JN (2000) Differences in regulation of pH(i) in large (>/=10 nuclei) and small (</=5 nuclei) osteoclasts. Am J Physiol Cell Physiol 279:C751–C761

Lingrel JB, Orlowski J, Shull MM et al (1990) Molecular genetics of Na, K-ATPase. Prog Nucleic Acid Res Mol Biol 38:37–89

Liu Y, Edwards RH (1997) The role of vesicular transport proteins in synaptic transmission and neural degeneration. Annu Rev Neurosci 20:125–156

Meinertzhagen IA, O'Neil SD (1991) Synaptic organization of columnar elements in the lamina of the wild type in *Drosophila melanogaster*. J Comp Neurol 305:232–263

Meinertzhagen IA, Pyza E (1999) Neurotransmitter regulation of circadian structural changes in the fly's visual system. Microsc Res Tech 45:96–105

Meinertzhagen IA, Sorra KE (2001) Synaptic organization in the fly's optic lamina: few cells, many synapses and divergent microcircuits. Prog Brain Res 131:53–69

Morise T, Okamato S, Ikeda M, Takeda R (1989) The possible role of endogenous digitalis-like substance in the regulation of circadian changes in urinary electrolyte excretion in man. Endocrinol Jpn 36:845–850

Moriyama Y, Futai M (1990) H^+-ATPase, a primary pump for accumulation of neurotransmitters, is a major constituent of brain synaptic vesicles. Biochem Biophys Res Commun 173:443–448

Moriyama Y, Maeda M, Futai M (1992) The role of V-ATPase in neuronal and endocrine systems. J Exp Biol 172:171–178

Nelson N (1992) The vacuolar H(+)-ATPase-one of the most fundamental ion pumps in nature. J Exp Biol 172:19–27

Nishiguchi E, Ozono S, Takakuwa Y et al (1995) Mechanism of the change in shape of human erythrocytes induced by lidocaine. Cell Struct Funct 20:71–79

Palmer MJ, Hull C, Vigh J et al (2003) Synaptic cleft acidification and modulation of short-term depression by exocytosed protons in retinal bipolar cells. J Neurosci 23:11332–11341

Pappas CA, Rioult MG, Ransom BR (1996) Octanol, a gap junction uncoupling agent, changes intracellular [H+] in rat astrocytes. Glia 16:7–15

Paul SM, Ternet M, Salvaterra PM et al (2003) The Na+/K+ ATPase is required for septate junction function and epithelial tube-size control in the *Drosophila* tracheal system. Development 130:4963–4974

Paul SM, Palladino MJ, Beitel GJ (2007) A pump-independent function of the Na, K-ATPase is required for epithelial junction function and tracheal tube-size control. Development 134:147–155

Picot M, Cusumano P, Klarsfeld A, Ueda R, Rouyer F (2007) Light activates output from evening neurons and inhibits output from morning neurons in the Drosophila circadian clock. PLos Biol 5:e315

Pyza E (2010) Circadian rhythms in the fly's visual systems. In: Darlene AD (ed) Encyclopedia of the eye, vol 1. Academic Press, Oxford, pp 302–311

Pyza E, Cymborowski B (2001) Circadian rhythms in behaviour and in the visual system of the blowfly *Calliphora vicina*. J Insect Physiol 47:897–904
Pyza E, Górska-Andrzejak J (2004) Involvement of glial cells in rhythmic size changes in neurons of the housefly's visual system. J Neurobiol 59:205–215
Pyza E, Górska-Andrzejak J (2008) External and internal inputs affecting plasticity of dendrites and axons of the fly's neurons. Acta Neurobiol Exp (Wars) 68:322–333
Pyza E, Meinertzhagen IA (1993) Daily and circadian rhythms of synaptic frequency in the first visual neuropile of the housefly's (*Musca domestica* L.) optic lobe. Proc Biol Sci 254:97–105
Pyza E, Meinertzhagen IA (1995) Monopolar cell axons in the first optic neuropil of the housefly, *Musca domestica* L., undergo daily fluctuations in diameter that have a circadian basis. J Neurosci 15:407–418
Pyza E, Meinertzhagen IA (1996) Neurotransmitters regulate rhythmic size changes amongst cells in the fly's optic lobe. J Comp Physiol A 178:33–45
Pyza E, Meinertzhagen IA (1997) Circadian rhythms in screening pigment and invaginating organelles in photoreceptor terminals of the housefly's first optic neuropile. J Neurobiol 32:517–529
Pyza E, Meinertzhagen IA (1999) Daily rhythmic changes of cell size and shape in the first optic neuropil in *Drosophila melanogaster*. J Neurobiol 40:77–88
Pyza E, Borycz J, Giebultowicz JM et al (2004) Involvement of V-ATPase in the regulation of cell size in the fly's visual system. J Insect Physiol 50:985–994
Senatorov VV, Hu B (1997) Differential Na+-K+-ATPase activity in rat lemniscal and non-lemniscal auditory thalami. J Physiol 502:387–395
Sheeba V (2008) The Drosophila melanogaster circadian pacemaker circuit. J Genet 87:485–493
Siwicki KK, Eastman C, Petersen G et al (1988) Antibodies to the *period* gene product of *Drosophila* reveal diverse tissue distribution and rhythmic changes in the visual system. Neuron 1:141–150
Strausfeld NJ (1976) Atlas of an insect brain. Springer, Berlin
Strausfeld NJ, Nässel DR (1980) Neuroarchitectures serving compound eyes of Crustacea and insects. In: Autrum H (ed) Handbook of sensory physiology, vol VII/6B: comparative physiology and evolution of vision in invertebrates. Springer, Berlin, pp 1–132
Stuart AE (1999) From fruit flies to barnacles histamine is the neurotransmitter of arthropod photoreceptors. Neuron 22:431–433
Suh J, Jackson FR (2007) *Drosophila* ebony activity is required in glia for the circadian regulation of locomotor activity. Neuron 55:435–447
Sun B, Salvaterra PM (1995a) Characterization of nervana, a *Drosophila melanogaster* neuron-specific glycoprotein antigen recognized by anti-horseradish peroxidase antibodies. J Neurochem 65:434–443
Sun B, Salvaterra PM (1995b) Two Drosophila nervous system antigens, Nervana 1 and 2, are homologous to the beta subunit of Na(+), K(+)-ATPase. Proc Natl Acad Sci U S A 92:5396–53400
Sun B, Wang W, Salvaterra PM (1998) Functional analysis and tissue-specific expression of Drosophila Na+, K+-ATPase subunits. J Neurochem 71:142–151
Sweadner KJ (1989) Isozymes of the Na^+/K^+-ATPase. Biochim Biophys Acta 988:185–220
Takahashi KI, Copenhagen DR (1996) Modulation of neuronal function by intracellular pH. Neurosci Res 24:109–116
Tataroglu O, Emery P (2015) The molecular ticks of the Drosophila circadian clock. Curr Opin Insect Sci 7:51–57
Wang H-Y, Huang R-C (2004) Diurnal modulation of the Na+/K+-ATPase and spontaneous firing in the rat retinorecipient clock neurons. J Neurophysiol 92:2295–2301
Wang Y, Telfer WH (1998) Cyclic-AMP-induced water uptake in a moth ovary: inhibition by bafilomycin and anthracene-9-carboxylic acid. J Exp Biol 201:1627–1635
Wang G, Schmied R, Ebner F, Korth M (1993) Intracellular sodium activity and its regulation in guinea pig atrial myocardium. J Physiol 465:73–84

Weber P, Kula-Eversole E, Pyza E (2009) Circadian control of dendrite morphology in the visual system of *Drosophila melanogaster*. PLoS One 4, e4290. doi:10.1371/journal.pone.0004290

Wieczorek H, Gerber G, Harvey WR et al (2000) Structure and regulation of insect plasma membrane H(+)V-ATPase. J Exp Biol 203:127–135

Xiao YT, Xiang LX, Shao JZ (2008) Vacuolar H(+)-ATPase. Int J Biochem Cell Biol 40:2002–2006

Xu P, Sun B, Salvaterra PM (1999) Organization and transcriptional regulation of *Drosophila* Na (+), K(+)-ATPase beta subunit genes: *Nrv*1 and *Nrv*2. Gene 236:303–313

Yamagata SK, Parsons SM (1989) Cholinergic synaptic vesicles contain a V-type and a P-type ATPase. J Neurochem 53:1354–1362

Zerr DM, Hall JC, Rosbash M et al (1990) Circadian fluctuations of period protein immunoreactivity in the CNS and the visual system of *Drosophila*. J Neurosci 10:2749–2762

Zheng L, de Polavieja GG, Wolfram V et al (2006) Feedback network controls photoreceptor output at the layer of first visual synapses in *Drosophila*. J Gen Physiol 127:495–510

Zimmermann B, Dames P, Walz B et al (2003) Distribution and serotonin-induced activation of vacuolar-type H^+-ATPase in the salivary glands of the blowfly *Calliphora vicina*. J Exp Biol 206:1867–1876

Chapter 15
The Circadian Clock in the Mammalian Kidney

Kristen Solocinski and Michelle L. Gumz

Abstract Healthy circadian rhythms are important for maintaining overall health. Several core clock genes, including Bmal1, Per, CLOCK, and Cry encode transcription factors that regulate gene expression in the kidney and in nearly all other organs and cell types. Modulation of clock genes can cause major physiological effects. Loss of any of the core clock genes in mice results in significant changes in blood pressure, indicating that the molecular clock is critical for regulation of blood pressure. The kidney regulates electrolyte and volume balance and is thus an important regulator of blood pressure. Several lines of evidence suggest a role for the kidney clock in blood pressure regulation.

Many aspects of renal function, including glomerular filtration rate and electrolyte excretion, are known to vary with a circadian rhythm. Multiple studies have demonstrated that the kidney is sensitive to food and light cues, consistent with a role for circadian rhythms in the regulation of renal function. In the kidney, clock genes are rhythmically expressed and thousands of genes that contribute to renal function are subject to transcriptional regulation by the core clock proteins. Indeed, several key circadian genes oscillate even in the fetal kidney. It is clear that the circadian clock is an important regulator of renal function and that a better understanding of how it functions can open up new avenues for the treatment of kidney disease and hypertension.

Keywords Renal • Sodium balance • Homeostasis • Electrolyte • Blood pressure • Per1 • Cry • Bmal1

K. Solocinski • M.L. Gumz (✉)
Department of Medicine, University of Florida, 1600 SW Archer Rd, Box 100224, Gainesville, FL 32610, USA

Department of Biochemistry and Molecular Biology, University of Florida, 1600 SW Archer Rd, Box 100224, Gainesville, FL 32610, USA
e-mail: Michelle.Gumz@medicine.ufl.edu

K.A. Hyndman, T.L. Pannabecker (eds.), *Sodium and Water Homeostasis*,
Physiology in Health and Disease, DOI 10.1007/978-1-4939-3213-9_15

15.1 Introduction

Since the formation of the earth, the sun has risen and set with a predictable rhythmicity giving the world day and night. This daily, circadian rhythm has been adapted by life in the form of rhythmic expression of genes and behaviors around a ~24 h day. Indeed, life forms from archaebacteria to humans have been shown to display circadian rhythms [reviewed in Richards and Gumz (2012)]. These rhythms were first noted by the French astronomer de Mairan three centuries ago with his observation that the opening and closing of the heliotrope plant occurred even in the absence of sunlight. Over a century ago, Vogel first reported circadian fluctuations in urine volume in humans. Twenty years ago, Dr. Joseph Takahashi and colleagues identified the CLOCK gene, one of the master regulators of the circadian clock in mammals (Vitaterna et al. 1994). Since then, there have been numerous advances in the field of circadian biology.

The circadian rhythms of the body are controlled by a master clock in the brain, which helps entrain numerous peripheral clocks throughout the body (Dibner et al. 2010). The master clock is located in the suprachiasmatic nucleus (SCN) in the hypothalamus of the brain. This clock is entrained by light signals that enter through the retina and are subsequently relayed to the SCN. These signals are transmitted to peripheral clocks via neuronal and humoral signals.

Through transcription-based feedback loops, a core group of clock genes forms a Transcription Translation Oscillating Loop (TTOL), which is the mechanism for the function of the circadian clock. Briefly, circadian proteins CLOCK and Bmal1 form a heterodimer capable of binding to E-box elements in the promoters of circadian target genes (Fig. 15.1). When this heterodimer binds to E-box elements of Cryptochrome 1 and 2 (Cry1 and Cry2) and Period 1 and 2 (Per1 and Per2), Per and Cry transcription is activated in the positive arm of the loop. Then Per and Cry interact and inhibit CLOCK and Bmal1 activity, forming the negative arm of the loop and inhibiting their own transcription. CLOCK/Bmal1 also binds to E-box elements in the retinoid-related orphan receptor (ROR) and REV-ERBα promoters. ROR and REV-ERBα in turn mediate opposing action on Bmal1 transcription.

15.2 Physiological Functions of the Kidney

The kidney is an important organ in the body because it not only removes waste but it is responsible for maintaining fluid and ion homeostasis. This is critical because even slight changes in ion balance can have deleterious effects. For example, alteration of sodium reabsorption can result in large changes in blood volume with subsequent effects on blood pressure. The kidney achieves blood filtration through the nephron, or functional unit of the kidney (Fig. 15.2). An adult human kidney contains about one million nephrons. Blood filtration starts when blood in capillaries comes into contact with the nephron via the glomerulus. Filtrate then

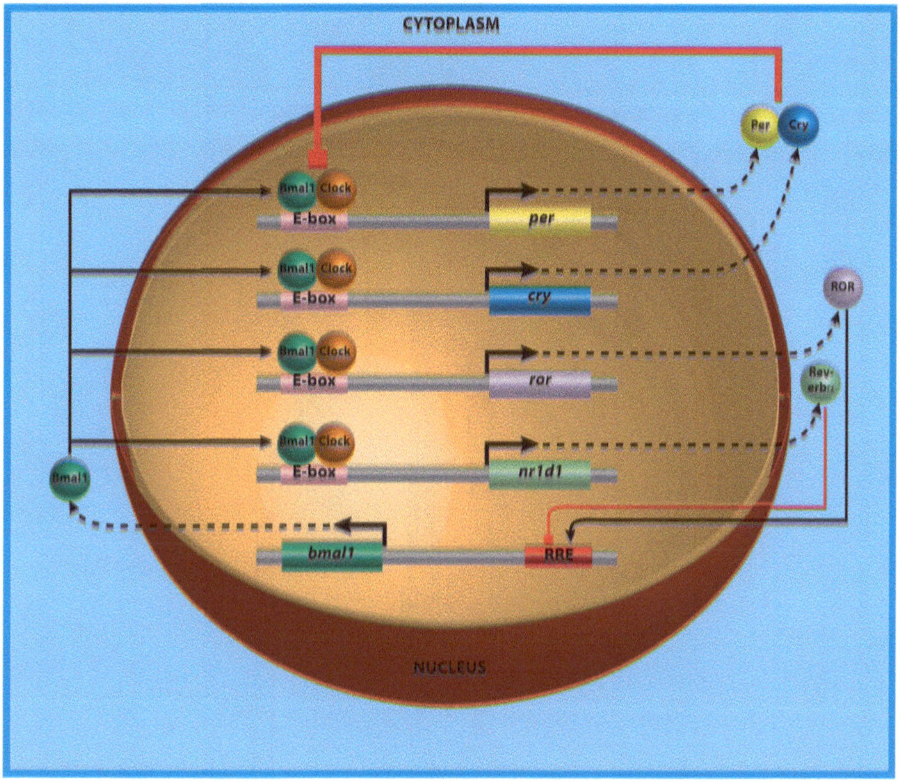

Fig. 15.1 Transcription-translation oscillating loop. Bmal1 and CLOCK heterodimerize and bind to E-box elements on target genes to increase levels of transcription. They make up the positive arm of the loop while the other two core components of the clock, Cry and Per proteins, make up the negative loop. Once translated, Cry and Per inhibit the actions of CLOCK and Bmal1. Other integral parts of the clock machinery, ROR and Rev-erbα (encoded by the *nr1d1* gene), also participate in the ~24 h oscillation of activation and inhibition of target gene transcription

passes into the proximal tubule (PT) where the bulk of sodium reabsorption occurs. The cells in this segment contain the sodium-glucose-like transporter 1 (SGLT1), which reabsorbs sodium and glucose and the sodium-hydrogen exchanger 3 (NHE3), which reabsorbs sodium and excretes hydrogen ions into the filtrate. Filtrate then passes into the Loop of Henle where active sodium reabsorption only occurs in the thick ascending loop (TAL). About 4 % of sodium reabsorption occurs in the next segment of the nephron, the distal convoluted tubule (DCT). The cells in this segment have the sodium-chloride co-transporter (NCC), which reabsorbs both sodium and chloride ions from the filtrate.

As filtrate enters the last part of the nephron, the collecting duct (CD), the final 4 % of sodium reabsorption from the filtrate occurs. This is a highly regulated process in order to avoid large changes in sodium reabsorption and blood pressure. All cell types in each segment of the nephron (PT, TAL, DCT, CD) have a

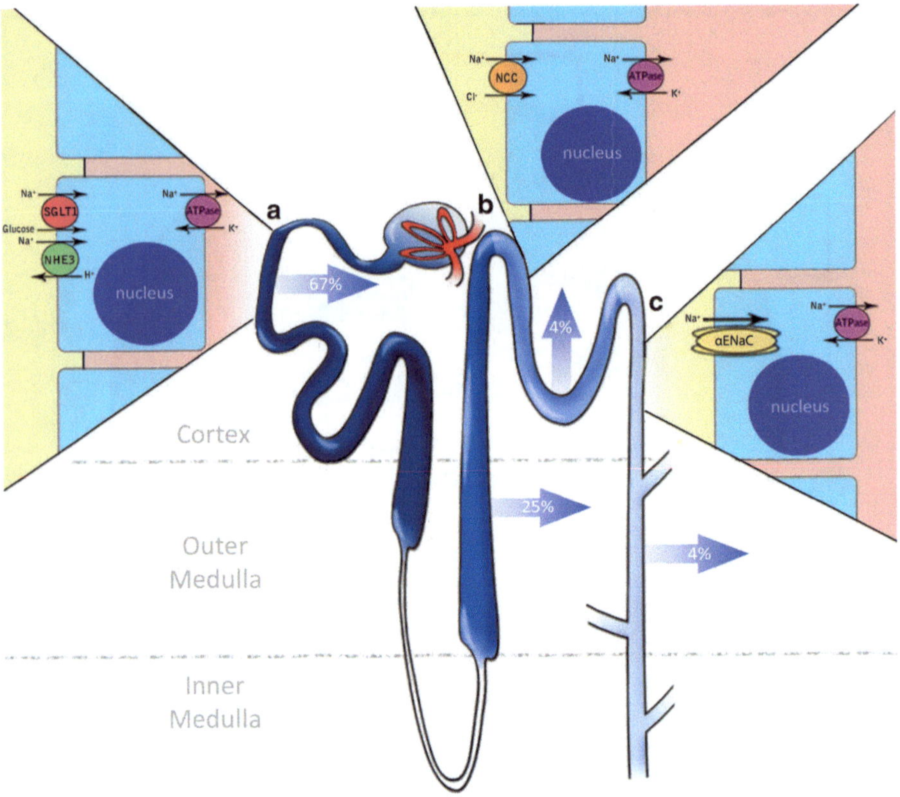

Fig. 15.2 Model of a renal nephron and representative cell types. (This illustration is focused on mechanisms of sodium reabsorption that have been linked to the circadian clock.) Blood enters the nephron at the glomerulus and is filtered as it passes through the tubule segments. Filtrate first passes through (**a**) the proximal convoluted tubule (PT) followed by the Loop of Henle (cell diagram not shown), through (**b**) the distal convoluted tubule (DCT) and (**c**) finally the collecting duct (CD). Sodium reabsorption decreases as filtrate moves along the nephron as indicated by *arrows* with approximate percentages of sodium reabsorption indicated. Each portion of the nephron is characterized by different sodium transporters, shown in the representative cell models. Different sections of the nephron are also found in different regions of the kidney, indicated by the *dashed lines* to separate cortex, outer medulla and inner medulla

sodium-potassium Na^+, K^+-ATPase on the basolateral side, which pumps sodium from the cells into the blood. In the principal cells of the collecting duct, the epithelial sodium channel (ENaC) facilitates the reabsorption of sodium from the filtrate. ENaC is regulated at many levels, one of which is by the hormone aldosterone. Aldosterone causes the increased transcription of the alpha subunit of ENaC via binding to mineralocorticoid receptors found inside the cell and subsequent translocation into the nucleus and binding to hormone response elements. The Gumz lab found that one of the core clock genes, Per1, actually regulates aldosterone thus showing that sodium reabsorption in the collecting duct of the kidney is at least partially under circadian control (Richards et al. 2013b).

Potassium is another electrolyte excreted in urine that has been shown to be regulated by the circadian clock as well (Gumz and Rabinowitz 2013). Numerous studies have shown that potassium excretion oscillates in a circadian rhythm with sodium in multiple species. Early microarray evidence accrued by Firsov et al. suggested that potassium transporters in the nephron are expressed in a circadian pattern (Zuber et al. 2009; Nikolaeva et al. 2012). Like sodium, potassium also seems to play a role in blood pressure. Not only does it appear that potassium excretion patterns are altered in disease states such as chronic kidney disease, potassium supplementation may be useful in treating hypertension (Zicha et al. 2011; Kanbay et al. 2013). It has been shown that potassium supplementation leads to increased sodium excretion and therefore decreased blood pressure. Potassium may also be beneficial in restoring a night time dip in blood pressure, lack of which is associated with higher risk of cardiovascular disease.

15.2.1 Evidence for a "Kidney Clock"

Like many other organs in the body, the kidney undergoes circadian fluctuations with respect to a number of aspects. In 1933, Robert Manchester reported on the diurnal rhythms of urine, sodium and potassium and showed that they peaked in the morning and reached a minimum at night (Manchester 1933). On a physiological level, many processes related to kidney function and blood pressure oscillate over the course of a day including glomerular filtration rate (GFR) (Koopman et al. 1989), sodium excretion and renal blood flow (Pons et al. 1996). At a transcriptional level, expression of clock genes including CLOCK, Bmal1, Cry1, Cry2, Per1 and Per2 oscillate with a ~24 h rhythm (Reppert and Weaver 2002).

The core clock proteins regulate the expression of up to 50 % of expressed genes in the body (Pizarro et al. 2013). The function of these clock genes is different depending on the tissue in which they are present. Wu et al. showed that a 30 min feeding stimulus was sufficient to not only significantly decrease transcript levels of the circadian gene Dbp but also to shift the peak of expression by 4 h in the heart but not in the kidney (Wu et al. 2012). The same study also showed decreased mRNA levels of Bmal1, Cry1, Per1 and Per2 in the heart with 30 min feeding stimulus while only Per1 mRNA was significantly decreased in the kidney.

Clock genes are ubiquitously expressed and dysregulation or loss of expression can cause profound physiological changes. In 2005, Okamura et al. provided the first evidence for the circadian control of a renal gene (Saifur Rohman et al. 2005). They showed that the Na^+, H^+ exchanger NHE3 is expressed in the membrane in a circadian manner and furthermore is regulated by CLOCK:Bmal1 heterodimers. This occurs in a similar fashion to that mentioned for aldosterone regulation. The CLOCK:Bmal1 heterodimers bind to E-box elements in the NHE3 gene to regulate its transcription. This group also showed that the circadian expression of NHE3 is severely blunted in Cry1/2 knockout (KO) mice, further supporting the notion of circadian control of this gene.

In 2006, Schibler et al. demonstrated the effects of knocking out three circadian transcription factors in mice (Dbp/Lef/Tcf triple KO) (Gachon et al. 2006). While single or double KOs of these factors do not produce much of a phenotype, triple KO mice usually do not live past 1 year. This may be due to the fact that the three factors have very conserved amino acid sequences and are able to compensate for the loss of one another. The triple KO caused changes in mRNA expression of numerous genes, many of which are related to drug metabolism or transport. Importantly, this group later showed that the Dbp/Lef/Tcf triple KO mice exhibited a phenotype of low blood pressure, reduced aldosterone levels and cardiac hypertrophy, providing strong evidence for the role of circadian proteins in cardiovascular physiology (Wang et al. 2010).

15.2.2 Circadian Clock Proteins and Blood Pressure

In 2007, Curtis et al. described the blood pressure phenotype of the Bmal1 KO mouse (Curtis et al. 2007). While the wild type mice have normal circadian variations in heart rate and blood pressure, KOs do not show these variations. The KO mice are also hypotensive, meaning they have low blood pressure, compared to the wild type mice. These results provided the first evidence that loss of a single circadian protein can dramatically alter blood pressure.

Per1 KO mice exhibit a lower blood pressure phenotype compared to wild type mice (Stow et al. 2012). This may be to due to increased renal endothelin 1 (ET-1, encoded by the *Edn1* gene) as Per1 KOs have increased levels of ET-1 in the medulla of the kidney. While ET-1 is known as a potent vasoconstrictor in the vasculature, ET-1 acts to decrease blood pressure via inhibition of ENaC in the renal collecting duct (Lynch et al. 2013; Bugaj et al. 2008). The subsequent reduction in sodium reabsorption results in a decrease in blood volume, leading to decreased blood pressure.

As previously stated, Per2 is part of the negative arm of the Transcription Translation Loop of the circadian clock mechanism. Vukolic et al. characterized the Per2 mutant mouse, which contains a mutated, not knocked out, Per2 protein (Vukolic et al. 2010). These mutant Per2 proteins are shortened, missing possible dimerization sites. This group found that the mutant mice have a higher heart rate during the light period than wild type mice but have similar locomotor activity. The Per2 mutant mice also had shorter circadian periods of heart rate, mean arterial pressure and locomotion as compared to wild type mice. The mutant mice also had smaller variations in heart rate and blood pressure between light and dark. Examples of normal and aberrant circadian oscillation are described in Fig. 15.3.

CLOCK KO mice are hypotensive and have mild diabetes insipidus but have normal 24 h blood pressure rhythms (Zuber et al. 2009). CLOCK KO mice also have altered circadian expression patterns of 20-HETE (20-hydroxyeicosatetraenoic acid), which is a regulator of blood pressure (Nikolaeva et al. 2012).

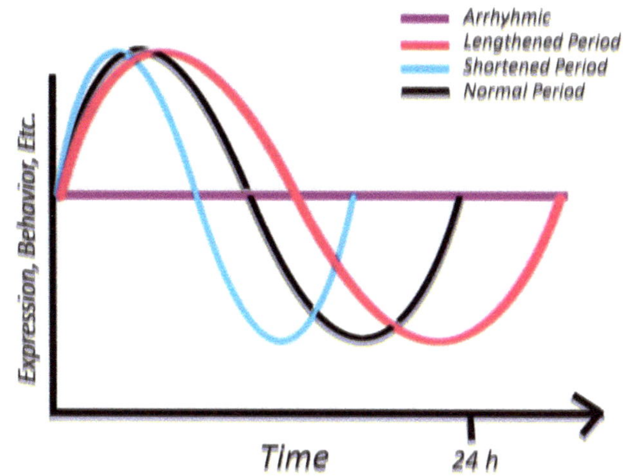

Fig. 15.3 Normal and aberrant circadian oscillation. A typical circadian cycle is depicted with the signal for gene expression or behavior on the y-axis and a 24 h period of time on the x-axis. The *black curve* represents a "normal" circadian period of ~24 h. In *blue* is a shortened period and in *magenta* is a lengthened period. A completely arrhythmic pattern is represented in the *purple line*

20-HETE can lead to both increases and decreases in blood pressure through actions on the kidney. If acting on preglomerular arterioles, the vasoconstrictive action of 20-HETE raises blood pressure (Nikolaeva et al. 2012). However, it is also capable of blocking channels and transporters that reabsorb sodium in the thick ascending limb and proximal tubule of the nephron, which leads to a decrease in blood pressure. Therefore, Firsov et al. have proposed the dysregulation of 20-HETE to be a contributing factor to the blood pressure phenotype of CLOCK KO mice. In 2014, Firsov elaborated on the role of Bmal1 by producing a mouse that has Bmal1 knocked out in renin-producing cells in the kidney (Tokonami et al. 2014). These Bmal1 mutant mice have decreased plasma aldosterone compared to control mice and also have significantly lower blood pressure compared to controls.

15.2.3 Development of the Kidney Clock

It has been known for decades that the fetal SCN is synchronized by the mother. After birth, maternal care helps to keep the clock entrained. Eventually, the neonatal SCN matures enough to take over its job as the central clock (Ohta et al. 2002). By using mice with luciferase-tagged Per2, Nishide et al. looked at the change in expression of Per2 in embryos to adult mice (Nishide et al. 2014). They showed that before birth, peak circadian rhythms of Per2 between the liver, kidney and lung were similar. However, after birth, these peaks changed between tissues. Furthermore, they showed that there is a phase shift in Per2 expression in lung and kidney from embryonic day 20 (E20) to adult mice. While there were shifts in these tissues, they found no significant change in the rhythms of the SCN, supporting the idea that it is not the sole source of entrainment for peripheral clocks.

One possible explanation for this shift in circadian expression irrespective of the SCN is that between 2 and 3 weeks after birth, pups switch from drinking milk to eating chow (Nishide et al. 2014). This change in nutrition composition may cause some alteration of digestive organs including the kidney. In addition, the pups are not fully mature after birth. Organs are still growing and changing which may also account for changes in circadian patterns due to loss or introduction of new factors because of the loss or change of cell phenotypes.

To probe even earlier stages of development, Dolatshad et al. looked at 10 day old embryos in conjunction with maternal rhythms (Dolatshad et al. 2010). They found that many core clock genes (CLOCK, Bmal1, Cry1 and Per2) were expressed as early as E10 but found little evidence that this expression was oscillating in a circadian manner up to E18. The maternal tissues, however, did express high levels of circadian oscillation. When embryonic tissues (heart, liver and kidney) were cultured, they did show circadian expression of the clock genes. While the authors determined that this rhythm was an artifact of the tissue culture process, it shows that the tissues are capable of being synchronized but something may prevent that from occurring. It is also possible that individual cells do have rhythmic expression but they are not synchronized within a tissue either due to interference by rapid mitosis of embryonic cells or some other factor.

Meszaros et al. explored this further by looking at clock genes specifically in rat embryonic kidneys (Meszaros et al. 2014). They saw that at E20 many clock (CLOCK, Per2) and clock-controlled genes (ENaC, Sgk1 and NHE3) exhibited circadian patterns of expressions whose amplitude was significantly increased later in life. At 1 week postpartum, even more genes were observed to display circadian patterns of expression including Bmal1, Cry1, Cry2, and Per1. These investigators also showed that by only allowing pups to nurse during the dark (active) period, their peak expression of CLOCK and Bmal1 was shifted by 12 h. This demonstrates how feeding can entrain clocks since normally the pups would feed during the mother's inactive period during the day. By changing the feeding time, the phase of expression of clock genes was shifted as well. This may occur because the SCN is not fully developed in rats until postnatal day 10 (Sladek et al. 2004) so nutrient (as opposed to light) cues entrain the peripheral clock in the kidney.

15.2.4 Effect of Food and Light Cues on the Kidney Clock

As previously mentioned, one of the main entrainment signals for the peripheral clocks is feeding time. While this has been established for some time, Oike et al. explored the effect of certain diets, particularly high salt diets in mice. It has already been shown that the type of diet can alter gene expression to adapt to new needs (Ferraris 2001). These authors looked at how high salt diets can affect circadian gene expression and found that in the liver, kidney and lung high salt diets decrease expression of Bmal1 while increasing expression of Dbp over time.

While food cues are significant in the entrainment of peripheral clocks, it is important to remember that the central clock, which sends signals to the peripheral clocks, is entrained primarily by light. Wu et al. investigated this by looking at the effects of only reversing the light/dark (LD) cycle, restricting feeding to the daytime (the inactive period of the rat) or reversing both the LD cycle and feeding time in rat kidneys (Wu et al. 2010). LD reversal did not alter the expression pattern of Bmal1, Cry1, CLOCK or Per2 but did delay the peak expression of Per1 by 4 h. In addition, the expression levels of Per1, Cry1, CLOCK and Bmal1 were altered only with the reversal of the LD cycle.

Feeding time reversal caused peak expression shifts of 8–12 h for CLOCK, Cry1 and Bmal1 after 7 days while causing 4 h shifts in Per1 and Per2. Combining LD and feeding time reversal resulted in a total reversal of the circadian expression of all five genes studied within 7 days. This evidence points to the cumulative effects of light and food cues regarding the entrainment of peripheral clocks.

15.2.5 Circadian Regulators Dec1 and Dec2 in the Kidney

In addition to Per and Cry homologs, Dec1 and Dec2 are also negative regulators of the clock transcription translation loop. They are expressed in a circadian rhythm in most tissues. Recently, Wu et al. studied their expression in the kidney and the heart of rats (Wu et al. 2011). Both Dec1 and Dec2 showed a pattern of daily oscillation in the heart while only Dec2 showed a similar oscillation in the kidney. By restricting feeding to daytime, as opposed to the normal night time feeding of rats, they showed that in the heart, Dec1 and Dec2 expression was shifted about 8 h. However, in the kidney, Dec2 expression was shifted 4 h and Dec1 expression became rhythmic.

They next reversed feeding and the light/dark schedule for the rats. After 7 days, peak expression was shifted for Dec1 in the heart and Dec1 and Dec2 in the heart and kidney. Expression of Dec1 in the kidney did not change significantly. Noshiro et al. showed that Dec1 exhibits a circadian rhythm of expression in mice, indicating that this clock protein acts in tissue- and species-specific ways (Noshiro et al. 2005). Along with the fact that Dec1 and Dec2 have similar DNA binding domains, this evidence suggests that Dec1 may be redundant in the kidney. This is one of many indications of the importance of maintaining circadian clock function.

15.2.6 Per1 in the Kidney

A role for Per1 in the regulation of renal function was first conceived when it was identified as a novel aldosterone target gene in a murine model of the inner medullary collecting duct (IMCD) (Gumz et al. 2003). Per1 was subsequently shown to be induced by aldosterone treatment in vivo and to mediate downstream action on the well-known aldosterone target gene, the alpha subunit of the renal epithelial sodium channel (αENaC) (Gumz et al. 2009).

In order for Per1 to get into the nucleus and affect its target genes, it has to be phosphorylated. This occurs through the actions of Casein Kinase 1 isoforms δ/ε (CK1δ/ε) (Takano et al. 2004). The Gumz lab showed that a CK1δ/ε inhibitor (PF670462) does inhibit the entry of Per1 into the nucleus in $mpkCCD_{c14}$ cells, which are a model of the murine CCD (Richards et al. 2012). This inhibition therefore prevents the interaction of Per1 with the E-box element from the promoter of the αENaC gene (*Scnn1a*). Furthermore, inhibition of Per1 nuclear entry decreases basal *Scnn1a* expression, indicating that Per1 not only increases ENaC transcription but is also involved in basal regulation of expression. This effect was observed at the protein level, as cells treated with the inhibitor show a 60 % decrease in αENaC membrane protein levels.

In addition to regulating *Scnn1a*, the Gumz lab investigated other genes that may be regulated by Per1 as well (Stow et al. 2012). Using $mpkCCD_{c14}$ cells, Per1 was knocked down using siRNA. Per1 knockdown caused changes in mRNA expression of four genes associated with sodium transport. Fxyd5 mRNA levels decreased in response to Per1 knockdown whereas Ube2e3, Cav-1 and Edn1 mRNA levels all increased. Fxyd5 increases the activity of the Na, K-ATPase, which pumps sodium reabsorbed by ENaC on the apical side of cells through the basolateral side back into the blood (Lubarski et al. 2005). Ube2e3 is an E3 ubiquitin ligase, a class of enzymes that are known to ubiquitinate the ENaC subunits, thus targeting these proteins for proteasomal degradation (Debonneville and Staub 2004). This action removes ENaC from the membrane, thereby decreasing sodium reabsorption. Loss of Per1 increases the mRNA levels of this ligase, which could lead to higher protein levels and increased ENaC degradation. Cav-1 is a lipid raft protein that has been shown to participate in removal of ENaC from the membrane (Lee et al. 2009). Finally, ET-1 causes blockade of ENaC through decreasing the open probability of the channels via a mechanism involving the ETB receptor and nitric oxide (Bugaj et al. 2008; Gallego and Ling 1996). In order to investigate the role of Per1 in the regulation of these ENaC-modulating proteins, Per1 expression was knocked down using siRNA in $mpkCCD_{c14}$ cells. ET-1 mRNA expression increased nearly four-fold after Per1 knockdown (Stow et al. 2012). After Per1 knockdown, mRNA and membrane αENaC protein levels were reduced (Richards et al. 2012). Cav-1 membrane protein levels were also increased, as were ET-1 protein levels. These results support a role for Per1 in the coordinate regulation of a number of genes that contribute to the regulation of ENaC in the kidney (Fig. 15.4).

Additional studies with Per1 KO mice showed that higher ET-1 protein levels were present in the inner medulla and cortex of the kidney compared to wild type mice (Stow et al. 2012). In wild type and Per1 KO mice, ET-1 levels were higher at noon (during their inactive period) when blood pressure is lower, showing an inverse relationship between blood pressure and renal ET-1 levels. This is consistent with the established action of ET-1 to inhibit ENaC activity, decreasing sodium reabsorption and therefore causing a decrease in blood pressure as well. Indeed, these Per1 KO mice exhibited an 18 mmHg decrease in mean arterial pressure

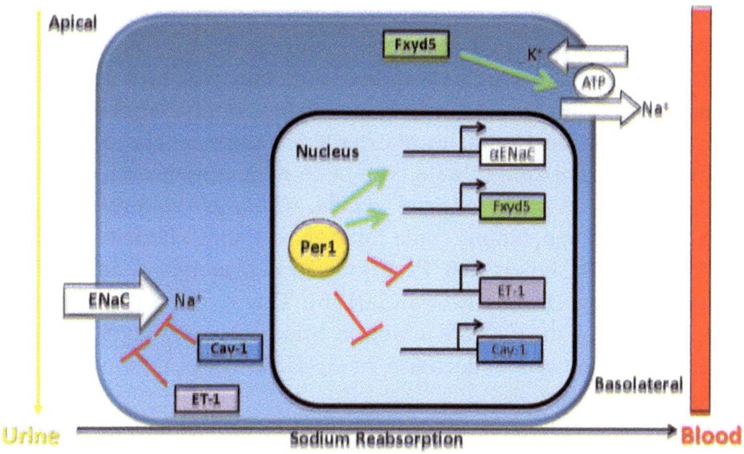

Fig. 15.4 Model for the action of Per1 in principal cells of the renal CD. Sodium derived from the filtrate enters a principal cell in the collecting duct through the epithelial sodium channel (ENaC). Sodium is exported from the cell through the action of the Na^+, K^+-ATPase and passes into the blood. ENaC is negatively regulated at the membrane by caveolin-1 (Cav-1) and endothelin (ET-1) while the Na^+, K^+-ATPase is positively regulated by Fxyd5. The genes encoding these proteins are regulated on a transcriptional level in the nucleus by Per1. Per1 activates transcription of the α subunit of ENaC (αENaC) and Fxyd5 but negatively regulates transcription of ET-1 and Cav-1. In other words, Per1 inhibits inhibitors of sodium reabsorption and activates activators of sodium reabsorption

compared to wild type mice. The contribution of ET-1 to this phenotype has not been directly tested.

Aldosterone regulates αENaC expression and Per1 appears to play a role in this regulation. Interaction of Per1 with E-box response elements in the αENaC promoter increases in aldosterone treated mpkCCD$_{c14}$ cells (Richards et al. 2013c). Aldosterone treatment also resulted in increased binding of RNA pol II to the ENaC promoter, which is indicative of increased transcription. This regulation is also evidenced by the fact that Per1 knockdown decreases the aldosterone-mediated induction of αENaC expression (Gumz et al. 2009). This effect is paralleled with use of the CK1δ/ε inhibitor (Richards et al. 2012). In mpkCCD$_{c14}$ cells, mRNA expression of αENaC is increased approximately 6× after addition of aldosterone compared to vehicle treated cells. In the presence of the CK1δ/ε inhibitor, αENaC mRNA levels drop below vehicle levels. With the CK1δ/ε inhibitor and aldosterone, mRNA levels increase about 1.5× compared to 6× with aldosterone treatment alone. Importantly, CK1δ/ε inhibitor treatment resulted in decreased ENaC activity in mpkCCD$_{c14}$ cells and in A6 cells, a model of the amphibian kidney.

In addition to aldosterone regulating Per1, Per1 appears to regulate aldosterone levels. The Gumz lab investigated this possibility using Per1 heterozygous mice, which have an approximate 50 % reduction in Per1 expression (Richards et al. 2013a). Like all mice, they are nocturnal and are active at night. Wild type and Per1 heterozygous mice were investigated at noon and midnight, the midpoint of their rest and active phases, respectively. Per1 heterozygous mice do not exhibit the normal increase in plasma aldosterone during their active phase like wild type mice and their plasma aldosterone levels are lower than wild type mice as well. This may be due to decreased 3β-HSD expression, which was observed in Per1 heterozygous mice in a pattern similar to plasma aldosterone levels (decreased at both noon and midnight compared to wild type mice with a blunted circadian pattern).

3β-HSD is produced in the adrenal glands, which are also responsible for the production of aldosterone, among other hormones. Using a human adrenal cell line (NCI-H295R), it was shown that Per1 knockdown causes a 58 % decrease in 3β-HSD mRNA levels (Richards et al. 2013b). This result was confirmed in vivo in wild type mice using the CK1δ/ε inhibitor. Mice injected with the inhibitor every 12 h for 2.5 days showed a decrease in 3β-HSD mRNA of around 60–70 % compared to wild type mice.

In contrast to Per1 heterozygous mice, Cry1/2 KO mice have increased levels of 3β-HSD and an accompanying increase in plasma aldosterone levels (Doi et al. 2010). Indeed Per1 and Cry1/2 appear to exhibit opposing actions on target gene expression (Richards et al. 2013a). Per1 causes repression of Cry2 and knockdown of Per1 with shRNA in mpkCCD$_{c14}$ cells does indeed result in increased Cry2 protein levels. This result was supported by the evidence that nuclear blockade of Per1 by the CK1δ/ε inhibitor increases cytosolic and nuclear Cry2 levels in AML12 (mouse liver) cells.

Similar results were found in mpkCCD$_{c14}$ cells and AML12 cells with regard to the effect of Per1 and Cry2 on Per1 target genes (Richards et al. 2013a). Per1 mRNA expression was decreased with Per1 siRNA and increased with Cry2 siRNA. CLOCK mRNA expression also decreased with Per1 siRNA and increased with Cry2 siRNA. Cry2 mRNA expression increased with Per1 siRNA while it decreased with Cry2 siRNA. These results are consistent with the hypothesis that Per1 suppresses Cry2 in the Transcription-Translation Oscillating Loop, which inhibits Cry2 from suppressing CLOCK/Bmal1. Further support for the opposing actions of Per and Cry comes from the observation that Per1/Cry1 KO mice have a normalization of the phenotypes observed in single Per1 and Cry1 KO mice (Oster et al. 2003). Oster et al. showed that while Cry1 KO mice have a shortened circadian period (of about 22.5 h) (van der Horst et al. 1999), knocking out Per1 in Cry1 KO mice restores the normal circadian period of about 24 h.

Per1 also regulates the NaCl cotransporter (NCC) in the kidney (Richards et al. 2014). The thiazide sensitive NCC is expressed in the apical membrane of cells in the distal convoluted tubule (DCT) of the kidney where it mediates sodium and chloride entry into the cell. Using a model of the DCT, mDCT15 cells, the Gumz lab demonstrated that either Per1 knockdown or treatment with the CK1δ/ε inhibitor resulted in decreased NCC expression. NCC mRNA levels were decreased in the renal cortex of Per1 heterozygous mice compared to wild type controls. Likewise, NCC expression was reduced in CK1δ/ε inhibitor-treated wild type mice compared to vehicle treated controls. Importantly, CK1δ/ε inhibitor treatment resulted in decreased NCC activity in mDCT15 cells, supporting a role for Per1 in the regulation of a key sodium transporter in the kidney.

15.3 Role of the Kidney Clock in the Regulation of Arterial Blood Pressure

The main function of the kidney is to filter wastes out of blood and maintain homeostasis of fluid and ions in the body. This maintenance helps maintain blood pressure within a healthy range. Hypertension, or high blood pressure, can cause a myriad of problems in the body including stroke, heart failure and renal disease. It is the main risk factor for developing cardiovascular disease, which is the cause of death of about one third of Americans every year (Go et al. 2014).

There have been numerous studies which show a correlation between the kidney and blood pressure. Grisk et al. (2002) transplanted kidneys from normotensive rats into spontaneously hypertensive rats (SHRs). This resulted in a decrease in MAP in over 50 mmHg compared to sham operated SHRs. When normotensive rats received a kidney from SHRs, the MAP increased by around 40 mmHg. These findings give strong evidence for the importance of the kidney in maintaining healthy blood pressure.

Almost all forms of Mendelian (inherited) hypertension can be traced to some defect in the Renin-Angiotenson-Aldosterone System (RAAS) (Lifton et al. 2001). Briefly, when low blood pressure is sensed, the juxtaglomerular cells release the enzyme renin. In the blood stream, renin converts angiotensinogen to angiotensin I which is then converted to angiotensin II by the angiotensin-converting enzyme (ACE). Angiotensin II (AngII) is itself a vasoconstrictor, which causes an increase in blood pressure due to decreased volume for blood to occupy. AngII also stimulates aldosterone secretion, which as previously stated, leads to an increase in αENaC transcription. Increased ENaC in the cell membrane allows for increased sodium reabsorption back into the blood and a consequent increase in blood pressure. There is increasing evidence for an intrarenal RAAS system as well as a whole-body system (Moon 2013). This further implicates the kidney in the development of hypertension.

Further evidence for the kidney's role in blood pressure maintenance has been provided by Coffman and colleagues [reviewed in Crowley et al. (2007)]. Dahl and Heine showed that a hypertensive phenotype tracks with the kidney by cross-transplantation studies using rats (Dahl and Heine 1975). They cross-transplanted kidneys from hypertension-resistant rats into hypertension-prone rats and found that it resulted in a significant drop in blood pressure. This phenomenon has been observed in human kidney transplant recipients as well (Curtis et al. 1983).

AngII mediates its actions primarily through the angiotensin receptor type 1 (AT_1R) (Timmermans et al. 1993). In mice, there are two subtypes of this receptor (A and B) and it is widely accepted that type A ($AT_{1A}R$) is the human homologue. Activation of AT_1 by AngII causes release of aldosterone and vasoconstriction in the kidney, both of which lead to an increase in blood pressure [reviewed in Crowley et al. (2007)]. Coffman et al. also did cross-transplantation studies using four groups of mice: WT, Systemic KOs having $AT_{1A}R$ only in the kidney, Kidney KOs having $AT_{1A}R$ everywhere but the kidney and finally Total KOs lacking

$AT_{1A}R$ everywhere (Crowley et al. 2007). These mice were generated by transplanting kidneys from either WT or $AT_{1A}R$ KO animals to respective recipients. For example, the Systemic KO was achieved through transplantation of a kidney from a WT animal into an $AT_{1A}R$ KO animal. The Kidney KO group had a significantly lower MAP compared to WT mice, demonstrating the impact the kidney has on blood pressure. It is important to note that the systemic KO animals had almost identical MAP values and similar circadian variation in MAP, demonstrating the importance of the systemic tissues in determining blood pressure as well.

Hypertension treatments such as ACE inhibitors aim to interrupt the RAAS system and decrease the amount of AngII being made, therefore reducing the hypertensive effects of the RAAS. Another possible target for hypertension treatment was observed in Cry1/Cry2 KO mice. These mice lacking integral circadian clock genes have salt-sensitive hypertension and increased plasma aldosterone (Doi et al. 2010). Microarray analysis identified an enzyme in the aldosterone synthesis pathway (Hsd3b6/3β-HSD) as having increased expression in these mice. This enzyme is involved in the conversion of pregnenolone to progesterone, which is an early step in aldosterone synthesis, and its expression is restricted specifically to zona glomerulosa cells of the adrenal gland. This enzyme could be a possible target to treat hypertension.

15.4 Role for Chronotherapy in Hypertension Treatment

Another proposed way of treating hypertension is chronotherapy. This is the administration of medicines with respect to time, primarily circadian time. Many hypertension medications are taken in the morning, mostly for convenience. However, a recently completed study on the effects of chronotherapy has shown that taking at least one hypertension medication at night not only significantly increases its effectiveness but also may increase its longevity. The MAPEC study also showed that nighttime hypertension medication dosing increased nocturnal dipping by about 30 % which, as previously mentioned, is associated with decreased risk of cardiovascular disease (Hermida et al. 2010; Kario and Shimada 2004).

15.5 Perspectives

In conclusion, it is quite apparent that the circadian clock is important for maintaining overall health and that modulation of clock genes can cause major physiological effects. Loss of any of the core clock genes in mice results in significant changes in blood pressure, indicating that these genes are important in its regulation. The kidney is also an important regulator of blood pressure and has been shown to display circadian rhythms of clock gene expression in addition to having specific genes under control of clock proteins themselves. This is important because it implicates the circadian clock in the proper function of the kidney which functions in a circadian manner as evidenced by the circadian rhythm of sodium excretion, renal blood flow and urine volume excretion. This further shows that blood pressure is at least partially under circadian control and provides new ideas for targets of hypertension treatments. It also sheds light on the fact that timing of treatment is important with regard to drug efficacy and patient well-being. It is clear that the circadian clock is an important regulator of renal function and that a better understanding of how it functions can lead to better treatments for kidney-related diseases.

References

Bugaj V, Pochynyuk O, Mironova E, Vandewalle A, Medina JL, Stockand JD (2008) Regulation of the epithelial Na+ channel by endothelin-1 in rat collecting duct. Am J Physiol Renal Physiol 295(4):F1063–F1070

Crowley SD, Gurley SB, Coffman TM (2007) AT(1) receptors and control of blood pressure: the kidney and more. Trends Cardiovasc Med 17(1):30–34. doi:10.1016/j.tcm.2006.11.002

Curtis JJ, Luke RG, Dustan HP et al (1983) Remission of essential hypertension after renal transplantation. N Engl J Med 309(17):1009–1015. doi:10.1056/NEJM198310273091702

Curtis AM, Cheng Y, Kapoor S, Reilly D, Price TS, Fitzgerald GA (2007) Circadian variation of blood pressure and the vascular response to asynchronous stress. Proc Natl Acad Sci U S A 104 (9):3450–3455

Dahl LK, Heine M (1975) Primary role of renal homografts in setting chronic blood pressure levels in rats. Circ Res 36(6):692–696

Debonneville C, Staub O (2004) Participation of the ubiquitin-conjugating enzyme UBE2E3 in Nedd4-2-dependent regulation of the epithelial Na+ channel. Mol Cell Biol 24(6):2397–2409

Dibner C, Schibler U, Albrecht U (2010) The mammalian circadian timing system: organization and coordination of central and peripheral clocks. Annu Rev Physiol 72:517–549. doi:10.1146/annurev-physiol-021909-135821

Doi M, Takahashi Y, Komatsu R et al (2010) Salt-sensitive hypertension in circadian clock-deficient Cry-null mice involves dysregulated adrenal Hsd3b6. Nat Med 16(1):67–74

Dolatshad H, Cary AJ, Davis FC (2010) Differential expression of the circadian clock in maternal and embryonic tissues of mice. PLoS One 5(3), e9855. doi:10.1371/journal.pone.0009855

Ferraris RP (2001) Dietary and developmental regulation of intestinal sugar transport. Biochem J 360(Pt 2):265–276

Gachon F, Olela FF, Schaad O, Descombes P, Schibler U (2006) The circadian PAR-domain basic leucine zipper transcription factors DBP, TEF, and HLF modulate basal and inducible xenobiotic detoxification. Cell Metab 4(1):25–36

Gallego MS, Ling BN (1996) Regulation of amiloride-sensitive Na+ channels by endothelin-1 in distal nephron cells. Am J Physiol 271(2 Pt 2):F451–F460

Go AS, Mozaffarian D, Roger VL et al (2014) Executive summary: heart disease and stroke statistics--2014 update: a report from the American Heart Association. Circulation 129 (3):399–410. doi:10.1161/01.cir.0000442015.53336.12

Grisk O, Kloting I, Exner J et al (2002) Long-term arterial pressure in spontaneously hypertensive rats is set by the kidney. J Hypertens 20(1):131–138

Gumz ML, Rabinowitz L (2013) Role of circadian rhythms in potassium homeostasis. Semin Nephrol 33(3):229–236. doi:10.1016/j.semnephrol.2013.04.003

Gumz ML, Popp MP, Wingo CS, Cain BD (2003) Early transcriptional effects of aldosterone in a mouse inner medullary collecting duct cell line. Am J Physiol Renal Physiol 285(4):F664–F673. doi:10.1152/ajprenal.00353.2002

Gumz ML, Stow LR, Lynch IJ et al (2009) The circadian clock protein Period 1 regulates expression of the renal epithelial sodium channel in mice. J Clin Invest 119(8):2423–2434. doi:10.1172/JCI36908

Hermida RC, Ayala DE, Mojon A, Fernandez JR (2010) Influence of circadian time of hypertension treatment on cardiovascular risk: results of the MAPEC study. Chronobiol Int 27 (8):1629–1651. doi:10.3109/07420528.2010.510230

Kanbay M, Bayram Y, Solak Y, Sanders PW (2013) Dietary potassium: a key mediator of the cardiovascular response to dietary sodium chloride. J Am Soc Hypertens 7(5):395–400. doi:10.1016/j.jash.2013.04.009

Kario K, Shimada K (2004) Risers and extreme-dippers of nocturnal blood pressure in hypertension: antihypertensive strategy for nocturnal blood pressure. Clin Exp Hypertens 26 (2):177–189

Koopman MG, Koomen GC, Krediet RT, de Moor EA, Hoek FJ, Arisz L (1989) Circadian rhythm of glomerular filtration rate in normal individuals. Clin Sci (Lond) 77(1):105–111

Lee IH, Campbell CR, Song SH et al (2009) The activity of the epithelial sodium channels is regulated by caveolin-1 via a Nedd4-2-dependent mechanism. J Biol Chem 284 (19):12663–12669

Lifton RP, Gharavi AG, Geller DS (2001) Molecular mechanisms of human hypertension. Cell 104(4):545–556

Lubarski I, Pihakaski-Maunsbach K, Karlish SJ, Maunsbach AB, Garty H (2005) Interaction with the Na, K-ATPase and tissue distribution of FXYD5 (related to ion channel). J Biol Chem 280 (45):37717–37724. doi:10.1074/jbc.M506397200, M506397200 [pii]

Lynch IJ, Welch AK, Kohan DE, Cain BD, Wingo CS (2013) Endothelin-1 inhibits sodium reabsorption by ET(A) and ET(B) receptors in the mouse cortical collecting duct. Am J Physiol Renal Physiol 305(4):F568–F573. doi:10.1152/ajprenal.00613.2012

Manchester RC (1933) The diurnal rhythm in water and mineral exchange. J Clin Invest 12 (6):995–1008

Meszaros K, Pruess L, Szabo AJ, Gondan M, Ritz E, Schaefer F (2014) Development of the circadian clockwork in the kidney. Kidney Int. doi:10.1038/ki.2014.199

Moon JY (2013) Recent update of renin-angiotensin-aldosterone system in the pathogenesis of hypertension. Electrolyte Blood Press 11(2):41–45. doi:10.5049/EBP.2013.11.2.41

Nikolaeva S, Pradervand S, Centeno G et al (2012) The circadian clock modulates renal sodium handling. J Am Soc Nephrol 23(6):1019–1026. doi:10.1681/ASN.2011080842

Nishide SY, Hashimoto K, Nishio T, Honma K, Honma S (2014) Organ-specific development characterizes circadian clock gene Per2 expression in rats. Am J Physiol Regul Integr Comp Physiol 306(1):R67–R74. doi:10.1152/ajpregu.00063.2013

Noshiro M, Furukawa M, Honma S et al (2005) Tissue-specific disruption of rhythmic expression of Dec1 and Dec2 in clock mutant mice. J Biol Rhythms 20(5):404–418

Ohta H, Honma S, Abe H, Honma K (2002) Effects of nursing mothers on rPer1 and rPer2 circadian expressions in the neonatal rat suprachiasmatic nuclei vary with developmental stage. Eur J Neurosci 15(12):1953–1960

Oster H, Baeriswyl S, Van Der Horst GT, Albrecht U (2003) Loss of circadian rhythmicity in aging mPer1−/−mCry2−/− mutant mice. Genes Dev 17(11):1366–1379. doi:10.1101/gad.256103

Pizarro A, Hayer K, Lahens NF, Hogenesch JB (2013) CircaDB: a database of mammalian circadian gene expression profiles. Nucleic Acids Res 41(Database issue):D1009–D1013. doi:10.1093/nar/gks1161

Pons M, Forpomes O, Espagnet S, Cambar J (1996) Relationship between circadian changes in renal hemodynamics and circadian changes in urinary glycosaminoglycan excretion in normal rats. Chronobiol Int 13(5):349–358

Reppert SM, Weaver DR (2002) Coordination of circadian timing in mammals. Nature 418 (6901):935–941

Richards J, Gumz ML (2012) Advances in understanding the peripheral circadian clocks. FASEB J 26(9):3602–3613. doi:10.1096/fj.12-203554

Richards J, Greenlee MM, Jeffers LA et al (2012) Inhibition of alphaENaC expression and ENaC activity following blockade of the circadian clock-regulatory kinases CK1delta/epsilon. Am J Physiol Renal Physiol 303(7):F918–F927. doi:10.1152/ajprenal.00678.2011

Richards J, All S, Skopis G et al (2013a) Opposing actions of Per1 and Cry2 in the regulation of Per1 target gene expression in the liver and kidney. Am J Physiol Regul Integr Comp Physiol 305(7):R735–R747. doi:10.1152/ajpregu.00195.2013

Richards J, Cheng KY, All S et al (2013b) A role for the circadian clock protein Per1 in the regulation of aldosterone levels and renal Na+ retention. Am J Physiol Renal Physiol 305(12): F1697–F1704. doi:10.1152/ajprenal.00472.2013

Richards J, Jeffers LA, All SC, Cheng KY, Gumz ML (2013c) Role of Per1 and the mineralocorticoid receptor in the coordinate regulation of alphaENaC in renal cortical collecting duct cells. Front Physiol 4:253. doi:10.3389/fphys.2013.00253

Richards J, Ko B, All S, Cheng KY, Hoover RS, Gumz ML (2014) A role for the circadian clock protein Per1 in the regulation of the NaCl co-transporter (NCC) and the with-no-lysine kinase (WNK) cascade in mouse distal convoluted tubule cells. J Biol Chem 289(17):11791–11806. doi:10.1074/jbc.M113.531095

Saifur Rohman M, Emoto N, Nonaka H et al (2005) Circadian clock genes directly regulate expression of the Na(+)/H(+) exchanger NHE3 in the kidney. Kidney Int 67(4):1410–1419

Sladek M, Sumova A, Kovacikova Z, Bendova Z, Laurinova K, Illnerova H (2004) Insight into molecular core clock mechanism of embryonic and early postnatal rat suprachiasmatic nucleus. Proc Natl Acad Sci U S A 101(16):6231–6236. doi:10.1073/pnas.0401149101

Stow LR, Richards J, Cheng KY et al (2012) The circadian protein period 1 contributes to blood pressure control and coordinately regulates renal sodium transport genes. Hypertension 59 (6):1151–1156. doi:10.1161/HYPERTENSIONAHA.112.190892

Takano A, Isojima Y, Nagai K (2004) Identification of mPer1 phosphorylation sites responsible for the nuclear entry. J Biol Chem 279(31):32578–32585. doi:10.1074/jbc.M403433200

Timmermans PB, Wong PC, Chiu AT et al (1993) Angiotensin II receptors and angiotensin II receptor antagonists. Pharmacol Rev 45(2):205–251

Tokonami N, Mordasini D, Pradervand S et al (2014) Local renal circadian clocks control fluid-electrolyte homeostasis and BP. J Am Soc Nephrol 25(7):1430–1439. doi:10.1681/ASN.2013060641

van der Horst GT, Muijtjens M, Kobayashi K et al (1999) Mammalian Cry1 and Cry2 are essential for maintenance of circadian rhythms. Nature 398(6728):627–630. doi:10.1038/19323

Vitaterna MH, King DP, Chang AM et al (1994) Mutagenesis and mapping of a mouse gene, Clock, essential for circadian behavior. Science 264(5159):719–725

Vukolic A, Antic V, Van Vliet BN, Yang Z, Albrecht U, Montani JP (2010) Role of mutation of the circadian clock gene Per2 in cardiovascular circadian rhythms. Am J Physiol Regul Integr Comp Physiol 298(3):R627–R634

Wang Q, Maillard M, Schibler U, Burnier M, Gachon F (2010) Cardiac hypertrophy, low blood pressure, and low aldosterone levels in mice devoid of the three circadian PAR bZip transcription factors DBP, HLF, and TEF. Am J Physiol Regul Integr Comp Physiol 299(4):R1013–R1019. doi:10.1152/ajpregu.00241.2010, ajpregu.00241.2010 [pii]

Wu T, Ni Y, Dong Y et al (2010) Regulation of circadian gene expression in the kidney by light and food cues in rats. Am J Physiol Regul Integr Comp Physiol 298(3):R635–R641

Wu T, Ni Y, Zhuge F et al (2011) Significant dissociation of expression patterns of the basic helix-loop-helix transcription factors Dec1 and Dec2 in rat kidney. J Exp Biol 214(Pt 8):1257–1263. doi:10.1242/jeb.052100

Wu T, Fu O, Yao L, Sun L, Zhuge F, Fu Z (2012) Differential responses of peripheral circadian clocks to a short-term feeding stimulus. Mol Biol Rep 39(10):9783–9789. doi:10.1007/s11033-012-1844-0

Zicha J, Dobesova Z, Behuliak M, Kunes J, Vaneckova I (2011) Preventive dietary potassium supplementation in young salt-sensitive Dahl rats attenuates development of salt hypertension by decreasing sympathetic vasoconstriction. Acta Physiol 202(1):29–38. doi:10.1111/j.1748-1716.2010.02248.x

Zuber AM, Centeno G, Pradervand S et al (2009) Molecular clock is involved in predictive circadian adjustment of renal function. Proc Natl Acad Sci U S A 106(38):16523–16528

Index

K.A. Hyndman, T.L. Pannabecker (eds.), *Sodium and Water Homeostasis*,
Physiology in Health and Disease, DOI 10.1007/978-1-4939-3213-9

MIX
Papier aus verantwortungsvollen Quellen
Paper from responsible sources
FSC® C105338

If you have any concerns about our products,
you can contact us on
ProductSafety@springernature.com

In case Publisher is established outside the EU,
the EU authorized representative is:
Springer Nature Customer Service Center GmbH
Europaplatz 3, 69115 Heidelberg, Germany

Printed by Libri Plureos GmbH
in Hamburg, Germany